The Physics of Structural Phase Transitions

Second Edition

T0234330

Springer
New York
Berlin
Heidelberg
Hong Kong
London
Milan
Paris
Tokyo

Minoru Fujimoto

The Physics of Structural Phase Transitions

Second Edition

With 95 Figures

Springer

Minoru Fujimoto
Department of Physics
University of Guelph
Guelph, Ontario
Canada, N1H 6C7

PACS: 64.70

Fujimoto, Minoru.
The physics of structural phase transitions / Minoru Fujimoto.–[2nd ed.].
p. cm.
Includes bibliographical references.

1. Phase transformations (Statistical physics) 2. Crystals. 3. Lattice dynamics. I. Title.
QC175.16.P5F85 2003
530.4′14–dc21

ISBN 978-1-4419-2349-3 e-ISBN 978-0-387-26833-0 Printed on acid-free paper.

© 2010 Springer Science+Business Media, Inc.
All rights reserved. This work may not be translated or copied in whole or in part without the written permission of the publisher (Springer Science+Business Media, Inc., 233 Spring Street, New York, NY 10013, USA), except for brief excerpts in connection with reviews or scholarly analysis. Use in connection with any form of information storage and retrieval, electronic adaptation, computer software, or by similar or dissimilar methodology now known or hereafter developed is forbidden.
The use in this publication of trade names, trademarks, service marks, and similar terms, even if they are not identified as such, is not to be taken as an expression of opinion as to whether or not they are subject to proprietary rights.

9 8 7 6 5 4 3 2 1

springeronline.com

To the memory of Professor M. Takéwaki
who inspired me with fantasy in
thermodynamics

Preface to the Second Edition

In the first edition, I discussed physical principles for structural phase transitions with applications to representative crystals. Although published nearly 6 years ago, the subject matter is so fundamental in solid states and I am convinced that this book should be revised in a textbook form to introduce the principles beyond the traditional theory of ideal crystals.

Solid-state physics of perfect crystals is well established, and lattice imperfections are treated as minor perturbations. The basic theories are adequate for most problems in stable crystals, whereas in real systems, disrupted translational symmetry plays a fundamental role, as revealed particularly in spontaneous structural changes. In their monograph *Dynamical Theory of Crystal Lattices*, Born and Huang have pointed out that a long-wave excitation of the lattice is essential in anisotropic crystals under internal or external stresses, although their theory had never been tested until recent experiments where neutron scattering and magnetic resonance anomalies were interpreted with the long-wave approximation. Also, the timescale of observations is significant for slow processes during structural changes, whereas such a timescale is usually regarded as infinity in statistical mechanics, and the traditional theory has failed to explain transition anomalies. Although emphasized in the first edition, I have revised the whole text in the spirit of Born and Huang for logical introduction of these principles to structural phase transitions. Dealing with thermodynamics of stressed crystals, the content of this edition will hopefully be a supplement to their original treatise on lattice dynamics in light of new experimental evidence.

We realize that in practical crystals, a collective excitation plays a significant role in the ordering process in conjunction with lattice imperfections, being characterized by a propagating mode with the amplitude and phase. Such internal variables are essential for the thermodynamic description of crystals under stresses, for which I wish to establish the logical foundation, instead of a presumptive explanation.

Constituting a basic theme in this book, the collective motion of dynamical variables is mathematically a nonlinear problem, where the idea of *solitons* casts light on the concept of local fields, in expressing the intrinsic mechanism of distant order involved in the collective motion in a wide range of temperature. While rather primitive at the present stage, I believe that this method leads us in a correct direction for nonlinear processes, along which structural phase transitions can be elucidated in further detail. I have therefore spent a considerable number of pages to discuss the basic mathematics for nonlinear physics.

I thank Professor E. J. Samuelsen for correcting my error in the first edition regarding the discovery of the central peak.

Mississauga, Ontario M. Fujimoto
September 2003

Preface to the First Edition

Structural phase transitions constitute a fascinating subject in solid state physics, where the problem related to lattice stability is a difficult one, but challenging to statistical principles for equilibrium thermodynamics. Guided by the Landau theory and the soft mode concept, many experimental studies have been performed on a variety of crystalline systems, while theoretical concepts acquired mainly from isotropic systems are imposed on structural changes in crystals. However, since the mean-field approximation has been inadequate for critical regions, existing theories need to be modified to deal with local inhomogeneity and incommensurate aspects, and which are discussed with the renormalization group theory in recent works. In contrast, there are many experimental results that are left unexplained, some of which are even necessary to be evaluated for their relevance to intrinsic occurrence. Under these circumstances, I felt that the basic concepts introduced early on need to be reviewed for better understanding of structural problems in crystals.

Phase transitions in crystals should, in principle, be the interplay between order variables and phonons. While it has not been seriously discussed so far, I have found that an idea similar to charge-density-wave condensates is significant for ordering phenomena in solids. I was therefore motivated to write this monograph, where basic concepts for structural phase transitions are reviewed in light of the Peierls idea. I have written this book for readers with basic knowledge of solid state physics at the level of *Introduction to Solid State Physics* by C. A. Kittel. In this monograph, the basic physics of continuous phase transitions is discussed, referring to experimental evidence, without being biased by existing theoretical models. Since many excellent review articles are available, this book is not another comprehensive review of experimental results. While emphasizing basic concepts, the content is by no means theoretical, and this book can be used as a textbook or reference material for extended discussions in solid state physics.

The book is divided into two parts for convenience. In Part One, I discuss basic elements for continuous structural changes to introduce the model of pseudospin condensates, and in Part Two various methods of investigation are discussed, thereby revealing properties of condensates. In Chapter 10, work on representative systems is summarized to conclude the discussion, where the results can be interpreted in light of fluctuating condensates.

I am enormously indebted to many of my colleagues who helped me in writing this book. I owe a great deal to S. Jerzak, J. Grindley, G. Leibrandt, D. E. Sullivan, H. –G. Unruh, G. Schaack, J. Stankowski, W. Windsch, A. Janner and E. de Boer for many constructive criticisms and encouragements. Among them, Professor Windsch took time to read through an early version of the manuscript, and gave me valuable comments and advice; Professor Unruh kindly provided me with photographs of discommensuration patterns in K_2ZnCl_4 systems; and Dr. Jerzak helped me to obtain information regarding $(NH_4)_2SO_4$ and $RbH_3(SeO_3)_2$, and to whom I express my special gratitude. Finally I thank my wife Haruko for her continuous encouragement during my writing, without which this book could not have been completed.

"It was like a huge wall!" said a blind man.
"Oh, no! It was like a big tree." said another blind man.
"You are both wrong! It was like a large fan!" said another.
Listening to these blind people, the Lord said, "Alas! None of you have seen the elephant!"

From East-Indian Folklore.

A Remark on Bracket Notations

Somewhat unconventional bracket notations are used in this monograph. While the notations $\langle Q\rangle$ and $\langle Q\rangle_s$ generally signify the spatial average of a distributed quantity Q over a crystal, the notation $\langle Q\rangle_t$ indicates the temporal average over the timescale t_o of observation.

In Chapters 8 and 9, the *bra* and *ket* of a vector quantity v, i.e. $\langle v|$ and $|v\rangle$, respectively, are used to express the corresponding row and column matrices in three-dimensional space to fascilitate matrix calculations. Although confusing at a glance with conventional notations in quantum theory, I do not think such use of brackets is of any inconvenience for discussions in this book.

Guelph, Ontario
April 1996

M. Fujimoto

Contents

Part II Experimental Studies

Part I

Basic Concepts

A structural phase transition can take place in a crystal when some distortion or reorientation in the *active groups* is collectively developed, which is characterized macroscopically by a change in lattice symmetry. Landau defined the *order parameter* in terms of irreducible representations of the symmetry element signifying the structural change, whereas the origin for phase transitions can be attributed to a physical change in the active group. Being considered as ordering phenomena in crystals, structural phase transitions should, in principle, be closely related to a spontaneous deformation in the lattice. We can therefore consider the interplay between active groups and their hosting lattice, which is responsible for a structural change at a specific thermodynamic condition. On the other hand, Cochran introduced the concept of soft phonons to deal with lattice stability, which was, nevertheless, deduced from two competing interactions of polar order variables in his model ionic crystal.

Although generally acceptable, there is still some confusion about these concepts when applied to structural problems as originally implied. Therefore, I have reconsidered their physical implications in practical crystals, so that critical anomalies observed by various experiments can be interpreted in terms of these interacting counterparts participating in phase transitions. It is also a significant fact that critical phenomena are so slow in timescale that observed results showed anomalies often conflicting with their thermodynamic interpretation. Generally, observed anomalies depend on the timescale of experiments, which is, nevertheless, considered as infinity in most statistical arguments based on the *ergodic* hypothesis. In reviewing thermodynamic concepts, we therefore pay specific attention to the timescale of observation, which is competitive with the characteristic time for critical fluctuations.

In Chapter 1, thermodynamic principles for isotropic media are reviewed for structural problems, whereas in Chapter 2, statistical concepts for ordering processes are reconsidered for typical order-disorder phenomena. In Chapter 3, classical *pseudospins* are proposed for binary structural transformations in crystals, where their anisotropic correlations in low dimensions are discussed for the singular behavior at transition points. The role played by soft phonons is discussed in Chapter 4, where the concept of *condensates* is introduced for the critical region, representing complexes of pseudospins and soft phonons. In Chapter 5, dynamics of condensates and their nonlinear character in the ordering process are discussed in relation to long-range order developing with decreasing temperature. The *soliton* is a promising concept for ordering processes, and hence the related mathematics is sketched in some detail, although the application to structural problems is still in its infancy at the present stage. Although constituting a recent topic of nonlinear physics, actual ordering processes are, by far, more complex than a simplified mathematical model can explain.

1

Thermodynamical Principles and the Landau Theory

1.1 Introduction

Basically phase transitions can be interpreted within the scope of thermodynamics, although precise knowledge of the transition mechanism is essential for critical regions. In most books of thermodynamics [1, 2, 3] phase equilibria in isotropic media are discussed at some length as simple examples, while structural phase transitions in crystals are complex and described only in sketchy manner [4, 5]. In nature, there are also many other types of phase transitions, e.g. conductor-to-insulator transitions, normal-to-superconducting phase transitions in metals, orientational ordering of macromolecules in nematic liquid crystals, and so on. Although depending on microscopic mechanisms in individual systems, Ehrenfest [6] classified phase transitions in terms of derivatives of the thermodynamical potential that exhibit discontinuous changes at transition temperatures T_c. The second-order phase transition, among others, characterized by a continuous change of the Gibbs potential at T_c is of particular interest, as the problem is related to a fundamental subject of lattice stability, if considering crystals within his classification scheme. In this chapter, we discuss a continuous phase transition in light of thermodynamical principles, although critical anomalies and a subsequent domain structure in an ordered phase cannot be elucidated properly, hence pertaining to an area beyond the limit of classical thermodynamics.

Landau [7] formulated a thermodynamical theory of continuous phase transitions in binary systems, which is sketched in Section 1.5. In his theory, a single variable called the *order parameter* emerges at T_c, signifying the ordered phase by its nonzero values that are related by inversion symmetry. He proposed that the variation of the Gibbs potential below T_c is expressed by a power series of the order parameter, implying that ordering is essentially a nonlinear process. Although well-accepted for a uniform phase, the order parameter should be redefined for anisotropic systems; in addition, critical anomalies cannot be explained by the Landau theory. The failure can partly be attributed to the fact that the theory ignores inhomogeneity in critical states due to distributed spontaneous strains in otherwise uniform crystals. Landau recognized such shortcomings in his abstract theory and suggested including spatial derivatives of the order parameter for an improved description of phase transitions. In such a revised Landau expansion, for example, an additional Lifshitz term composed of such derivatives can be responsible for lattice modulation. However, even in such a revised theory, it is still not clear if anomalies arise from a dynamical behavior of the order parameter.

Needless to say, phase transitions are phenomena in a macroscopic scale. In a noncritical phase away from T_c if sufficiently uniform, thermodynamical properties can be described by the ergodic average over distributed microscopic variables, representing the order parameter. In contrast, the critical region is dictated by short-range correlations among those variables in slow motion, for which the ensemble average is obviously inadequate. Whereas in a modified theory known as the Landau-Ginzburg theory [8] derivatives of the order parameter express the spatial inhomogeneity, critical anomalies cannot be fully explained due partly to time-dependent fluctuations. At the present stage where a reliable model has yet to be established for the transition mechanism, the thermodynamical approach still provides a first approximate step toward the problem. Experimentally, on the other hand, it is a prerequisite to identify the order variable in a given system in terms of constituent ions and molecules, whereas their behavior in anisotropic lattices needs to be visualized from observed results.

This chapter is devoted to reviewing relevant thermodynamical principles, thereby the primary account of phase transitions can be dealt with, although the Landau theory has only limited access to anisotropic systems. In view of the presence of many articles on liquid and magnetic systems, particularly an excellent monograph by Stanley [9], our discussion on isotropic systems here can be limited to minimum necessity.

1.2 Phase Equilibria in Isotropic Systems

Thermodynamical properties of an isotropic and chemically pure substance can be described by the Gibbs potential $G(p,T)$, where the pressure p and the temperature T are external variables representing the surroundings in equilibrium with it. It is noted that such a substance in equilibrium is *uniform*, as $G(p,T)$ is specified only by these variables. Conversely, however, as evident from liquid vapor equilibrium, the substance may not necessarily be homogeneous under a given p-T condition, thus, such a condensing system should be described with two potentials, G_1 and G_2, to represent these phases individually. In fact, these phases can coexist in equilibrium in a certain range of p and T, being maintained by exchanging heat and mass. Accordingly, these Gibbs potentials of coexisting phases should be involved in different internal mechanisms specified by the numbers of constituent particles N_1 and N_2 while p and T remain as common external variables. On the other hand, for a crystal, the structural detail should specify the Gibbs potential, although insignificant for thermal properties, as discussed later. Hence, the knowledge of isotropic equilibria provides a useful guideline for structural phase transitions.

The thermodynamical equilibrium under a given p-T condition is determined by minimizing the Gibbs potential. For a two-phase system, we minimize the total Gibbs function $G = G_1 + G_2$, where $G_1 = G_1(N_1, p, T)$ and $G_2 = G_2(N_2, p, T)$, namely

$$\mathrm{d}G = 0 \qquad \text{and} \qquad N_1 + N_2 = N = \text{constant}.$$

Therefore, the phase equilibrium can be specified by

$$\left(\frac{\partial G_1}{\partial N_1}\right)_{p,T} = \left(\frac{\partial G_2}{\partial N_2}\right)_{p,T}.$$

Here the derivative $\mu = (\partial G/\partial N)_{p,T}$ is called the *chemical potential*, which is the same as the Gibbs potential per particle. Using chemical potentials for the two phases, the equilibrium condition can be expressed as

$$\mu_1(p,T) = \mu_2(p,T), \tag{1.1}$$

indicating that p and T for the phase equilibria are not independent. As illustrated in Fig. 1.1, the two phases are represented graphically by areas separated by the curve given by (1.1), on which at all points (p,T), the phases are in equilibrium.

Comparing phase equilibria at two proximate temperatures T and $T + \delta T$ on the equilibrium line, we expect a pressure difference $\delta p = (\mathrm{d}p/\mathrm{d}T)\delta T$ between them, corresponding to the small temperature difference $\delta T \ll T$. The slope $\mathrm{d}p/\mathrm{d}T$ of the curve can be obtained from arbitrary variations δp and δT at a point (p,T), for which we consider that the chemical potential is continuous across the line in arbitrary manner, that is,

$$\delta\mu_1(p,T) = \delta\mu_2(p,T), \tag{1.2}$$

where

$$\delta\mu_1(p,T) = \left[\left(\frac{\partial \mu_1}{\partial T}\right)_p + \left(\frac{\partial \mu_1}{\partial p}\right)_T \left(\frac{\mathrm{d}p}{\mathrm{d}T}\right)\right]\delta T = \left[-s_1 + v_1\left(\frac{\mathrm{d}p}{\mathrm{d}T}\right)\right]\delta T$$

and

$$\delta\mu_2(p,T) = \left[-s_2 + v_2\left(\frac{\mathrm{d}p}{\mathrm{d}T}\right)\right]\delta T \quad \text{for } \delta p = 0.$$

Here, $s_\mathrm{i} = -(\partial\mu_\mathrm{i}/\partial T)_p$ and $v_\mathrm{i} = (\partial\mu_\mathrm{i}/\partial p)_T$ are specific entropies and volumes of the phases $\mathrm{i} = 1$ and 2, respectively. From (1.2) we can derive the Clausius-Clapayron relation

$$\mathrm{d}p/\mathrm{d}T = \frac{(s_1 - s_2)}{(v_1 - v_2)} = \frac{\Delta s}{\Delta v}, \tag{1.3}$$

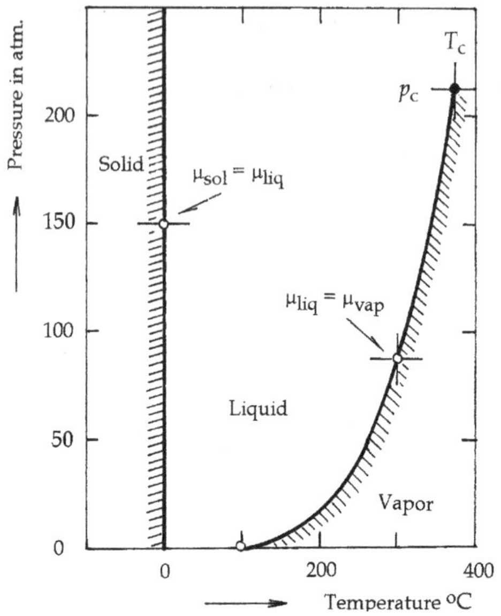

Fig. 1.1. A phase diagram of H_2O, where the chemical potentials μ_{sol}, μ_{liq} and μ_{vap} of ice, liquid water and vapor phases, respectively, are shown in the p-T plane. The phase boundary between ice and water is not exactly vertical, but with a large negative slope. The equilibrium line between water and vapor is terminated at the critical point (p_c, T_c).

where Δs and Δv signify the structure difference between phases 1 and 2; i.e. the finite entropy difference corresponds to the *latent heat* per particle $L = T\Delta s$, and the finite volume difference indicates a *packing* difference. Equation (1.3) determines the rate at which the equilibrium pressure varies with the equilibrium temperature in the p-T diagram.

As an example, the reciprocal rate $\mathrm{d}T/\mathrm{d}p$ determines the variation of the transition temperature with pressure, which is positive for liquid-vapor transitions because $v_{\mathrm{vapor}} \gg v_{\mathrm{liquid}}$, where the latent heat is always absorbed by the vapor. Also notable is that the boiling point of liquid rises with increasing pressure, whereas, applying (1.3) to a liquid-solid transition, the freezing temperature can either rise or fall, depending on the sign of Δv during solidification.

We can derive a useful expression for the vapor pressure of liquid by integrating (1.3). Ignoring very small v_{liquid} as compared with much larger v_{vapor}, the vapor pressure can be determined from the differential equation

$$\frac{\mathrm{d}v_{\mathrm{vapor}}}{\mathrm{d}T} = \frac{L}{Tv_{\mathrm{vapor}}} = \frac{Lp_{\mathrm{vapor}}}{\mathrm{k_B}T^2},$$

where the vapor is assumed to obey the ideal gas law, i.e. $v_{\mathrm{vapor}} = \mathrm{k_B}T/p_{\mathrm{vapor}}$, where $\mathrm{k_B}$ is the Boltzmann constant. Assuming, further, that L is independent of temperature, the above equation can be easily integrated and

$$p_{\text{vapor}} = p_{\text{o}} \exp\left(-\frac{L}{\text{k}_\text{B}T}\right),$$

where the constant of integration p_{o} corresponds to the pressure of an ideal gas, namely $p_{\text{vapor}} = p_{\text{o}}$, if $L = 0$. From this result, it is clear that for such an ideal vapor, the vapor pressure remains constant during isothermal condensation, providing a useful physical supplement to the van der Waals isotherm to delineate the mathematical conjecture in the equation of state (see Section 1.4).

1.3 Phase Diagrams and Metastable States

With the aid of a *phase diagram*, it is instructive to see how chemical potentials of two coexisting phases behave in the vicinity of their equilibrium. For a uniform substance, the Gibbs potential G can be used, but for two or more phases in equilibrium, chemical potentials are more convenient because of (1.1). Normally, the chemical potential μ is a continuous function of p and T, as shown by a smooth mathematical surface in the three-dimensional μ-p-T space of Fig. 1.2. Therefore, for liquid-vapor equilibrium, such surfaces of two phases should intersect in a curve, along which the two chemical potentials take an equal value. The two phases can generally coexist, whereas at arbitrary points other than those on the equilibrium line, only one of these phases with a lower value of μ can be stable.

For a simple isotropic substance like water, exhibiting three phases, i.e. solid, liquid and vapor, these phase surfaces may intersect in pair to give three equilibrium curves in the μ-p-T space. However, if a point lying on all three surfaces or eqilibrium lines can be found, these three phases can coexist at such a point called the *triple point* (Fig. 1.3). Usually, a phase diagram is drawn in two dimensions for convenience, e.g. with two variables p and T at a constant μ, corresponding to the three-dimensional μ-p-T surface projected on the p-T plane. Similar projections can also be obtained on the μ-p and μ-T planes, providing useful phase diagrams at constant T and at constant p conditions, respectively.

Although intersecting curves in phase diagrams represent accessible equilibrium states, it is important to realize that in practical systems, there are always so-called *metastable states*, which are represented, for example, by a point x on the extention of a constant μ-line in Fig. 1.3. Deviated from the vapor-liquid equilibrium curve, hence unstable thermodynamically, such a metastable state can often be observed as if it were stable. For instance, a vapor can be compressed to a pressure higher than the vapor pressure, if there are no appreciable *nuclei* for initiating condensation. Although rather vaguely defined, the "nuclei" expresses the presence of unavoidable impurities in practical systems, playing a significant role in condensation. Such a

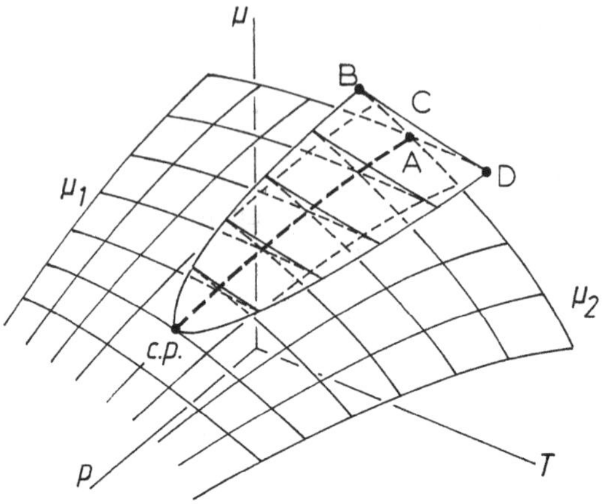

Fig. 1.2. Chemical potential surfaces $\mu_1(p,T)$ and $\mu_2(p,T)$ for two phases in equilibrium, that is represented by the intersection A...c.p. shown by the thick broken line, where c.p. is the critical point. The points B and D are possible metastable states at a constant p.

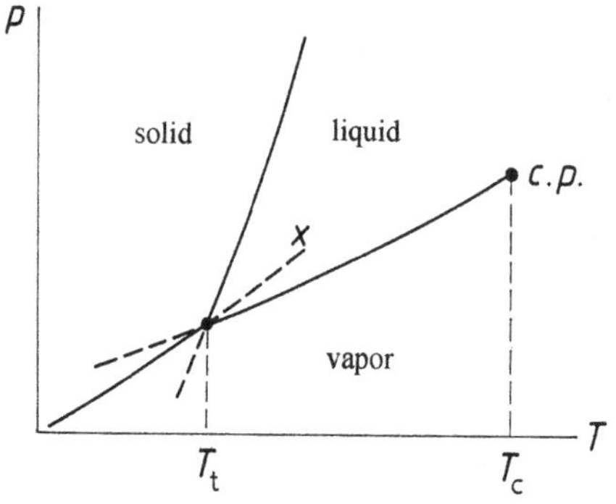

Fig. 1.3. The triple point T_t and the critical point T_c in a system of three phases. The point x on the extension of the solid vapor equilibrium line represents a supersaturated state.

metastable vapor, called *supersaturated*, is unstable against external disturbance like shock waves, resulting in sudden condensation.

The nature of a metastable state can be discussed with a phase diagram. Figure 1.4b shows a μ-T diagram, where μ-curves 1 and 2 are crossing at a temperature T_x while p is kept constant. It is noted that such a crossing point is uniquely specified by the chemical potential μ, whereas the transition between two phases is generally discontinuous in terms of the Gibbs potential G.

In Fig. 1.4b for a μ-p diagram, drawn as $\mu_2 < \mu_1$ for $T < T_x$, the state y on the μ_1-curve is unstable in this region. However, such a state y below T_x

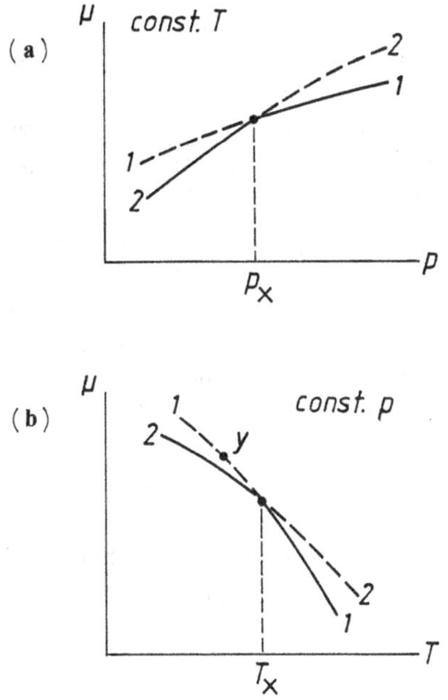

Fig. 1.4. Equilibrium of two phases: (a) a phase diagram at a constant T, where the intersection p_x is the equilibrium pressure, (b) a diagram at a constant p, where T_x is the equilibrium temperature.

may be observed as metastable if the temperature T_y is not much different from T_x. On the other hand, the phase 1 is stable at temperatures above T_x. When the phase 1 represents a vapor phase stable at high temperatures, it can be supersaturated when the temperature is lowered to below the boiling point. Conversely, liquid 2 can be *superheated* when heated upward from a lower temperature through T_x. In this case, superheated liquid is metastable above T_x.

The stability of a metastable phase depends on rather ill-defined "nuclei" existing in the given system. Although extrinsic in nature, such nuclei are essential for phase transitions to occur, which are thermodynamically irreversible. Impurities and lattice defects in a crystalline solid play a role similar to nuclei in a condensing system, being essential for domain formation in the ordering process.

The slope of a μ-curve in a μ-p diagram represents the specific volume $v = (\partial \mu/\partial p)_T$, which is always positive (Fig. 1.4a). Therefore, the stable phase can be specified by a smaller v in this case, i.e. $v_\mathrm{liquid} < v_\mathrm{vapor}$. In solid-liquid equilibrium, either phase can be stable, depending on which phase has a smaller specific volume. In vapor-liquid and vapor-solid transitions, in contrast, the vapor phase, as characterized by a larger v, is obviously stable at all temperatures above transitions. In a μ-T diagram, on the other hand, the slope of an equilibrium curve is determined by the specific entropy, $-s = (\partial\mu/\partial T)_p$, which is negative as shown in Fig. 1.4b. In this diagram, the stable phase is always specified by a larger entropy.

A transition from liquid to vapor begins to occur when v_{liquid} is increased by heating under a constant p. On the other hand, vapor starts to change to liquid if v_{vapor} decreases with increasing p under a constant T. The liquid in the former case needs to be further heated beyond the threshold for complete vaporization, whereas the vapor in the latter case be further pressurized untill all vapor molecules condense. Thus, two phases can coexist until one phase is completely transformed to the other. However, when the limit of $v_{\text{vapor}} \to v_{\text{liquid}}$ is achieved, the two phases cannot be distinguished, where the state of substance is called *critical.* The critical state can be specified by a point $(p_{\text{c}}, T_{\text{c}})$ in the p-T diagram, where p_{c} and T_{c} are referred to as critical pressure and critical temperature, respectively. It is noted at a critical point that the rate $\mathrm{d}p/\mathrm{d}T$ cannot be determined from the Clausius-Clapayron equation, and so the equilibrium curve must be terminated there, as illustrated in Fig. 1.1.

Microscopically however, v_{vapor} and v_{liquid} may not be equal at the critical point if molecular clusters of a finite size are responsible for initiating condensation. In this case, $v_{\text{vapor}} - v_{\text{liquid}} = \Delta v$ is not zero at the critical point, at which a latent work, $-p_{\text{c}}\Delta v$, is required for completing condensation. In this context, the transition cannot be continuous, although it can be considered approximately as second-order if Δv is negligible. In addition, the size of a molecular cluster is unknown, due perhaps to diverse nucleation processes. Nevertheless, it is evident from opalescent experiments that such liquid droplets can actually be observed at the threshold of condensation in some systems, where the droplet size should be of the order of the wavelength of scattered light. For details, interested readers are referred to Stanley's book [9].

In the above argument, the transition temperature T_x is not a unique parameter for isotropic phase transitions, as it depends also on the vapor pressure p. On the other hand, a phase transition in solids is normally observed under ambient atmospheric pressure around 1 atm.Hg, where the properties are virtually unchanged by p, and, hence, T_{c} is regarded as characteristic for a structural change. In some cases however, such a transition temperature T_{c} can be hypothetical if T_{c} appears higher than the melting point of the crystal, where the real transition may be observed under a pressure p higher than 1 atm.

1.4 The van der Waals Equation of State

Thermodynamical properties of a real gas can be described adequately by the van der Waals equation of state, which is capable of explaining the significant feature of a classical gas, namely condensation, at least qualitatively. Although approximate, it is instructive to see how the theory can deal with condensation phenomena as a first-order phase transition.

A real gas is distinct from an ideal gas obeying the Boyle-Charles law in that finite attractive molecular interactions and nonzero molecular volume are taken into account in deriving the equation of state. Qualitatively, attractive molecular forces should reduce the vapor pressure from that of an idealized gas, where such forces are completely ignored. Van der Waals considered molecular interactions as averaged over long ranges, which were expressed in the form proportional to $1/V^2$. Accordingly, the effective pressure is given by $p+a/V^2$, where p is the external pressure and a is a constant of the constituent molecule. Thus, we realize that in the van der Waals theory, molecular interactions are evaluated in the *mean-field* approximation. Further, the volume for molecular motion cannot be considered as equal to the container volume, but is one from which the total molecular volume should be subtracted. Gas molecules are very small objects in a large container, but the total molecular volume is not negligible particularly at a high density of condensing gas. He expressed the effective gas volume by $V-b$, where V is the container volume and b is another constant of the constituent.

The van der Waals equation is written for 1 mole of a gas as

$$\left(p+\frac{a}{V^2}\right)(V-b)=RT, \tag{1.4}$$

where $R=8.314$ joule/deg/mol is the gas constant. Values of the constants a and b are tabulated for representative gases in many standard books of thermodynamics (See e.g. ref 2). To discuss the general feature of van der Waals isotherms in a p-T diagram, we rewrite (1.4) in the algebraic form

$$V^3-\frac{b+RT}{p}V^2+\frac{a}{b}V+\frac{ab}{p}=0. \tag{1.4a}$$

This cubic equation has either one or three real roots for given external variables p and T. Figure 1.5 shows p-V curves at various temperatures, known as van der Waals isotherms, among which a particular one at $T=T_c$ has a point of inflection at (p_c,T_c) determined by a horizontal tangent. Mathematically, for all isotherms at temperatures above T_c, (1.4a) has only one real root V at a given p above p_c, whereas for those below T_c three real intersections V_1, V_2 and V_3 occur with a horizontal line of a given p below p_c. The isotherms for $T>T_c$ represent clearly uniform states signified by p and V, whereas those for $T<T_c$ may be interpreted for the condensing state consisting of two distinct vapor and liquid phases. However, we realize that there are some mathematical conjectures in (1.4a), conflicting with physical realities.

First, we notice that the isotherm, as represented by (1.4), will change continuously between the two categories when the temperature varies through T_c. The three roots below T_c become equal to V_c, when the critical point is approached from below. Therefore, in the limit of $T\to T_c$, (1.4b) should be

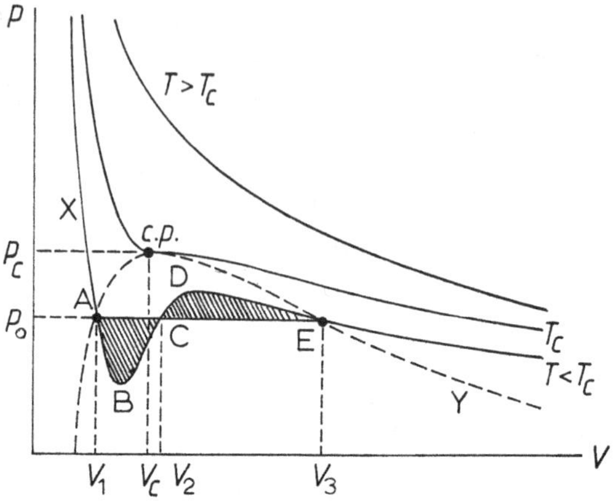

Fig. 1.5. Van der Waals' isotherms for vapor liquid equilibria in a p-T diagram. At T_c, liquid and vapor phases are in critical equilibrium, specified at c.p. by p_c and T_c. At temperatures below T_c, the two phases are represented by chemical potentials $\mu(A)$ and $\mu(E)$, respectively, coexisting at all points on the horizontal line AE at a constant vapor pressure p_o. The figure shows that p_o can be determined as area(ABC) = area(CDE).

written as

$$(V - V_\mathrm{c})^3 = 0,$$

indicating that

$$3V_\mathrm{c} = b + \frac{RT_\mathrm{c}}{p_\mathrm{c}}, \quad 3V_\mathrm{c}^2 = \frac{a}{p_\mathrm{c}} \quad \text{and} \quad V_\mathrm{c}^3 = \frac{ab}{p_\mathrm{c}}.$$

Therefore, critical values p_c, V_c and T_c are all determined by the molecular constants a and b; that is

$$p_\mathrm{c} = \frac{a}{27b^2}, \quad V_\mathrm{c} = 3b \quad \text{and} \quad T_\mathrm{c} = \frac{8a}{27b^2}. \tag{1.5}$$

Using these results, we can confirm that

$$\left(\frac{\partial p}{\partial V}\right)_{T=T_\mathrm{c}} = 0 \quad \text{and} \quad \left(\frac{\partial^2 p}{\partial V^2}\right)_{T=T_\mathrm{c}} = 0,$$

which are the requirements for the inflection point with horizontal tangent at $T = T_\mathrm{c}$.

It is realized that such a continuity of the van der Waals isotherm at T_c originates from the mean-field assumption for the molecular interaction, thereby the whole system is regarded as homogeneous. However, this is contradictory to the presence of two phases, i.e., liquid droplets coexisting with vapor at the threshold of condensation.

Second, the thermodynamical inequality

$$\left(\frac{\partial p}{\partial V}\right)_T < 0. \tag{1.6}$$

must be obeyed by all isothermal p-V curves, expressing the fact that the pressure should always decrease with increasing volume. However, in contradiction to the inequality (1.6), van der Waals isotherms below T_c indicate that the part marked BD in Fig. 1.5 has a positive slope, which is clearly a mathematical conjecture involved in the van der Waals theory.

Using the Clausius-Clapayron equation, we showed in Section 1.2 that the vapor pressure should remain constant during isothermal condensation. Assuming such a vapor pressure p_o, all real states between B and D should therefore be found on the straight horizontal line $p = p_o$ instead of the curved BD with a positive slope. On this part of an isotherm at T, all points on the straight line EA should represent equilibrium states of the vapor-liquid mixture at p_o and T, where the states E and A are interpreted as the beginning and final stages of condensation, and at a point P in-between we consider coexisting phases of the mixture.

When the vapor is compressed from a state Y along an isotherm below T_c, condensation begins to form liquid at E. On further compressing to P, more liquid is formed, coexisting with vapor under a constant p_o. At this stage, we consider that v_{liquid} is unchanged, whereas v_{vapor} changes as V is reduced from V_3 by compressing the vapor. Representing the two phases by Gibbs potentials G_{vapor} and G_{liquid}, we can write the following relation for the condensation process.

$$G_{\text{vapor}}(\mathrm{E}) = G_{\text{vapor}}(\mathrm{P}) + G_{\text{liquid}}(\mathrm{P}) - p_o(V_3 - V_\mathrm{P}),$$

where

$$\lim_{\mathrm{P}\to\mathrm{A}} G_{\text{vapor}}(\mathrm{P}) = 0 \quad \text{and} \quad \lim_{\mathrm{P}\to\mathrm{A}} G_{\text{liquid}}(\mathrm{P}) = G_{\text{liquid}}(\mathrm{A})$$

for isothermal compression at T, so that we have the relation

$$G_{\text{liquid}}(\mathrm{A}) = G_{\text{vapor}}(\mathrm{A}) + p_o(V_3 - V_1). \tag{1.7}$$

Assuming, on the other hand, that the van der Waals equation is acceptable over the entire region of isotherms for $T < T_c$, we may write a thermadynamical relation

$$G(p,T) = G(p_o,T) - \int_\mathrm{E}^\mathrm{P} \left(\frac{\partial G}{\partial V}\right)_T \mathrm{d}V = G_{\text{vapor}}(\mathrm{E}) - \int_\mathrm{E}^\mathrm{P} p\mathrm{d}V.$$

Therefore, in the limit of $\mathrm{P} \to \mathrm{A}$, we can write

$$G_{\text{liquid}}(\mathrm{A}) = G_{\text{vapor}}(\mathrm{E}) - \int_\mathrm{E}^\mathrm{A} p\mathrm{d}V = G_{\text{vapor}}(\mathrm{E}) - p_o(V_3 - V_1), \tag{1.8}$$

suggesting that the areas ABC and CDE above and below the horizontal line $p = p_o$ are considered as equal but with opposite signs. At $T = T_c$ in particular, $V_3 = V_1$ and, hence, $G_{liquid}(T_c) = G_{vapor}(T_c)$ from (1.8), confirming that the transition at T_c is continuous, whereas below T_c, transitions are discontinuous by the amount of work $p_o(V_3 - V_1)$. Using chemical potentials and numbers of molecules in these phases, the Gibbs functions are expressed as

$$G_{vapor}(P) = N_{vapor}(P)\mu_{vapor}(P) - p_o(V_3 - V_P)$$

and

$$G_{liquid}(P) = N_{liquid}(P)\mu_{liquid}(P).$$

It is noted that $N_{liquid} = N$ at state A, and $N_{vapor} = N$ at state E, so that $G_{liquid}(A) = N\mu_{liquid}(A)$ and $G_{vapor}(E) = N\mu_{vapor}(E)$. Hence, $\mu_{vapor}(E) = \mu_{liquid}(A)$ on the straight equilibrium line in Fig. 1.5, whereas the corresponding Gibbs potentials at E and A are discontinuous by $p_o(V_3 - V_1)$ on this straight isotherm. Known as the Maxwell *equal-area construction*, it is clear that the horizontal line $p = p_o$ has been drawn as consistent with the equilibrium condition (1.1) derived thermodynamically.

In the van der Waals equation, the molecular interaction in a gas phase is considered in the mean-field approximation. It is a logical assumption that gas molecules are in random motion, whereas molecules are in *near order* in the condensed phase, which is, however, a physical addendum to the equation of state for interpreting the equation of state. Also assumed is that the liquid phase is regarded as uniform, as specified by a parameter v_{liquid}. In this theory, the mixed phase is characterized in terms of $\Delta v = v_{vapor} - v_{liquid}$ at temperatures close to T_c, which may be regarded as the *order parameter* for the transition between these phases. However, these assumptions are clearly too simplified to deal with the "critical region," for which a more detailed description is obviously required. Nevertheless, even in this approximation, the corresponding specific density difference $\Delta\rho$, where $\rho = 1/v$, is more practical than Δv, and so we can define

$$\eta = \Delta\rho = \rho_{liquid} - \rho_{vapor} \approx \rho_{liquid} \tag{1.9}$$

as the order parameter. As remarked, such a definition is valid for a uniform state, and, hence, acceptable in the mean-field accuracy for the mixed state below T_c. Being essential for phase transitions in general, the order parameter as such is usually defined as

$$\eta = \rho_{order} - \rho_{disorder} \approx \rho_{order}$$

and it is appropriate to consider the liquid phase as ordered and the vapor phase as disordered.

For an isotropic liquid phase, the Gibbs potential can be expressed as a function of the order parameter η, i.e., $G = G(\eta)$. The order parameter defined as (1.9) is an *internal variable* for a condensed phase, signifying molecular interactions. Nevertheless, the Gibbs potential of a condensing liquid is determined by the surrounding vapor in equilibrium, where the order parameter represents the response of the system to heat transfer from the surroundings and to an external compression. In the former case, such an η is a thermodynamic variable, as determined by external T, whereas in the latter it is a mechanical variable, implying that the liquid condenses as the vapor is compressed at p. Also, it is noted that η may not necessarily be defined as positive for binary order in solids while defined as a positive variable in (1.9).

1.5 Second-Order Phase Transitions and the Landau Theory

1.5.1 The Ehrenfest Classification

In Fig. 1.4b, equilibrium between different phases under a constant p condition was illustrated in a μ-T plane. Here, two chemical potential curves, $\mu_1(T)$ and $\mu_2(T)$, intersect at a temperature T_x, at which the slopes are generally unequal, i.e., $(\partial\mu_1/\partial T)_p \neq \partial\mu_2/\partial T)_p$. Such a discontinuity in the slope at T_x corresponds to a finite change in entropy $\Delta s = s_1 - s_2$ at T_x, signifying a latent heat $T_x\Delta s$ for the transition. On the other hand, in Fig. 1.4a where two μ-curves intersect at a point p_x under a constant T, the discontinuity in the slope $(\partial\mu/\partial p)_T$ at T_x characterizes the transition in the μ-p diagram. In this case, an amount of external work $-p_x\Delta v$ is required for a mass transfer between the two phases, as discussed for the van der Waals theory. Figure 1.6 shows the behavior of such isotherms in the μ-p diagram when the critical region is approached from below T_x. Characterized by discontinuous first-order derivatives $(\partial\mu/\partial p)_T$, transitions at pressure p_o are called the *first order*, whereas at the critical point p_c, the transition is continuous, where the two phases cannot be distinguished by the first derivatives. However, the

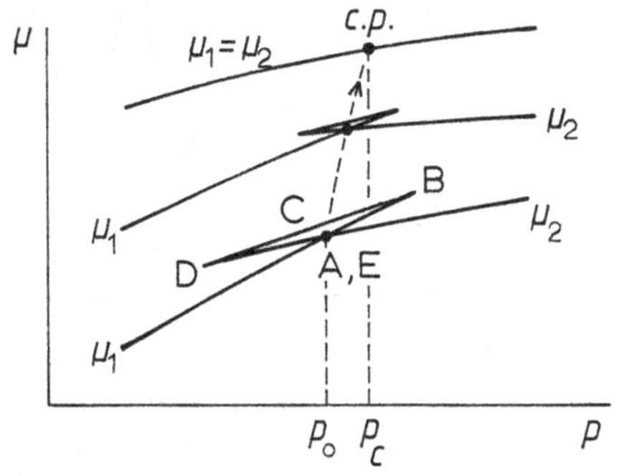

Fig. 1.6. Two-phase equilibria in the μ-p plane at different temperatures, showing the behavior of the intersection A (= E in Fig. 1.5) between $\mu_1(p)$ and $\mu_2(p)$, when c.p. is approached, indicating that transitions are discontinuous, except at c.p.

two phases can be distinguished in principle if higher-than-first-order derivatives of Gibbs potentials are unequal at p_c. Ehrenfest called such transitions as *second-* and *higher*-order phase transitions, depending on the order of the lowest nonvanishing derivative.

Figure 1.7a illustrates such a transition in a μ-T diagram that is signified by a common tangent at T_c. However, if the curvatures at T_c are unequal, one of these phases can always be more stable than the other on both sides of T_c, hence representing no phase transitions. On the other hand, if we consider a system of a "single" phase at all temperatures, instead of two, we can show that Ehrenfest's criterion for the second-order phase transition can be fulfilled mathematically. In such a single-phase system, the transition should be spontaneous and characterized by a discontinuous change in curvature at T_c, i.e., $\Delta(\partial^2\mu/\partial T^2)_{p,T=T_c} \neq 0$, and $\Delta(\partial\mu/\partial T)_{p,T=T_c} = 0$. However, for such a system, the number of constituents is constant and insignificant, and so the

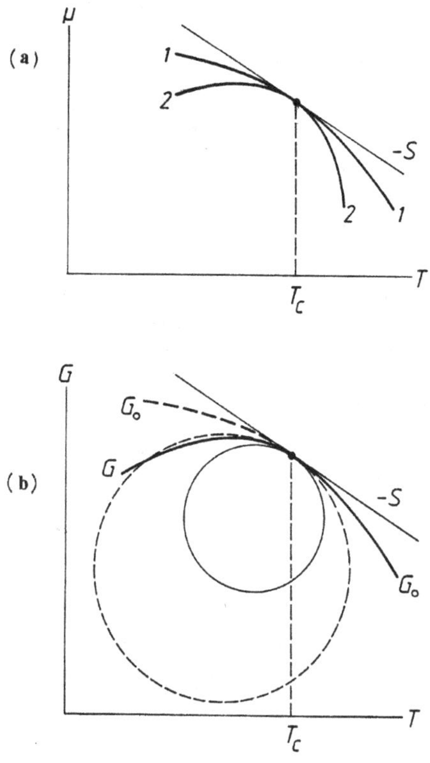

Fig. 1.7. The second-order phase transition illustrated in the μ-T diagram (a) when interpreted as two phases are in equilibrium, and in the G-T diagram (b) when interpreted as a continuous change in the entropy S with a discontinuous curvature at T_c, as indicated by two circles of different radii.

transition can be described in terms of the Gibbs function; that is

$$\Delta\left(\frac{\partial G}{\partial T}\right)_p = 0 \quad \text{and} \quad \Delta\left(\frac{\partial^2 G}{\partial T^2}\right)_p \neq 0 \quad \text{at} \quad T = T_c. \tag{1.10}$$

Denoting the Gibbs potential under a constant p as $G_o(T)$ for $T > T_c$, the second-order transition to $G(T)$ on lowering temperature can be expressed as

$$G(T) = G_o(T_c) - \Delta G(T_c - T).$$

where

$$\Delta G(T_c - T) = G_o(T_c) - G(T) = \tfrac{1}{2}(\partial^2 \Delta G/\partial T^2)_{T=T_c}(T_c - T)^2 + \ldots . \tag{1.11}$$

Here, the first-order term of $T_c - T$ is absent in the expansion because $\Delta(\partial G/\partial T)_p = 0$ at $T = T_c$, where the leading term proportional to $(T_c - T)^2$ represents a change of the curvature at T_c that corresponds physically to discontinuity in the heat capacity ΔC_p.

Although the above argument is acceptable for interpreting the second-order phase transition, observed anomalies in the critical region are by no means as simple as the above simple theory can explain. In addition, although higher-than-second-order terms may dominate the expansion of (1.11), there is no supporting evidence from practical systems so far reported in the literature. In this context, we only need to consider second-order transitions. Also, it is interesting to note that metastable states do not exist in second-order phase transitions, as characterized experimentally by the absence of *hysteresis*.

1.5.2 The Landau Theory

Typically, the second-order phase transition can be found in systems undergoing *order-disorder* phase transitions, for which the order parameter η is characterized by an *inversion* mechanism. Although defined as a scalar in an isotropic system, η, as such, can be a vector quantity if related to the directional displacement of active species in the crystalline system. Thermodynamic properties as a whole do not reflect the domain structure that arises from the intrinsic inversion mechanism, although they can respond to an applied field or stress. For example, ordered states in a binary system can be twofold under no external action due to inversion or reflection symmetry, for which thermodynamic properties represented by the Gibbs potential are invariant by inversion or reflection of η in the mirror plane; that is,

$$G(\eta) = G(-\eta). \tag{1.12}$$

Landau further postulated for a binary system that the Gibbs potential can be expanded into an infinite power series of η, which is expressed as

$$G(\eta) = G_o + \tfrac{1}{2}A\eta^2 + \tfrac{1}{4}B\eta^4 + \tfrac{1}{6}C\eta^6 + \ldots, \tag{1.13}$$

where $G_o = G(T_c)$ and the coefficients A, B, C ... are normally smooth functions of temperature. Here, it is noted that odd-power terms are not included in the expansion, because of the invariance requirement (1.12).

Emerging at T_c, the magnitude of η is infinitesimal at temperatures close to T_c, and the series expansion in (1.13) can be truncated at the quartic term $\frac{1}{4}B\eta^4$ for the critical region without losing accuracy. Therefore, at near T_c, the order parameter can be determined by minimizing the truncated Gibbs potential

$$G(\eta) = G_o + \tfrac{1}{2}A\eta^2 + \tfrac{1}{4}B\eta^4 \tag{1.12a}$$

In this case, the order parameter in thermal equilibrium can be determined from

$$\frac{\partial G}{\partial \eta} = A\eta + B\eta^3 = \eta(A + B\eta^2) = 0,$$

yielding simple solutions; i.e., either

$$\eta = 0, \tag{1.14a}$$

or

$$\eta = \pm\left(-\frac{A}{B}\right)^{1/2} \tag{1.14b}$$

near the critical temperature. The solution $\eta = 0$ of (1.14a) is the value of the order parameter in the disordered state above T_c, where the minimum of $G(\eta)$ occurs at $\eta = 0$, as characterized by $A > 0$ and $B = 0$ at T_c. On the other hand, the other solution (1.14b) can be real if $A < 0$ and $B > 0$ and assigned to the ordered phase below T_c.

In the disordered phase, the Gibbs potential is expressed by $G(\eta) = \frac{1}{2}A\eta^2$, where the equilibrium is given by $\eta = 0$, giving stability against fluctuating order, which is warranted by a positive A. In contrast, the factor A changes to a negative value below T_c from positive above T_c, so that the equilbrium is no longer at $\eta = 0$, shifting to $\eta_o = \pm(-A/B)^{1/2}$ determined with a positive potential $\frac{1}{4}B\eta^4$ that emerges at T_c. As will be discussed later, such a quartic potential is related to correlations among microscopic order variables.

For the coefficient A changing signs at T_c, Landau has given the expression

$$A = A'(T - T_c) \quad \text{where} \quad A' > 0, \tag{1.15}$$

giving $A > 0$ and $A < 0$ for $T > T_c$ and $T > T_c$. On the other hand, the coefficient B is zero and positive in the ranges above and below T_c, respectively. Accordingly, the equilibrium value of the order parameter η_o is expressed as

$$\eta_o = 0 \quad \text{for } T > T_c, \tag{1.16a}$$

and

$$\eta_o = \pm\{(A'/B)(T_c - T)\}^{1/2} \quad \text{for } T < T_c. \tag{1.16b}$$

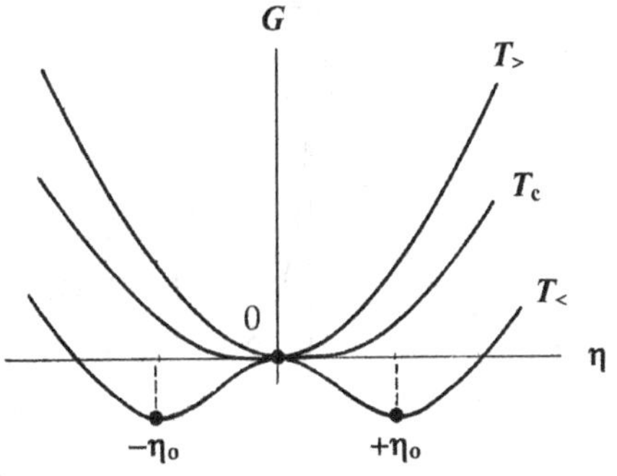

Fig. 1.8. Changes of the Gibbs potential $G(\eta)$ in the vicinity of T_c: from a parabolic above T_c to a double-well potential below T_c, where the equilibrium is specified by $\eta = 0$ and $\eta = \pm\eta_o$, respectively.

In Fig. 1.8, such Gibbs potential curves are sketched as a function of η in the vicinity of T_c, showing that the equilibrium is at $\eta = 0$ above T_c, but shifting continuously to $\pm\eta_o$ with decreasing temperature below T_c.

If a system is regarded macroscopically as homogeneous, the internal interactions can be evaluated with the mean-field approximation. Nevertheless, as a consequence of binary symmetry (1.12), the ordered phase is spontaneously separated into two *domains* below T_c of equal volume in most cases, behaving like two phases of opposite polarizations. Although properties of domains, as represented by the Gibbs potentials $G(\pm\eta_o)$ are identical, we cannot explain thermodynamically how an intermingling state of opposite sublattices occurs as alternative cases. In addition, it is significant that these oppositely polarized domains are transformable from one to another by an external field or stress. As discussed later for a magnetic system, such a transformation is analogous to vapor-liquid transitions that occur by compressing or decompressing the vapor. Such a conversion is first order and *irreversible*, as interactions with *extrinsic* agents such as lattice defects cause energy loss during a hysteresis cycle. When the order parameter is a *responsive* variable to an external action, the domain conversion can be utilized for testing theories in single-domain samples.

The parabolic temperature-dependence expressed by (1.16b) is a consequence of the mean-field hypothesis. In the critical region, the observed temperature dependence of η_o was not quite as described by (1.16b) but often close to parabolic toward T_c. Nevertheless, it is a usual practice to employ the thermodynamical approach where observed deviations from mean-field predictions can be analyzed with an empirical temperature dependence

$$\eta_o \propto (T_c - T)^\beta.$$

Here, the exponent β represents a deviation from the mean-field value $\frac{1}{2}$. In contemporary theory [10], the value of β is attributed to the dimensionality of order variables, although its physical implication is not clear. Practically,

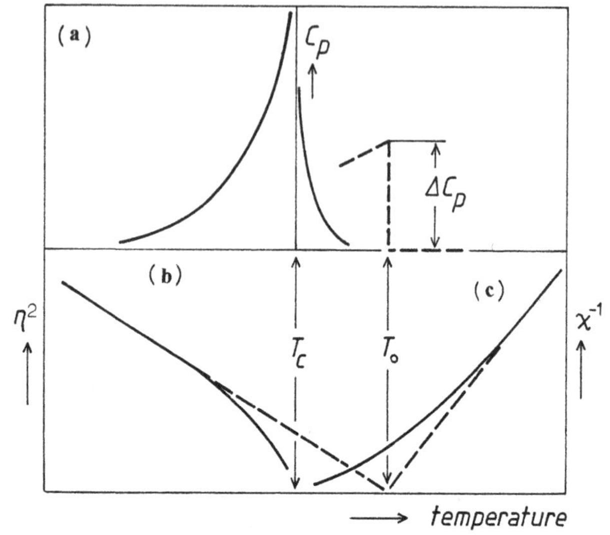

Fig. 1.9. Critical anomalies in a second-order phase transition observed (a) in the specific heat C_p (λ anomaly), (b) the squared order parameter η^2 (a deviation from a parabolic $\eta(T_c - T)$, and (c) the susceptibility (a Curie-Weiss anomaly). Here, the broken lines indicate predictions by the mean-field theory.

a curve for η_o^2 plotted against T in the range below T_c shows a significant deviation from a straight line for $\beta = \frac{1}{2}$, intersecting the base line $\eta_o = 0$ at a temperature T_o that is substantially higher than known value of T_c, as illustrated in Fig. 1.9b. In practice, the deviation $T_c - T_o = \Delta T$ is considered as a measure for criticality in the region close to T_c.

In the Landau theory, the entropy and heat capacity of an ordering system can be calculated by the relation $S = -(\partial G/\partial T)_p$, which is continuous at T_c, namely

$$S(T > T_c) = S(T_c) \quad \text{and} \quad S(T < T_c) = S(T_c) + \frac{A'^2}{2B}(T_c - T).$$

Therefore, in the limit of $T \to T_c$, we have $S(T > T_c) = S(T < T_c)$. On the other hand, the heat capacity is twofold at T_c, as seen from

$$C_p(T_c) = -\left[T\left(\frac{\partial S}{\partial T}\right)_p\right]_{T=T_c} = 0 \quad \text{whereas} \quad C_p(T_c) = -\frac{A'^2 T_c}{2B},$$

as calculated with the above S for $T > T_c$ and for $T < T_c$, respectively. Therefore, the discontinuity in the specific heat is given by $\Delta C_p = A'^2 T_c/B$ at T_c in the mean-field approximation.

However, such a discontinuity ΔC_p as shown in Fig. 1.9a has never been observed in any continuous phase transitions, indicating that the Landau theory fails in the critical region. Figure 1.10 is a typically observed heat-capacity

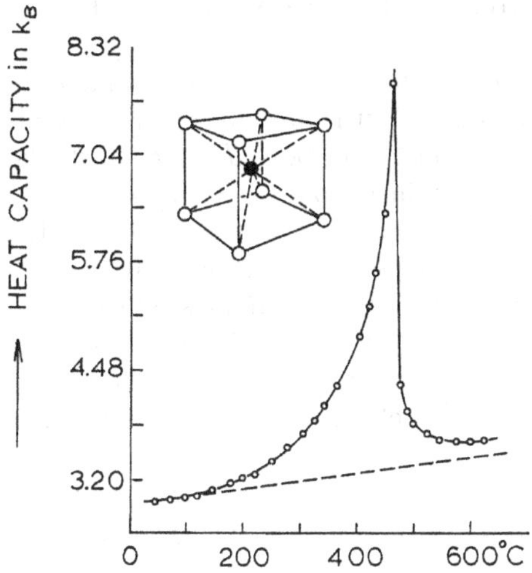

Fig. 1.10. The heat capacity of β-brass, showing a typical λ anomaly in the vicinity of T_c. (From F. C. Nix and W. Shockley, Rev. Mod. Phys. 10, 1 (1938).)

curve, characterized by a sharp rise, as T_c is approached from above, followed by a gradual decrease with decreasing temperature. Resembling the Greek letter λ, such a specific heat curve is considered as characteristic of order-disorder systems, e.g., β-brass [11]. However, lacking of a proper explanation, observed λ-anomalies can also be expressed by another set of exponents;

$$C_p(T > T_c) \propto (T - T_c)^{-\alpha} \quad \text{and} \quad C_p(T < T_c) \propto (T_c - T)^{-\alpha'}.$$

These empirical exponents, α, α′, and β together with others for response functions, constitute the basis of the scaling theory.

The thermodynamical theory with the truncated Gibbs potential is valid for a weakly ordered state except in the critical region. Although the quartic term is essential below T_c, its implication is limited to mean-field accuracy. The transition is continuous at T_o in the mean-field theory, but, practically, it is not as simple as implied by the Landau theory. In the noncritical region below T_c, the long-range order dominates the ordering process, where it is nonlinear, as expressed by higher terms in the Landau expansion (1.13). In practice, the presence of anomalies make it uncertain whether such a transition is continuous or discontinuous. However, as Blinc and Zeks [12] have pointed out, the Landau expansion truncated at the term $C\eta^6$, for example, may lead to a first-order transition if $B < 0$ and $C > 0$. Apart from the physical interpretation of these coefficients, their argument is not inconsistent with the present one. If the order parameter emerges abruptly at T_c with a finite magnitude, the transition should be of first order in any case.

1.6 Susceptibilities and the Weiss Field

Depending on the physical nature, the order parameter may interact with an applied field or stress F. If such an external F is applied, the crystal can be ordered to some extent even above T_c, for which the Gibbs potential is no longer invariant under inversion, i.e. $G(\eta) \neq G(-\eta)$, as signified by the interaction term $\mp\alpha\eta F$, where α is a positive constant. In this case domain volumes can be unequal, depending on the strength of F, and the transition between two domains cannot be continuous. On the other hand, if $F = 0$, the function $G(\eta)$ is invariant for inversion and the transition is continuous, where domains for $\pm\eta$ occupy exactly each one-half of the crystal volume. In this section we discuss the effect of a weak F for a small value of η in the region close to T_c.

These $\pm\eta$ can be assigned to domains under F, which are approximately related by inversion as in the Landau theory, and we can consider the susceptibility for expressing the linear response of domains to the applied F. On the other hand, for a finite magnitude of η, Weiss considered that the internal field F_{int} is essential for the singular behavior of the susceptibility, from which the transition temperature T_c can be determined.

1.6.1 Susceptibility of an Order Parameter

In the presence of F, we can write the Gibbs potentials for two domains as

$$G(\pm\eta) = G_o + \tfrac{1}{2}A\eta^2 + \tfrac{1}{4}B\eta^4 \mp \alpha\eta F, \tag{1.17}$$

which is truncated at η^4 for small η, considering F as a sufficiently weak perturbation. Obviously, the system of dipolar η is forced to be ordered by F to some extent, even though no correlations can be effective above T_c. Accordingly, the transition becomes *diffuse* in a system under F, where no clear-cut transition temperature can be found. Nevertheless, the equilibrium under a given F can be determined by minimizing $G(\pm\eta)$ of (1.17) with respect to η; namely the equation

$$\frac{\partial G(\pm\eta)}{\partial\eta} = A(\pm\eta) + B(\pm\eta)^3 \mp \alpha F = 0$$

should be solved for the equilibrium value of η. Ignoring further the term $B\eta^3$ for a small η, we can immediately obtain expressions for the response at temperatures very close to T_c; that is,

$$\chi = \alpha\eta/F = (\alpha^2/A')/(T - T_c) \quad \text{and} \quad \chi(\alpha^2/A')/(T_c - T) \tag{1.18}$$

for $T > T_c$ and $T < T_c$, respectively, which are known as the Curie-Weiss law. The susceptibility χ goes to infinity as T_c is approached, which should be observed with a very small applied field $F \approx 0$, thereby identifying the transition as second order. On the other hand, the responses below and above T_c should be different in principle, because of the correlation term B below T_c that was ignored in the above derivation of (1.18). In addition, in the critical region, the observed χ does not obey the Curie-Weiss law, as shown in Fig. 1.9c, where the failure of the mean-field approximation is evident. Empirically observed $1/\chi$ can be expressed by critical exponents γ and γ' as

$$1/\chi \propto (T - T_c)^{\gamma} \quad \text{and} \quad 1/\chi \propto (T_c - T)^{\gamma'},$$

in the regions above and below T_c, respectively. The exponents γ and γ' are both equal to 1 in the mean-field approximation.

1.6.2 The Weiss Field in a Ferromagnetic Domain

A real gas condenses at temperatures below T_c, which is related to the molecular constants a and b, according to the van der Waals theory. The transformation between vapor and liquid at a constant $T < T_c$ can be performed by an external work (1.7). For a binary phase transition, the Landau theory gives only an approximate description, leaving the origin of the transition temperature T_c unexplained. For a magnetic system, Weiss introduced the concept of *molecular field* to express the average magnetic interactions in a crystal, which was considered as responsible for a singular behavior of the susceptibility. He postulated the presence of an internal field expressed by

$$\boldsymbol{B}_{\text{int}} = \lambda \boldsymbol{M} \tag{1.19a}$$

in a uniformly magnetized magnet, where the magnetization $\boldsymbol{M}$ is the order parameter, and λ is called the Weiss constant. Later Heisenberg proposed theoretically that microscopic magnetic interactions originate from a quantum-mechanical exchange mechanism between adjacent magnetic ions in the crystal. According to his theory, such magnetic interactions are expressed as the correlation energy $-\sum_{\text{i}}\sum_{\text{j}} J_{\text{ij}}\boldsymbol{s}_{\text{i}}.\boldsymbol{s}_{\text{j}}$, where J_{ij} is the exchange integral between two ions i and j, representing the magnitude of correlation between the spins $\boldsymbol{s}_{\text{i}}$ and $\boldsymbol{s}_{\text{j}}$. Writing this expression as $-\sum_{\text{i}} \boldsymbol{s}_{\text{i}}.(\sum_{\text{j}} J_{\text{ij}}\boldsymbol{s}_{\text{j}})$, the sum $\sum_{\text{j}} J_{\text{ij}}\boldsymbol{s}_{\text{j}}$ can be interpreted as the instantaneous local field $\boldsymbol{B}_{\text{int}}(\text{i})$ at the spin $\boldsymbol{s}_{\text{i}}$. We can therefore define the spatial average $\langle \boldsymbol{B}_{\text{int}}(\text{i})\rangle_{\text{s}} = \langle \sum_{\text{i}}(\sum_{\text{j}} J_{\text{ij}}\boldsymbol{s}_{\text{j}})/\text{N}\rangle_{\text{s}} = \lambda\langle \boldsymbol{M}_{\text{i}}\rangle_{\text{s}}$, where N is the total number of spins, as the internal magnetic field $\boldsymbol{B}_{\text{int}}$ in the mean-field approximation. Using the macroscopic magnetization $\boldsymbol{M}$ defined by $\langle \boldsymbol{M}_{\text{i}}\rangle_{\text{s}}$, the Weiss field can be expressed as $\boldsymbol{B}_{\text{int}} = \lambda \boldsymbol{M}$. In fact, this is the expression of $\boldsymbol{B}_{\text{int}}$ including the long-range contributions, although the Heisenberg formula is originally for spin correlations in short ranges. The internal field $\boldsymbol{B}_{\text{int}}$ can be expressed more generally by a power series of $\boldsymbol{M}$ as in the Landau theory.

In the presence of an applied field $\boldsymbol{B}_{\text{o}}$, the spin $\boldsymbol{s}_{\text{i}}$ is considered to be in the effective field $\boldsymbol{B}_{\text{o}} + \boldsymbol{B}_{\text{int}}$, so that the Weiss relation (1.19a) can be modified as

$$\boldsymbol{M} = \chi_{\text{o}}(\boldsymbol{B}_{\text{o}} + \boldsymbol{B}_{\text{int}}) \tag{1.19b}$$

where χ_o is the paramagnetic susceptibility that obeys the Curie law

$$\chi_o = C/T \qquad (C \text{ is the Curie constant}),$$

giving the basic response from uncorrelated spins. Nevertheless, combining these relations, the magnetic susceptibility can be expressed by

$$\chi_o = M/B_o = C/(T - C\lambda) = C/(T - T_o), \tag{1.20}$$

which is the Curie-Weiss law for $T > T_o$, where $T_o = C\lambda$ represents the transition temperature in the mean-field application. Thus, the linear Weiss field (1.19a) may be considered as responsible for the singular behavior of magnetization $\boldsymbol{M}$ at $T = T_o$.

Although logical, the Weiss field $\boldsymbol{B}_{\text{int}}$ defined by (1.19a) and (1.19b) may remain as a conjecture, unless substantiated in experiments. In ferroelectric crystals, an internal electric field can be considered by analogy, which was in fact detected using polar molecular probes. In this context, the Weiss field should be considered as real in a magnetized crystal. As related to a nonzero constant λ, T_o can be interpreted as related to a minimum spin cluster combined with short-range correlations, which is analogous to a liquid droplet for initial condensation of a gas. In Chapter 3, we will consider that second-order phase transitions are initiated with such ordered clusters formed with minimum correlations.

It is significant that the Weiss fields, $+\boldsymbol{B}_{\text{int}}$ and $-\boldsymbol{B}_{\text{int}}$, are related to $+\boldsymbol{M}$ and $-\boldsymbol{M}$ in opposite domains, respectively, whereas the external field $\boldsymbol{B}_o$ is applied to the whole crystal. Therefore, the Gibbs potentials of these domains in $\boldsymbol{B}_o$ can be written as

$$G_+ = G_o - (+\boldsymbol{M}).(+\boldsymbol{B}_{\text{int}} + \boldsymbol{B}_o) \quad \text{and} \quad G_- = G_o - (-\boldsymbol{M}).(-\boldsymbol{B}_{\text{int}} + \boldsymbol{B}_o),$$

where G_o is the potential for $\boldsymbol{B}_o = 0$ and, hence,

$$G_+ - G_- = -2\boldsymbol{M}.\boldsymbol{B}_o. \tag{1.21}$$

Analogous to (1.8) for condensation, relation (1.21) describes a "phase transition" between two domains of magnetization $\boldsymbol{M}$, which can be performed by the external work $-\boldsymbol{M}.\boldsymbol{B}_o$. Depending on $\boldsymbol{M}$, a magnetized body can therefore become to a single domain under a sufficiently strong applied field $\boldsymbol{B}_o$. In fact, (1.21) expresses the empirical fact that the total energy for domain transformation consists of $-\boldsymbol{M}.\boldsymbol{B}_o$ minus $-(-\boldsymbol{M}).\boldsymbol{B}_o$ on these domains, if assuming that no energies are required to move domain walls. It is significant that domains can be switched in *soft* magnets by applying an external field $\boldsymbol{B}_o$, although difficult in *hard* (*permanent*) magnets where domains are locked in the crystal by lattice defects. At the critical temperature T_c, the transition is thermodynamically second-order because $G_+ = G_-$ if $\boldsymbol{B}_o = 0$. In contrast, for $T < T_c$ the domain conversion is a first-order transition, as shown in Fig. 1.11 for an idealized ferromagnet.

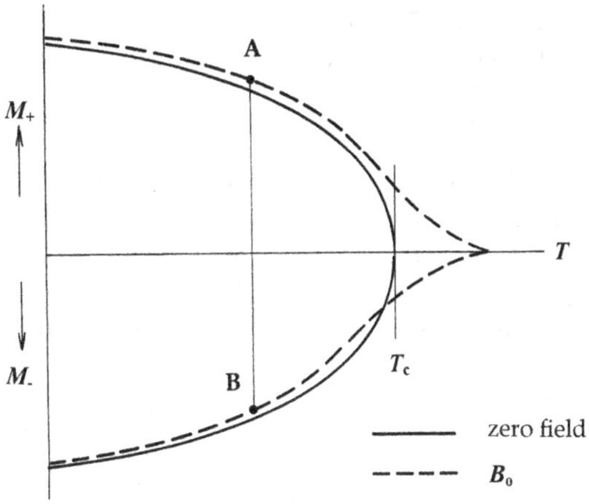

Fig. 1.11. Comparison of magneitization curves of a soft ferromagnet without an external field and with a weak applied field B_o. The transition is generally sharp with $B_o = 0$, but becomes diffuse with $B_o \neq 0$. Varying B_o magnetic domains behave like two phases in a first-order equilibrium, converting between A and B.

1.7 Critical Anomalies, Beyond Classical Thermodynamics

In the critical region of a second-order phase transition, observed anomalies cannot be explained by thermodynamical principles, as indicated by appreciable deviations from the mean-field theory in all experimental results. We can attribute these anomalies to correlated order variables [13], in spite of unknown dynamical origin for fluctuations. However, at this stage, it may be worth speculating about what could be responsible for critical anomalies before proceeding with arguments beyond the limit of classical thermodynamics.

Corresponding to thermodynamic equilibrium at the minimum of the Gibbs potential, it is logical to consider internal fluctuations that may arise from inside the system in equilibrium with the surroundings at given p and T. It is noted that the order parameter η at parabolic minima of the Gibbs potential $G(\eta)$ can be subjected to a harmonic motion rather than random fluctuations. In addition, such a change in η should be described in terms of a space-time variation in the lattice structure, where ΔG can be expressed as a function of $\delta\eta = \eta - (0, \pm\eta_o)$. For a small deviation $\delta\eta$, we can write

$$\Delta G(\delta\eta) = G(\eta) - G(0) = \tfrac{1}{2}A\delta\eta^2, \quad A = A'(T - T_o) \qquad \text{for } T > T_o,$$

and

$$\Delta G(\delta\eta \pm \eta_o) = \tfrac{1}{2}A(\delta\eta \pm \eta_o)^2, \quad A = A'(T_o - T) \qquad \text{for } T < T_o.$$

Corresponding to these excitations ΔG, the kinetic energy $\frac{1}{2}m(\mathrm{d}\delta\eta/\mathrm{d}t)^2$ can be considered, where m is the effective mass. Therefore, we may define the characteristic frequency by $\varpi = (A/m)^{1/2}$ of the oscillator. Using Landau's expression for A, the frequency of such harmonic fluctuations can be expressed as

$$\varpi_{>} \propto (T - T_o)^{1/2} \quad \text{and} \quad \varpi_{<} \propto (T_o - T)^{1/2}, \tag{1.22}$$

for $T > T_o$ and $T < T_o$, signifying the phase transition by these softening frequencies as $\varpi_{>}$ and $\varpi_{<}$, respectively. In fact, such fluctuations can be related to the interaction with the lattice undergoing a symmetry change, for which Cochran has proposed *soft modes* (See Chapter 4). Consider that the average of ΔG over a short timescale t_o of observation can be detected; that is

$$\langle \Delta G \rangle_t = t_o^{-1} \int_o^{t_o} \Delta G dt, \tag{1.23}$$

is nonzero and detectable if $\varpi t_o < 1$, and otherwise vanishes for a long t_o. The softening frequency ϖ can become very low, when T_o is approached, so that the characterstic time $\tau = 2\pi/\varpi$ of fluctuations becomes competitive with t_o, giving rise to detectable $\langle \Delta G \rangle_t$ [14]. Such a nonzero avarage $\langle \Delta G \rangle_t$ can be considered for critical anomalies, in which the spatial profile of fluctuations should be explicit. Consequently, the crystal appears to be inhomogeneous, when the temporal fluctuations are slowed down.

Landau [7] recognized such nature of critical fluctuations and described the spatial inhomogeneity in terms of distributed Gibbs potentials,

$$\langle \Delta G \rangle_t = \sum\nolimits_i \langle \Delta g_i \rangle_t,$$

which represents the average of local $\langle \Delta g_i \rangle_t$ at a position i. At each position i, he considered a small but sufficiently large volume to consider a meaningful macroscopic average, thereby the substance is regarded as inhomogeneous, being specified by local pressure p_i and temperature T_i. For a nonzero Δg_i, these p_i and T_i should be different from p and T of the surroundings, and we can write the relation

$$\Delta g_i = g_i(p_i, T_i) - g_i(p, T) \geq -(T_i - T)\Delta s_i + (p_i - p)\Delta v_i,$$

where Δs_i is a change in the local entropy and Δv_i is the corresponding small change in volume at i. Here, the inequality expresses that such a process for local entropy production is *irreversible.* Judging from an observed symmetry change at the transition, a lattice excitation should be involved in such an intrinsic mechanism for entropy production.

At this point, it should be emphasized that critical fluctuations in the second-order phase transitions are not in the same category as random thermodynamic fluctuations that are caused generally by fluctuating external variables, Δp and ΔT. In normal states of a crystal, such thermodynamic fluctuations are so small that its thermal properties are well described by *ergodic* averages, whereas critical fluctuations, in contrast, should be associated with the spatial deformation of the lattice, which is not ergodic, thereby making the system mechanically inhomogeneous. According to recent investigations, critical fluctuations are by no means of random type but are found to be sinusoidal in character.

1.8 Remarks on Critical Exponents

In Landau's thermodynamical interpretation, critical anomalies can be expressed by the spatial average of distributed Gibbs potentials locally deviated from equilibrium. Signified by a long timescale τ, the critical fluctuation is so slow in the region close to T_c that the spatial profile becomes explicit if observed in the timescale $t_o \leq \tau$. Under the circumstances, it is not possible to describe the anomalies in thermodynamical terms, because they are related to the deformed lattice in the critical region, as will be discussed later in Chapter 3. Nevertheless, observed anomalies of η, χ and C_p are empirically analyzed in thermodynamic terms with critical exponents on $\Delta T = T_c - T$, as expressed by

$$\eta \propto (\Delta T)^{\beta},$$
$$\chi_{>} \propto (-\Delta T)^{-\gamma}, \quad \chi_{<} \propto (\Delta T)^{-\gamma'},$$

and

$$C_{p>} \propto (-\Delta T)^{-\alpha}, \quad C_{p<} \propto (\Delta T)^{-\alpha'}.$$

These exponential expressions are intended to deal with anomalies within the framework of thermodynamics, for which no rigorous justification can be made, hence remaining hypothetical. By hypothetical, we mean that a leading mechanism is to be sought for prevailing phase transitions. Although the physical implication is not clear, these formula are simple enough to express deviations from mean-field predictions with these critical exponents. Nevertheless, there are some universal relations among these exponents in various systems, constituting the basis of the *scaling theory* for phase transitions.

In the scaling theory, the spatial inhomogeneity is scaled down to renormalized units by grouping microscopic variables at lattice sites, so that the dimensionality of ordering can be taken into account in principle, similar to the short-range interactions for clusters in a given anisotropic crystals (See Chapter 3). By doing so, the system can be regarded as quasi-uniform, allowing to describe phase transitions thermodynamically. On the other hand, we consider collective modes of variables that arise from ordered clusters in short ranges, whereas the scaling approach appears to be a little too naïve to deal with anisotropic correlations in practical crystals. Instead of relying on the mathematical hypothesis, we prefer to consider the crystallographic model of short-range correlations as an appropriate alternative in our discussions on structural phase transitions. The scaling theory constitutes a highly topical theoretical objective in modern statistical physics, although we do not discuss it in this monograph.

2

Order Variables, Their Correlations and Statistics: the Mean-Field Theory

2.1 Order Variables

In Chapter 1, we defined the order parameter as a macroscopic variable that signifies phase transitions. Originating from microscopic variables σ_m attached to ions or molecules *active* at lattice sites m, their *ensemble average* can be considered to represent the macroscopic order parameter η. Needless to say, such an average is meaningful only if the system is regarded as sufficiently uniform. In addition, if varying as functions of space-time coordinates at a long wavelength, such variables σ_m may not be subjected to statistical averaging in an inhomogeneous state of crystals.

In a disordered phase above T_c, these σ_m are generally in fast random motion, so that the time average $\langle\sigma_m\rangle_t$ vanishes at each lattice site. In contrast, below T_c, these variables are in slow correlated motion, where $\langle\sigma_m\rangle_t$ averaged over the timescale of observation may take a variety of values distributed over the crystal. Furthermore, at temperatures close to T_c, the crystal is not fully ordered and topologically inhomogeneous, as illustrated in Fig. 2.1, leading to either domains or a sublattice structure with decreasing temperature. Therefore, the ensemble average is valid only if calculated at least for such a subsystem, instead of the whole crystal. Needless to say, observed results should be so interpreted as related to the observing condition.

Although the active group should be identified in a given system, the variable σ_m is not always evident from the chemical formula or unit-cell structure, except for a few simple cases. Pending identification of σ_m, as is often the case, one has to investigate their dynamical behavior in the critical region. For structural phase transitions, these variables σ_m in collective motion play a significant role, constituting a main objective in our studies. We shall hereafter call microscopic σ_m the *order variable* to distinguish it from the corresponding macroscopic order parameter η.

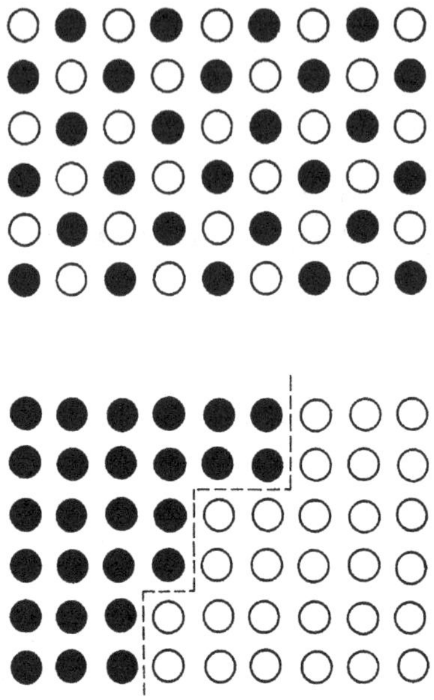

Fig. 2.1. Schematic ordered phases of a binary system in two dimensions, where states of constituents are shown by open and filled circles: (a) ordered phase consisting of intermingling sublattices, and (b) ordered phase of two "opposite" domains. Domain boundaries are shown by broken lines.

In the statistical approach, the relation between η and σ_m may be written in the spatial average

$$\eta = \sum_m \langle \sigma_m \rangle_t / N \tag{2.1a}$$

of $\langle \sigma_m \rangle_t$ in a subsystem of N lattice sites, provided that σ_m are uncorrelated or only weakly correlated. On the other hand, when locally correlated, the time average should be first calculated for a cluster of correlated σ_m, which is then averaged over the lattice space in the subsystem; namely,

$$\eta = \langle \eta_i \rangle_s = \sum_i \eta_i / N' \quad \text{where} \quad \eta_i = \left\langle \sum_{\text{cluster}} \sigma_m \right\rangle_t , \tag{2.1b}$$

and N' the number of clusters. Although unspecified, if such correlated clusters are predominant, the state of the subsystem is regarded as thermodynamically uniform, as postulated in the renormalization group theory. Nevertheless, the crystal is generally inhomogeneous during the ordering process, for which we need precise knowledge of correlated σ_m in collective motion.

Ordering processes in crystals are considerably more complex than in isotropic media, since the strained lattice plays a hidden role [15]. The collective mode of σ_m varies at a slow rate, as inferred from critical anomalies observed in experiments with different timescales. Processes in solid states are generally slow, but timescales of observation are not as seriously considered as in the critical region. In this chapter, we review existing statistical theories on binary systems, which are discussed in light of a slow variation. In solid-states, values of $\langle \sigma_{\mathrm{m}} \rangle_{\mathrm{t}}$ are usually calculated by using probabilities at sites m, which are in fact a valid concept in fast processes, where the timescale is assumed as *infinity* with the ergodic hypothesis that is the basis for statistical theories of random processes.

2.2 Probabilities, Short- and Long-Range Correlations, and the Mean-Field Approximation

2.2.1 Probabilities

In a binary alloy AB such as Cu-Zn (β-brass), spontaneous atomic ordering takes place as the temperature is lowered through T_{c}, due to diffusive atomic rearrangement among lattice sites. If such a rearrangement rate is sufficiently fast, as compared with the timescale of observation, we can interpret σ_{m} as defined by the relation

$$\sigma_{\mathrm{m}} = p_{\mathrm{m}}(\mathrm{A}) - p_{\mathrm{m}}(\mathrm{B}) \tag{2.2}$$

and

$$p_{\mathrm{m}}(\mathrm{A}) + p_{\mathrm{m}}(\mathrm{B}) = 1, \tag{2.3}$$

where $p_{\mathrm{m}}(\mathrm{A})$ and $p_{\mathrm{m}}(\mathrm{B})$ represent *probabilities* for the site m to be occupied by an atom A and by B, respectively. Subject to realistic observations, we are, in fact, uncertain whether the rearrangement process occurs at a sufficiently fast rate. Considering thermal rearrangements of individual atoms, the process can be sufficiently fast, but very slow if they are in collective motion. Nevertheless, sufficiently fast rearrangements and a long timescale of observation are assumed for traditional statistical theories to support the probably concept.

In a disordered phase where atoms are uncorrelated, these probabilities can take only two values, either 1 or 0: for example, if the site m is occupied by an atom A, $p_{\mathrm{m}}(\mathrm{A}) = 1$, otherwise $= 0$. In this case, we can also write $p_{\mathrm{m}}(\mathrm{B}) = 0$ or 1, referring to an atom B at a site m. These probabilities in a disordered state are independent of m if all lattice sites are occupied by either A or B, and no vacant sites in the crystal. Below T_{c}, on the other hand, due to atomic correlations, different sites m and n are not independently occupied, for which the probabilities $p_{\mathrm{m}}(\mathrm{A})$ and $p_{\mathrm{m}}(\mathrm{B})$ related by (2.2) and (2.3) can take virtually any continuous values between 1 and 0, because of various atomic arrangements in the neighboring sites around the site m. Accordingly, these probabilities and variable σ_{m} can be considered as continuous functions of space-time coordinates, which are called *classical variables* to distinguish them from quantum-mechanical variables such as *spins*. It is noted that quantum-mechanical variables are characterized by discrete values if uncorrelated, although behaving like classical variables if they are heavily correlated.

Signified by such probabilities, order variables σ_m in slow motion are virtually quasi-static, being distributed over lattice sites, although varying at a sufficiently fast rate for time averaging. The order parameter η can then be calculated by (2.1a) as a spatial average called the *mean-field average*. Needless to say, such a mean-field average is meaningful only if the spatial *variance* for distribution is sufficiently small. The validity of such an order parameter is evaluated by the *correlation function* defined by

$$\Gamma(r_{mn}) = \langle(\sigma_m - \eta)(\sigma_n - \eta)\rangle = \langle\sigma_m\sigma_n\rangle - \eta^2\delta_{mn}, \tag{2.4}$$

where r_{mn} is the distance between σ_m and σ_n, and for the last expression we have used the relations $\langle\sigma_m\rangle = \langle\sigma_n\rangle = \eta$. Here, δ_{mn} is the Kronecker delta, whose value is 1 for $m = n$, and otherwise it is 0 for $m \neq n$. For complete disorder, $\Gamma(r_{mn}) = 0$ for all pairs for $m \neq n$, meaning that $\sigma_m\sigma_n = 0$ for no correlations.

On the other hand, the correlation function $\Gamma(r_{mn})$ is nonzero for all pairs in an ordering process, where the product $\sigma_m\sigma_n$ should be significant. Therefore, the correlation energy can be expressed as proportional to $\sigma_m\sigma_n$; that is

$$E_{mn} = -J_{mn}\sigma_m\sigma_n, \tag{2.5}$$

where the coefficient J_{mn} is a function of the distance r_{mn}, representing the magnitude of correlation between σ_m and σ_n. For convenience, the negative sign attached to (2.5) expresses the stable arrangement of σ_m and σ_n, which are correlated at a lower energy. Although assumed for a correlated pair, (2.5) can also be derived directly for a simple system with short-range energies, as shown next.

Writing interaction energies between two atoms at sites m and n as $\varepsilon_{AB}(m,n)$, $\varepsilon_{AA}(m,n)$ and so forth, the short-range interaction energy E_m for σ_m can be expressed in terms of probabilities defined by (2.2) and (2.3), namely

$$\begin{aligned} E_m = \sum\nolimits_n E_{mn} = \sum\nolimits_n [&p_m(A)\varepsilon_{AA}(m,n)p_n(A) + p_m(B)\varepsilon_{BB}(m,n)p_n(B) \\ &+p_m(A)\varepsilon_{AB}(m,n)p_n(B) + p_m(B)\varepsilon_{BA}(m,n)p_n(A)], \end{aligned} \tag{2.6}$$

Considering that only interactions between nearest neighbors are essential, (2.6) can be simplified, particularly for a cubic lattice where r_{mn} are all equal, and hence the site specification (m, n) can be omitted from ε_{AB}, ε_{AA} and ε_{BB} in (2.6). Further, using (2.2) and (2.3), these probabilities can be replaced by order variables σ_m and σ_n; namely

$$p_m(A) = \tfrac{1}{2}(1 + \sigma_m), \quad p_m(B) = \tfrac{1}{2}(1 - \sigma_m)$$

and

$$p_n(B) = \tfrac{1}{2}(1 + \sigma_n), \quad p_n(A) = \tfrac{1}{2}(1 - \sigma_n).$$

Substituting these relations in $E_{\rm mn}$, we have

$$\begin{aligned} E_{\rm mn} &= \tfrac{1}{2}(2\varepsilon_{\rm AB} + \varepsilon_{\rm A} + \varepsilon_{\rm B}) \\ &\quad + \tfrac{1}{4}(\varepsilon_{\rm AA} - \varepsilon_{\rm BB})(\sigma_{\rm m} + \sigma_{\rm n}) \\ &\quad + \tfrac{1}{4}(2\varepsilon_{\rm AB} - \varepsilon_{\rm AA} - \varepsilon_{\rm BB})\sigma_{\rm m}\sigma_{\rm n} \\ &= \text{const.} - K(\sigma_{\rm m} + \sigma_{\rm n}) - J\sigma_{\rm m}\sigma_{\rm n}, \end{aligned} \tag{2.7}$$

where

$$K = \tfrac{1}{4}(\varepsilon_{\rm AA} - \varepsilon_{\rm BB}) \quad \text{and} \quad J = \tfrac{1}{4}(\varepsilon_{\rm AA} + \varepsilon_{\rm BB} - 2\varepsilon_{\rm AB}).$$

Here, the parameter J represents the magnitude of binary correlations with the nearest neighbors, corresponding to $J_{\rm mn}$ in (2.5). On the other hand, the factor K can be zero, if $\varepsilon_{\rm AA} = \varepsilon_{\rm BB}$, as in most binary systems, while $K \approx 0$ for alloys of similar atoms for which $\varepsilon_{\rm AA} \approx \varepsilon_{\rm BB}$, and hence the term of K is generally insignificant. The first constant term in (2.7) is independent of order variables, and hence insignificant. In this way, the equation (2.7) has been confirmed as essentially the same as (2.5).

2.2.2 The Concept of a Mean Field

Order variables $\sigma_{\rm m}$ in crystals were defined as statistical variables, using occupation probabilities at lattice sites, as given by (2.2). Correlations among these variables at lattice sites are basically molecular interactions in short-ranges, which can be interpreted in terms of probabilities for *like* or *unlike* arrangements of atoms, although their ranges are not specified in (2.7). Nevertheless, writing

$$E_{\rm m} = -\sigma_{\rm m}\left(\sum\nolimits_{\rm n} J_{\rm mn}\sigma_{\rm n}\right),$$

the quantity $\sum_{\rm n} J_{\rm mn}\sigma_{\rm n} = F_{\rm m}$ can be considered as the internal field acting on $\sigma_{\rm m}$ when summed over effective ranges of $r_{\rm mn}$, i.e. $E_{\rm m} = -\sigma_{\rm m}F_{\rm m}$. Taking distances $r_{\rm mn}$ for the nearest and next-nearest neighbors, the correlation energy $E_{\rm m}$ is called the *short-range interaction energy* at m. Statistically, we may proceed to calculate the time average $\langle F_{\rm m}\rangle_{\rm t}$ as the effective local field at site m below $T_{\rm c}$. Then, the spatial average of these $\langle F_{\rm m}\rangle_{\rm t}$ can be calculated over the whole subsystem to obtain the effective macroscopic field. In the mean-field approximation, such a *long-range average* can be considered as a meaningful quantity for a system that obeys thermodynamic principles. An ordered system can therefore be characterized by the presence of such a macroscopic internal field $F = \langle F_{\rm m}\rangle_{\rm s} = \sum_{\rm m}\langle F_{\rm m}\rangle_{\rm t}/\mathbf{N}$ in mean-field approximation.

We can also express such probabilities by long-range averages, that is

$$p(\text{A}) = \langle p_{\rm m}(\text{A})\rangle_{\rm s} \quad \text{and} \quad p(\text{B}) = \langle p_{\rm m}(\text{B})\rangle_{\rm s},$$

for which the relation $p(\text{A}) + p(\text{B}) = 1$ holds at all sites. For a binary system, the order parameter can be defined for the two subsystems as

$$\eta_1 = \eta = p(\text{A}) - p(\text{B})$$

and

$$\eta_2 = -\eta = p(\mathrm{B}) - p(\mathrm{A}).$$

It is noted in general that $1 \geq \eta_1 \geq 0$ and $0 \geq \eta_2 \geq -1$, where $1 \geq p(\mathrm{A})$, $p(\mathrm{B}) \geq 0$. For complete disorder, $\eta_1 = \eta_2 = 0$, and so $p(\mathrm{A}) = p(\mathrm{B}) = \frac{1}{2}$. On the other hand, for complete order, $\eta_1 = 1$ and $\eta_2 = -1$, which correspond to $p(\mathrm{A}) = 1$, $p(\mathrm{B}) = 0$ and $p(\mathrm{B}) = 1$, $p(\mathrm{A}) = 0$, respectively.

If long-range correlations are significant, the binary ordering in an alloy AB can simply be interpreted in terms of average probabilities $p(\mathrm{A})$ and $p(\mathrm{B})$ in the mean-field accuracy. If $J > 0$, two attracting *unlike* atoms at shortest distances lower the interaction energy by $-J$, whereas repelling *like* pairs are unstable by the amount $+J$ in the same domain. On the other hand, two intermingling sublattices can be stabilized in anti-ordered crystal. In the former case, considering only nearest neighbors, the average number of interacting A-B and B-A pairs can be expressed by $2\mathbf{N}zp(\mathrm{A})p(\mathrm{B})$, where z is the number of lattice sites in the shortest distance and $\mathbf{N} = \frac{1}{2}N$ is the total number of sites in each domain. Therefore, the number of unlike pairs and the corresponding interaction energy are given by

$$N_{\mathrm{AB}} = 2\mathbf{N}p(\mathrm{A})p(\mathrm{B}) = \tfrac{1}{2}Nz(1-\eta^2)$$

and the total ordering energy is

$$E = E_1 + E_2 = \text{const.} + 2J\{\tfrac{1}{2}Nz(1-\eta^2)\} = \text{const.} + \tfrac{1}{2}NzJ(1-\eta^2), \quad (2.8)$$

which are consistent with $N_{\mathrm{AB}} = \frac{1}{2}Nz$ and $E = \text{const.} + \frac{1}{2}NzJ$ in the disordered state for $\eta = 0$. It is interesting to note that in complete order, $\eta = \pm 1$ determined from $N_{\mathrm{AB}} = 0$, $E = \text{const.}$, and the energy difference between ordered and disordered states, i.e. $-\frac{1}{2}NzJ$, represents the amount of macroscopic energy lowered from the disordered state. During the ordering process, the energy resulted from partial order is therefore given by the η-dependent term in (2.8), i.e.

$$\Delta E = -\tfrac{1}{2}\mathbf{N}zJ\eta^2, \quad (2.9)$$

which is negative in both domains, representing the correlation energy averaged in the mean-field approximation. Bragg and Williams used (2.9) for their statistical theory of binary alloys, as outlined in Section 2.3.

In the mean-field approximation, mutual interaction energies are averaged in space of the entire system, which is represented by a single internal variable η. Although inadequate for the critical region, the dynamical response to an applied field or stress F can be estimated as due to the energy $-\alpha\eta F$, where α is a constant. By analogy with the Weiss field in a ferromagnet, we rewrite (2.9) by using the effective internal field F_{int} as

$$\Delta E = -\alpha\eta F_{\mathrm{int}},$$

where

$$F_{\mathrm{int}} = \left(\frac{NzJ}{2\alpha}\right)\eta \quad (2.10)$$

with the factor α used for adjusting units. Like the magnetic Weiss field $\boldsymbol{B}_{\text{int}} = \lambda \boldsymbol{M}$, the field F_{int} is not directly measurable under normal circumstances. However, it is significant that such F_{int} can be combined with an external field F, when dealing with the response of order variables to F. In this context, the Weiss field is not a mere theoretical concept, but representing a real internal field in an ordered phase. In fact, as will be discussed in Chapter 9, the internal electric field F_{int} in some ferroelectric crystals was detected by dipolar paramagnetic probes in magnetic resonance experiments.

2.3 Statistical Mechanics of an Order-Disorder Transition

The long-range order is a concept first introduced by Bragg and Williams in their statistical theory of binary alloys. They considered that thermal properties of a partially ordered alloy can be specified by the order parameter η and the macroscopic correlation energy $-E(\eta)$, postulating that the ordering system is a *canonical ensemble* governed by statistical principles. Statistically, correlated A-B pairs at the nearest-neighbor sites are responsible for such an ordered state, which occurs in a large number of combinations $g(\eta)$. Such a large "degeneracy" of the energy $-E(\eta)$ corresponds to the entropy $S(\eta) = \mathrm{k_B} \ln g(\eta)$ under a constant-volume condition, and we minimize the Helmholtz free energy $F(\eta) = E(\eta) - TS(\eta)$ to obtain the equilibrium value of η at a given temperature T.

In a single-domain crystal, in order for $\mathbf{N}$ lattice sites to be occupied by either A or B atoms with no vacancies, the combination number is given by

$$g(\eta) = \begin{pmatrix} \mathbf{N} \\ \mathbf{N}p(\mathrm{A}) \end{pmatrix} \begin{pmatrix} \mathbf{N} \\ \mathbf{N}p(\mathrm{B}) \end{pmatrix} = \begin{pmatrix} 1 \\ \frac{1}{2}(1+\eta) \end{pmatrix} \begin{pmatrix} 1 \\ \frac{1}{2}(1-\eta) \end{pmatrix}.$$

The free energy can be expressed with the partition function

$$Z(\eta) = Z(0) g(\eta) \exp(\tfrac{1}{2} N z J \eta^2 / \mathrm{k_B} T),$$

as

$$F(\eta) = -\mathrm{k_B} T \ln Z(\eta) = E(\eta) - \mathrm{k_B} T \ln g(\eta).$$

From the condition $(\partial F/\partial \eta)_{\mathrm{V}} = 0$, we obtain

$$\frac{\partial}{\partial \eta} \left\{ \ln Z(0) + \ln g(\eta) + \frac{1}{2} \frac{N z J \eta^2}{\mathrm{k_B} T} \right\} = 0.$$

Using the Stirling formula for a large N, the term $\ln g(\eta)$ can be evaluated approximately using

$$\frac{\partial \ln g(\eta)}{\partial \eta} = -\frac{N}{2} \ln \frac{1+\eta}{1-\eta}.$$

Hence, the equilibrium order parameter at T can be determined from the equation

$$\frac{zJ}{k_B T}\eta = \ln\frac{1+\eta}{1-\eta},$$

or

$$\eta = \tanh\frac{zJ\eta}{zk_B T}. \tag{2.11}$$

Equation (2.11) can be solved graphically by finding the intersection between the straight line

$$y = \frac{zJ}{2k_B T}\eta$$

and the curve

$$\eta = \tanh y$$

in the η-y plane, as illustrated in Fig. 2.2a. Here, we note that if $2k_B T/zJ \geq 1$, only the origin $\eta = 0$ is the intersection, whereas for $2k_B T/zJ < 1$, another intersection is found at η in the range $0 < \eta \leq 1$, representing partial order with the limit of $\eta \to 1$ as complete order. In this diagram, the transition

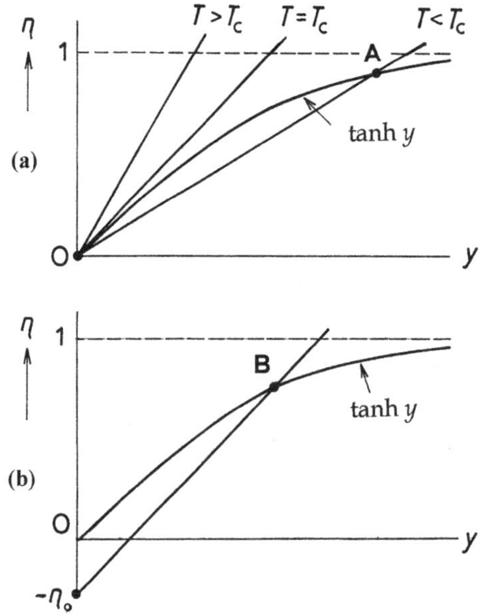

Fig. 2.2. (a) Graphical solutions for the spontaneous order parameter. The straight line $y = (zJ/2k_B T)\eta$ intersects the curve $\eta = \tanh y$ at a point A to give a real solution $\eta \neq 0$ if $T > T_c$, whereas the only intersection is $\eta = 0$, if $T < T_c$. (b) Graphical solutions for ferromagnetic order. The intersection B is always present between the straight line $y = (T/T_c)\eta - \eta_o$ and the curve $\eta = \tanh y$ below T_c, but no solution above T_c.

between disordered to ordered states can be specified as $2k_BT_c/zJ = 1$, where the unit slope at the origin gives the transition temperature T_c, namely

$$T_c = zJ/2k_B. \tag{2.12a}$$

Writing $y = (T/T_c)\eta$, from (2.11) we have

$$\frac{T}{T_c}\eta = \tanh^{-1}\eta \approx \eta + \eta^3/3$$

for a small η, from which an approximate relation

$$\eta^2 \approx \frac{3(T_c - T)}{T_c} \qquad \text{for } T < T_c \tag{2.12b}$$

can be derived. Hence for a small η, the order parameter shows a parabolic temperature-dependence, which is a consequence of the mean-field approximation.

The heat capacity for ordering can be calculated easily with the above results; that is

$$C_V = \frac{\partial E(\eta)}{\partial T} = \frac{dE}{d\eta}\cdot\frac{d\eta}{dT} = (-NzJ\eta)\left(-\frac{3}{2T_c\eta}\right) = 3Nk_B,$$

when T_c is approached closely from below. In the disordered phase, $C_V = 0$ as $\eta = 0$, and, hence, the discontinuity at T_c is $\Delta C_V = 3Nk_B$.

2.4 The Ising Model for Spin-Spin Correlations

In ferromagnetic crystals, internal magnetic interactions are quantum mechanical, and expressed by the Heisenberg exchange energy between spins $\boldsymbol{s}_m$ and $\boldsymbol{s}_n$, i.e.

$$\mathbf{H}_{mn} = -2J_{mn}\boldsymbol{s}_m.\boldsymbol{s}_n, \tag{2.13}$$

where J_{mn} is the exchange integral between unpaired electrons of magnetic ions (3d electrons in iron-group ions) at lattice sites m and n. In a *uniaxial* magnetic crystal characterized by the unique z axis, the spin vectors are primarily in *precession* at a constant frequency ω_m around the axis z, keeping the components s_{mz} constant all the time, provided that the spin-spin interactions due to the terms $s_{mx}s_{nx} + s_{my}s_{ny}$ in (2.13) are treated as perturbations. In this case, known as the *random phase approximation*, the spin-spin correlations are described by the time average $\langle \mathbf{H}_{mn}\rangle_t$ calculated over the timescale t_o of observation. Namely,

$$\langle \mathbf{H}_{mn}\rangle_t = -2J_{mn}[s_{mz}s_{nz} + \langle s_{mx}s_{nx} + s_{my}s_{ny}\rangle_t],$$

where the second term vanishes if $2\pi/\omega_{\mathrm{m}}$ and $2\pi/\omega_{\mathrm{n}}$ are both shorter than t_{o}. It is noted that if these *precessions* can be assumed at random in phase, we can write in the zero order

$$\langle \mathbf{H}_{\mathrm{mn}} \rangle_{\mathrm{t}} = -2J_{\mathrm{mn}} s_{\mathrm{mz}} s_{\mathrm{nz}}, \tag{2.14}$$

where only z components of spin vectors are significant. Known as the *Ising* model [16], (2.14) is essentially due to random phases in spin precessions, where s_{mz} can be related to probabilities for two quantum states $\pm\frac{1}{2}$ of a spin at site m, analogous to the classical binary variable for ordering. Representing occupation probabilities of the spin states, such an interpretation of spin components s_{mz} provides a useful classification of magnetic ordering in various types, antiferro-, ferri-, spiral- and other kinds of order. Although unspecified in the above, we have considered that the origin for a unique z-axis is generally attributed to the significant magnetic anisotropy in a given crystal.

Here, we have discussed the Ising model for a simplified ferromagnetic system, but the idea for such an Ising spin can be applied to other binary systems as well. For example, in binary alloys where diffusive atomic rearrangements are responsible for ordering, Ising's spins can be used for the statistical description. At a site m, the classical spin variable s_{mz} can be specified by the state

$$|m\rangle = a_{\mathrm{m}}|+\tfrac{1}{2}\rangle + b_{\mathrm{m}}|-\tfrac{1}{2}\rangle, \tag{2.15a}$$

where $|\pm\frac{1}{2}\rangle$ are the wavefunctions for an uncorrelated spin s_{m}, and the coefficients a_{m} and b_{m} are normalized as

$${a_{\mathrm{m}}}^2 + {b_{\mathrm{m}}}^2 = 1. \tag{2.15b}$$

In this case, ${a_{\mathrm{m}}}^2$ and ${b_{\mathrm{m}}}^2$ are interpreted as the probabilities for the site m to be occupied by $+\frac{1}{2}$ and $-\frac{1}{2}$ spins, i.e. $p(+\frac{1}{2})$ and $p(-\frac{1}{2})$, respectively. We can therefore define the order variable by

$$\sigma_{\mathrm{m}} = {a_{\mathrm{m}}}^2 - {b_{\mathrm{m}}}^2. \tag{2.15c}$$

Assuming nearest-neighbor interactions, the short-range interaction energy is expressed as

$$\begin{aligned} E_{\mathrm{m}} &= \sum\nolimits_{\mathrm{n}} \langle \mathrm{m,n}| < |\mathbf{H}_{\mathrm{mn}}\rangle_{\mathrm{t}}|\mathrm{m,n}\rangle \\ &= -2J\sum\nolimits_{\mathrm{n}} [{a_{\mathrm{m}}}^2{a_{\mathrm{n}}}^2\langle ++|s_{\mathrm{mz}}s_{\mathrm{nz}}|++\rangle + {b_{\mathrm{m}}}^2{b_{\mathrm{n}}}^2\langle --|s_{\mathrm{mz}}s_{\mathrm{nz}}|--\rangle \\ &\qquad + {a_{\mathrm{m}}}^2{b_{\mathrm{n}}}^2\langle +-|s_{\mathrm{mz}}s_{\mathrm{nz}}|+-\rangle + {b_{\mathrm{m}}}^2{a_{\mathrm{n}}}^2\langle -+|s_{\mathrm{mz}}s_{\mathrm{nz}}|-+\rangle], \end{aligned}$$

where we have considered $z = 8$ and $J_{\mathrm{mn}} = J$ for a cubic lattice. For spins $\frac{1}{2}$, these matrix elements are

$$\langle ++|s_{\mathrm{mz}}s_{\mathrm{nz}}|++\rangle = \langle --|s_{\mathrm{mz}}s_{\mathrm{nz}}|--\rangle = \tfrac{1}{4}$$

and

$$\langle + - |s_{mz}s_{nz}| + - \rangle = \langle - + |s_{mz}s_{nz}| - + \rangle = -\tfrac{1}{4},$$

hence

$$\begin{aligned} E_m &= -2J\sum_n [\tfrac{1}{4}(a_m{}^2a_n{}^2 + b_m{}^2b_n{}^2) - \tfrac{1}{4}(a_m{}^2b_n{}^2 + b_m{}^2a_n{}^2)] \\ &= -2J\sum_n [\tfrac{1}{4}(a_m{}^2 - b_m{}^2)(a_n{}^2 - b_n{}^2)] = -\tfrac{1}{2}\sigma_m \sum_n J\sigma_n. \end{aligned}$$

Replacing the factor $\frac{1}{2}J$ by $2J_{mn}$, we can write $E_{mn} = -J_{mn}\sigma_m\sigma_n$, which is identical to (2.5). In contrast to the spin s_{mz} with two states $|\pm\frac{1}{2}\rangle$, the binary order variable σ_m is characterized by two values of probabilities ± 1, and called the *pseudospin.*

Considering the sum $F_m = \sum_n J\sigma_n$ for the local field at a site m, and the spatial average $F_{int} = \langle F_m \rangle$ represents the the intermal field at all sites in the crystal in the mean-field approximation. Therefore,

$$\langle E_m \rangle = -\tfrac{1}{2}\mathrm{N}F_{int}\left\langle \sum_m \sigma_m \right\rangle = -\tfrac{1}{2}\mathrm{N}F_{int}\eta,$$

where

$$F_{int} = \left\langle J\sum_n \sigma_n \right\rangle = J\eta. \tag{2.16}$$

The internal ordering energy in each of two domains can then be expressed by the identical formula, that is

$$E_1(\eta) = E_2(-\eta) = \langle E_m \rangle = -\tfrac{1}{2}\mathrm{N}J\eta^2 = -\tfrac{1}{4}NJ\eta^2.$$

Accordingly, the internal energy of the whole crystal is

$$E = E_1(\eta) + E_2(-\eta) = -\tfrac{1}{2}NJ\eta^2.$$

In conventional notations, $s_{mz} = \frac{1}{2}\sigma_m$, the magnetic moment and the internal magnetic field are expressed as $\mu_m = g\beta s_{mz}$ and $B_{mz} = (2/g\beta)\langle F_m \rangle$, thereby writing $E = -MB_{int}$, where B_{int} is the Weiss field and M is the macroscopic magnetization. Although derived from magnetic spins with random phase approximation, the Ising spin σ_m defined by (2.15c) can be conveniently used to describe any binary correlations, whenever probabilities matter.

2.5 The Role of the Weiss Field in an Ordering Process

In the mean-field approximation, spontaneous ordering signified by η can be considered as induced by the Weiss internal field F_{int}, both η and F_{int} emerging at T_o, whose magnitudes increase with decreasing temperature. Although derived specifically for binary alloys, (2.11) can be modified with binary probabilities $p(\pm 1)$ for Ising's spin states ± 1 that can be expressed by the Boltzmann statistics, as shown below. Considering the internal field F_{int} of (2.10) where α is set equal to 1 for simplicity, (2.11) is written as

$$\eta = \tanh\left[\frac{zJ}{2k_BT}\eta\right] = \tanh\left(\frac{F_{int}}{k_BT}\right)$$
$$= \left[\exp\left(+\frac{F_{int}}{k_BT}\right) - \exp\left(-\frac{F_{int}}{k_BT}\right)\right] \Big/ Z,$$

where

$$Z = \exp\left(+\frac{F_{int}}{k_BT}\right) + \exp\left(-\frac{F_{int}}{k_BT}\right)$$

is the partition function for energies $\pm F_{int}$ of the Ising spin $\sigma = \pm 1$ in the field F_{int}. Hence, in the mean-field theory, the order parameter η is determined by the difference of the Boltzmann probabilities $p(+1) = Z^{-1}\exp(+F_{int}/k_BT)$ and $p(-1) = Z^{-1}\exp(-F_{int}/k_BT)$ for these states, i.e. $\eta = p(+1) - p(-1) = \langle\sigma_m\rangle_s$. It is notable that in the mean-field approximation the spatial average is determined by the thermal average of pseudospin energy in the internal field F_{int}, although F_{int} may remain as a conjecture unless supported by experimental evidence.

In a uniaxial ferromagnet, the order parameter is given by the average of pseudospins, i.e. $\eta = \langle\sigma_m\rangle_s$, and the internal energy is $-\frac{1}{2}\mathbf{N}zJ\eta^2$. Applying a magnetic field B_o, the internal energy in each domain can be written as

$$E_1(+\eta) = -\tfrac{1}{2}\mathbf{N}_1 zJ\eta^2 - \mathbf{N}_1(g\beta\eta)B_o$$

and

$$E_2(-\eta) = -\tfrac{1}{2}\mathbf{N}_2 zJ\eta^2 + \mathbf{N}_2(g\beta\eta)B_o,$$

where $\mathbf{N}_1$ and $\mathbf{N}_2$ are not equal to $\frac{1}{2}N$. Therefore, we can write the internal energies per order variable as $\varepsilon_+ = E_1(+\eta)/\mathbf{N}_1$ and $\varepsilon_- = E_2(-\eta)/\mathbf{N}_2$; that is,

$$\varepsilon_+ = -\tfrac{1}{2}zJ - g\beta B_o \quad \text{and} \quad \varepsilon_- = -\tfrac{1}{2}zJ + g\beta B_o.$$

We may consider the probabilities for these states as given by the Boltzmann statistics, i.e.

$$p(+1) = Z^{-1}\exp\left(-\frac{\varepsilon_+}{k_BT}\right) \quad \text{and} \quad p(-1) = Z^{-1}\exp\left(-\frac{\varepsilon_-}{k_BT}\right),$$

where

$$Z = \exp\left(-\frac{\varepsilon_+}{k_BT}\right) + \exp\left(-\frac{\varepsilon_-}{k_BT}\right).$$

Accordingly,

$$\frac{p(+1)}{p(-1)} = \exp\frac{zJ\eta + 2g\beta B_o}{k_BT} = \frac{1+\eta}{1-\eta},$$

where the last expression was derived from the definition of $p(\pm 1)$ in Section 2.2. From this relation, we obtain

$$\eta = \tanh\frac{\frac{1}{2}zJ\eta + g\beta B_o}{k_BT}, \tag{2.17}$$

which is the equation to be solved for η in the field B_o. Equation (2.17) can be solved for η in exactly the same manner as (2.11), by finding graphically the intersection between the straight line

$$y = \left(\frac{zJ}{2k_B T}\right)\eta + \frac{g\beta B_o}{k_B T}$$

and the curve $y = \tanh^{-1}\eta$. Writing $zJ/2k_B = T_c$ as defined in (2.12a), these are reexpressed as

$$\eta = \left(\frac{T}{T_c}\right)y - \frac{g\beta B_o}{k_B T} \quad \text{and} \quad \eta = \tanh y. \tag{2.18}$$

Figure 2.2b illustrates these intersecting lines in the η-y plane, where the straight one at $T = T_c$ intersects the η axis at $\eta_o = -g\beta B_o/k_B T_c$, which is numerically very small in practical cases. For example, for $\beta = 1$ Bohr's magneton, assuming $B_o = 3$ weber/m^2 at $T_c \sim 10^3$K in a typical ferromagnet, η_o is only of the order of 10^{-2}. It is noted that in the presence of B_o (2.18) has always a real solution at all temperatures, and hence there is no critical temperature, although η is singular at T_c if $B_o = 0$.

Above T_c, when the temperature is close to T_c, we can set $y \approx \eta$ and $T = T_c$ in (2.18), and obtain

$$\eta = \frac{T}{T_c}\eta - \frac{g\beta B_o}{k_B T_c},$$

which is the Curie-Weiss formula as written in the form

$$\chi = \frac{Ng\beta}{2B_o} \cdot \eta = \frac{C}{T - T_c}, \quad \text{where} \quad C = \frac{Ng^2\beta^2}{2k_B}.$$

While substantiated only in ferroelectric crystals, from the above discussion we have the reason to believe that the Weiss field is a real internal field in the mean-field accuracy in ordered magnetic crystals as well.

We have so far reviewed statistical theories of binary ordering with the mean-field approximation, which can be described in terms of the Ising spin σ_m located at a lattice site m. It is noted that such statistical variables defined at lattice points may not necessarily be periodic functions of the lattice if their correlations are insignificant. Furthermore, in the mean-field theory, all σ_m are represented by the spatial average $\langle\sigma_m\rangle_s$, making the crystal of a uniform substance, thereby their ordering appears to have nothing to do with the lattice structure. In this section, we have in fact shown that binary values of the $\langle\sigma_m\rangle_s$ are determined the Boltzmann probabilities, implying that the ordering is a thermal process at a given temperature. Although the role of lattice is implicit in the statistical argument, we realize that the thermal accessibility of these states should be attributed to random collisions with phonons, thereby contributing to the free energy as expressed by TdS under a constant-volume condition.

On the other hand, the failure of the mean-field theory in the critical region signifies that the ordering process is slow in the timescale of observation, where the probability cannot be a meaningful concept. Instead, in so-called *displacive* systems, collective displacements prevail in the critical region, as will be discussed in Chapter 3. Unlike the Ising spin σ_m for probabilities, such displacement vectors at lattice sites can violate translational symmetry of the lattice, generating strains if they are correlated *incommensurately* and, hence, responsible for a symmetry change at T_c. Below T_c, the free energy should, therefore, change by an internal *mechanical work* $\mathrm{d}W$, resulting in the ordered crystal with a deformed lattice.

In the lattice dynamical theory, it is known that in normal crystals, there are three independent acoustic modes at long wavelengths, in addition to a large number of high-frequency modes, representing thermal vibrations of the lattice. In this context, these lattice modes in different categories can be responsible for a change of the Gibbs free energy under given external variables p and T, where mechanical and thermal contributions are expressed by the terms $\mathrm{d}W$ and $T\mathrm{d}S$, respectively. We discuss such order variables in displacive systems in Chapter 3.

3

Collective Modes of Pseudospins in Displacive Crystals and the Born-Huang Theory

3.1 Displacive Crystals

In a stable crystal, constituent ions or molecules are arranged at regular lattice sites, whereas thermodynamic properties reflect on the lattice symmetry only implicitly. Represented by a unit cell, an idealized crystal is regarded as macroscopically *uniform*, provided that surfaces and lattice defects are insignificant. For such a "perfect" crystal, the thermodynamical Gibbs potential can be expressed primarily as a function of external variables, p and T, as in isotropic systems. On the other hand, a structural transition between crystalline phases is characterized by a symmetry change, which, however, does not constitute by itself the responsible transition mechanism. Many crystals undergoing phase transitions exhibit *reconstructive* structural changes, whereas in some phase transitions, active ions or parts of active groups displace their positions continuously from regular lattice sites. Thus, thermodynamically, reconstructive transitions are first order, whereas displacive ones are second order. Normally, in the former cases, symmetries in the two phases above and below T_c are not necessarily related, whereas in the latter continuous transitions, two phases are related, inviting theoretical interest in regard to structural instability.

In a displacive system, the structural change is signified by spontaneous linear or angular displacements of active ions or groups at the transition temperature T_c. At the transition threshold, the active groups begin to exhibit such displacements, whose magnitudes increase with decreasing temperature, as evidenced from diffuse X-ray diffraction [19] and magnetic resonance results [20], [21]. In a continuous phase transition, such displacements are believed to occur collectively, resulting in a change of macroscopic symmetry.

Perovskite crystals provide typical examples of displacive phase transitions. Given the chemical formula ABO_3, the unit cell in the *normal* phase consists of an octahedral complex ${BO_6}^{2-}$ surrounded by eight A^{2+} ions at corners of the cubic cell, as shown in Fig. 3.1a. Crystals of the perovskite family are rich in types of structural change, exhibiting a variety of displacement schemes. For example, in the ferroelectric phase transition of $BaTiO_3$ at

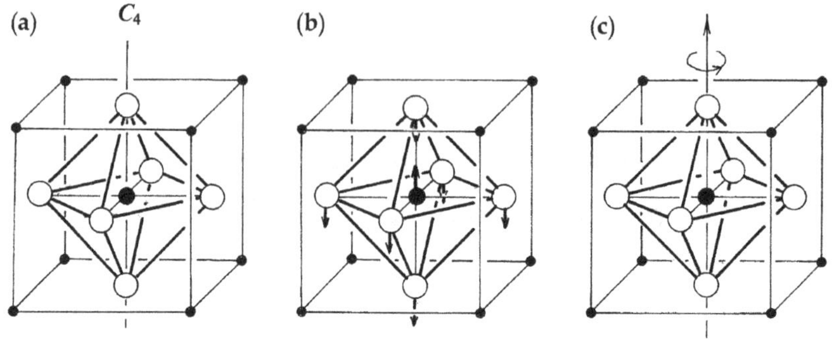

Fig. 3.1. Unit cells in the perovskite structure: (a) the normal phase, (b) linear ionic displacements along the C_4 axis in $BaTiO_3$, and (c) a rigid-body rotation of TiO_6^{2-} octahedra around the C_4 axis in $SrTiO_3$ crystals.

405K, the structural change is signified by an off-center displacement of the central Ti^{4+} ion along the C_4 axis parallel to one of the cubic axes (Fig. 3.1b). For the phase transition of $SrTiO_3$ at 105K, on the other hand, a rotational displacement of ${TiO_6}^{2-}$ around a C_4 axis is responsible (Fig. 2.1c).

Known as a binary transition, the low-temperature phase is characterized by two opposite directions of the order parameter, which are related by *inversion* or *reflection* on the *mirror plane* in the crystal. Accordingly, the corresponding microscopic order variable can be specified by $\sigma_{mz} = \pm 1$ of a classical vector $\boldsymbol{\sigma}_m$, when those at different lattice sites are not correlated. In the above examples, σ_{mz} represents linear displacements of Ti^{4+} along $\pm z$ directions in $BaTiO_3$, and small-angle rotations $\pm\delta\varphi$ around the z axis in $SrTiO_3$. Although defined as a scalar in the Landau theory, directional displacements as order variables are vectors in the lattice. For such a vector order variable $\boldsymbol{\sigma}_m$, a local potential $V(\boldsymbol{\sigma}_m)$ changes at T_c, whereas the corresponding change in the Gibbs potential ΔG is given by the spatial average $\langle V(\boldsymbol{\sigma}_m)\rangle_s$.

In the normal phase $T > T_c$, ΔG should be zero, whereas the Gibbs potential G is contributed by distributed local potential where the minimum is at $\boldsymbol{\sigma}_m = 0$; that is,

$$V_>(\boldsymbol{\sigma}_m) = \tfrac{1}{2}a_x{\sigma_{mx}}^2 + \tfrac{1}{2}a_y{\sigma_{my}}^2 + \tfrac{1}{2}a_z{\sigma_{mz}}^2 \qquad (3.1a)$$

with positive coefficients, providing the lattice with stability at $\boldsymbol{\sigma}_m = (\sigma_{mx}, \sigma_{my}, \sigma_{mz}) = 0$; hence, $\langle V_>(\boldsymbol{\sigma}_m)\rangle_t = 0$. Accordingly, the normal phase is characterized by $\langle \boldsymbol{\sigma}_m\rangle_t = 0$ and $\langle \boldsymbol{\sigma}_m.\boldsymbol{\sigma}_n\rangle_t = 0$. Here, it is noted that the local symmetry axes x, y and z are not necessarily the same as the lattice symmetry axes in general.

On the other hand, below T_c, for such a displacement $\boldsymbol{\sigma}_m$, the minimum of the local potential is no longer at the origin, but shifting to new positions say $\sigma_{mz} = \pm\sigma_o$ on the z axis, or at angles $\pm\delta\varphi_o$ of twist around z in the local

(a)

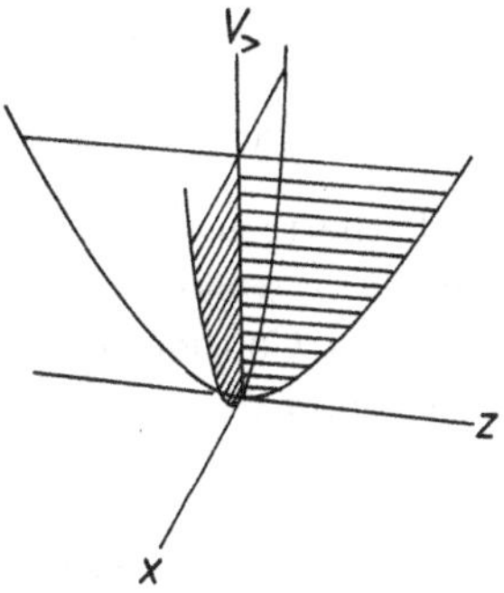

(b)

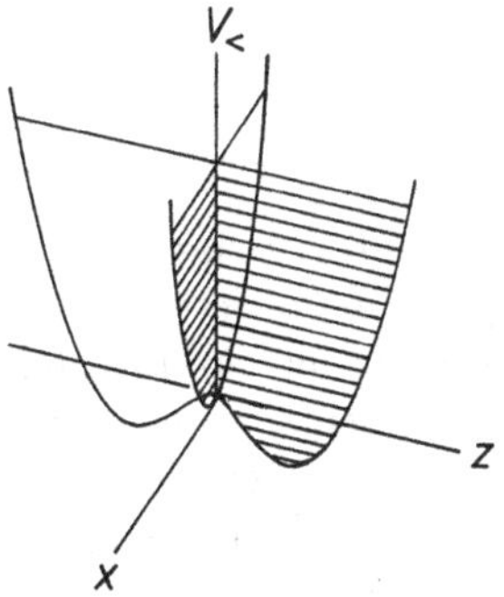

Fig. 3.2. Local crystalline potentials in a quasi-two-dimentional lattice. (a) a paraboloidal $V_>$ above T_c; (b) a double-well paraboloidal $V_<$ below T_c.

potential. Therefore, such a potential for $T < T_c$ can be written as

$$V_<(\boldsymbol{\sigma}_m) = \tfrac{1}{2}a_x{\sigma_{mx}}^2 + \tfrac{1}{2}a_y{\sigma_{my}}^2 + \tfrac{1}{2}a_z{\sigma_{mz}}^2 + \tfrac{1}{4}b_z{\sigma_{mz}}^4, \tag{3.1b}$$

where $a_z < 0$ and $b_z > 0$, whereas a_x and a_y are unchanged from (3.1a). Figures 3.2a and 3.2b illustrate such potentials $V_>$ and $V_<$ for $a_y = 0$, and there are two minima in $V_<$ at $\sigma_{mz} = \pm\sigma_o = \pm(-a_z/b_z)^{1/2}$, representing positions in static equilibrium, similar to curves in Fig. 1.8. Dynamically, we can therefore consider fluctuations in the vicinity of $\pm\sigma_o$ as well as between $+\sigma_o$ and $-\sigma_o$, while their physical origin is unspecified. The latter can essentially be quantum-mechanical tunneling, although characterized as a classical jumping over the central barrier in some cases. If these variables $\boldsymbol{\sigma}_m$ and $\boldsymbol{\sigma}_n$ at different sites m and n are correlated, we have the relation $\boldsymbol{\sigma}_m.\boldsymbol{\sigma}_n \neq 0$, resulting in collective displacements, where interactions between nearest neighbors are particularly significant. In a long timescale, we can define probabilities $p_m(+\sigma_o)$ and $p_m(-\sigma_o)$ for the variable $\boldsymbol{\sigma}_m$ to be at one of the minima $+\sigma_o$ or another $-\sigma_o$ by writing

$$\frac{\sigma_{mz}}{\sigma_o} = p_m(+\sigma_o) - p_m(-\sigma_o), \quad \text{where} \quad p_m(+\sigma_o) + p_m(-\sigma_o) = 1. \tag{3.2}$$

For uncorrelated pseudospins, we have

$$\langle\sigma_{mz}\rangle_t = 0 \quad \text{and} \quad \langle p(+\sigma_o)\rangle_s = \langle p(-\sigma_o)\rangle_s = \tfrac{1}{2}.$$

Further, when correlated, the transversal components σ_{mx} and σ_{my} can also be significant in this model for a classical motion of $\boldsymbol{\sigma}_m$ whose dimensionality in the crystal space matters. Considering a continuous motion in the xz plane, for instance, the direction of $\boldsymbol{\sigma}_m$ can be signified by an angle θ_m from the z axis, and so

$$\sigma_{mz} = \sigma_o \cos\theta_m \quad \text{and} \quad \sigma_{mx} = \sigma_o \sin\theta_m,$$

where

$$\cos\theta_m = p_m(+\sigma_o) - p_m(-\sigma_o) \quad \text{and} \quad \sin\theta_m = \{2p_m(+\sigma_o)p_m(-\sigma_o)\}^{1/2}.$$

In this description the angle θ_m varies from one site to another, indicating that these correlated $\boldsymbol{\sigma}_m$ are in collective motion in the lattice. Nevertheless it is noted that the local potentials are invariant for inversion of the classical pseudospin $\boldsymbol{\sigma}_m \rightarrow -\boldsymbol{\sigma}_m$. The displacive phase transition can be signified by the outset of nonzero correlations $\langle\boldsymbol{\sigma}_m.\boldsymbol{\sigma}_n\rangle_t$ for collective $\boldsymbol{\sigma}_m$ at T_c, at which the local potentials $V_<(\boldsymbol{\sigma}_m)$ and $V_<(\boldsymbol{\sigma}_n)$ begin to show a quartic anharmonicity as given by the last term in (3.1b), being responsible for the correlation between displacements at sites m and n.

However, depending on the timescale of observation, a slow classical displacement vector may not be subjected to thermal probabilities, for which the Ising spin is not an adequate model. By analogy with the Heisenberg exchange interaction, for correlations between such displacements, we postulate the Hamiltonian

$$\mathbf{H}_{mn} = -J_{mn}\boldsymbol{\sigma}_m.\boldsymbol{\sigma}_n \tag{3.3a}$$

where J_{mn} represents the correlation between classical pseudospins $\boldsymbol{\sigma}_m$ and $\boldsymbol{\sigma}_n$ at sites m and n. It is noted that all components of a displacement vector are significant for the correlation energy (3.3a), in contrast to an Ising spin that is signified only by its z component.

A classical pseudospin vector is often signified by its direction, while the amplitude is considered as a small constant at a given temperature. Writing $\boldsymbol{\sigma}_m = \sigma_o \boldsymbol{e}_m$ in this case, where $\boldsymbol{e}_m$ is the unit vector along $\boldsymbol{\sigma}_m$, the correlation energy (3.3a) can be expressed as

$$\mathbf{H}_{mn} = -\sigma_o{}^2 J_{mn}\boldsymbol{e}_m.\boldsymbol{e}_n. \tag{3.3b}$$

Here, although written by analogy with the exchange integral, the correlation parameter J_{mn} cannot be evaluated from the first principle. Nevertheless, it is temperature dependent in crystals, and there are some intervening ions and molecules between $\boldsymbol{\sigma}_m$ and $\boldsymbol{\sigma}_n$, making its evaluation difficult. For such $\boldsymbol{\sigma}_m$ representing active groups, the sign of J_{mn} is more significant than the magnitude for the binary arrangement in crystals; either in parallel or antiparallel.

In the following discussions, we postulate the equations (3.3a) and (3.3b) for pseudospin correlations responsible for structural phase transitions, leaving values of J_{mn} to empirical evaluation.

3.2 The Landau Criterion for Classical Fluctuations

Located at regular lattice points in a crystal, the displacive variables $\boldsymbol{\sigma}_m$ can be a periodic function of the lattice period. On the other hand, we must consider the fact that such distributed finite displacements $\boldsymbol{\sigma}_m$ violate the lattice symmetry locally, resulting in appreciable strains in the lattice. If occurring with a finite amplitude in a long timescale, the lattice should be distorted in principle, destabilizing the structure. Born and Huang [18] discussed strained crystals theoretically, although structural instability during phase transitions was not their primary concern. However, it is a serious matter for spontaneous structural changes and, therefore, we must discuss it in light of their theory (Section 3.4). In this section, before dealing with the theoretical implication, we discuss the Landau criterion [7] for a classical order variable, which is instructive for us to proceed to the problem of interactions between order variables and the hosting lattice in relation to the structural stability.

It is noted that in a distorted crystal, such dynamical displacements could take place at very low frequencies, being characterized as acoustic excitations at long wavelengths. If these waves become stationary, the crystal appears to be inhomogeneous due to unevenly distributed densities. Landau realized such spatial inhomogeneity in stressed crystals [7], which was described as consisting of many small volumes $\mathrm{d}V(\boldsymbol{r},t)$ at a position $\boldsymbol{r}$, where thermodynamic properties are deviated from equilibrium. In his description, the variation of classical order variables can arise from a density deviation defined by $\Delta\rho(\boldsymbol{r},t) = \mathrm{d}\sigma(\boldsymbol{r},t)/\mathrm{d}V$, which is determined by distributed thermodynamic probabilities $\Delta p(\boldsymbol{r},t) = p_o - p(\boldsymbol{r},t)$, where p_o is referred to the uniform distribution in the undistorted crystal. Consequently, a negative distributed entropy per volume can be considered for distributed probabilities $-\Delta p$ as expressed by the Boltzmann relation

$$\Delta s(\boldsymbol{r},t) = \mathrm{k_B} \ln\{p(\boldsymbol{r},t)/p_o\} = \mathrm{k_B} \ln[\{1 - \Delta p(\boldsymbol{r},t)\}/p_o].$$

Assuming $\Delta p/p_o \ll 1$, the decrease in entropy $\Delta s(\boldsymbol{r},t)$ is at most of the order of $\mathrm{k_B}$, providing a heat energy $T_c\Delta s = \mathrm{k_B}T_c$ transferred to the lattice at T_c, consequently causing an excitation $\Delta\varepsilon(\boldsymbol{r},t)$ in the crystal. We note at this stage that such an energy transfer process should be irreversible thermodynamically, hence we can write an inequality

$$\Delta\varepsilon(\boldsymbol{r},t) \le T_c\Delta s(\boldsymbol{r},t) = \mathrm{k_B}T_c. \tag{3.4}$$

Unlike in the normal phase, the lattice is thus excited under a critical condition due to displacive ordering. Further, it is noted that such a negative entropy production as caused by varying local volume $dV_i(\boldsymbol{r},t)$ should be associated with the work $-p_i dV_i$, which essentially originates from spontaneous local stresses in the lattice.

On the other hand, classically such an excitation energy $\Delta\varepsilon$ can be described as related to a decay of the collective order variable $\boldsymbol{\sigma}(\boldsymbol{r},t)$; that is,

$$\frac{\partial\boldsymbol{\sigma}}{\partial t} = -\frac{\boldsymbol{\sigma}}{\tau}, \tag{3.5}$$

implying that the ordering energy is relaxed presumably to the lattice excitation in this case. Nevertheless, such an energy transfer $\Delta\varepsilon$ is basically quantum mechanical, although the classical decay (3.5) should be interpreted in the limit of $\hbar \to 0$. Denoting the Hamiltonian density for the responsible interaction between $\boldsymbol{\sigma}$ and the lattice by $\mathbf{H}$, the quantum-mechanical order variable operator $\boldsymbol{\sigma}_{\mathrm{op}}$ should obey the Heisenberg equation

$$i\hbar\left\langle\frac{\partial\boldsymbol{\sigma}_{\mathrm{op}}}{\partial t}\right\rangle_{\mathrm{t}} = \langle[\mathbf{H},\boldsymbol{\sigma}_{\mathrm{op}}]\rangle_{\mathrm{t}}, \tag{3.6}$$

where the time derivative in the classical approach should be evaluated by the average over the relaxation time interval $\Delta\tau$, allowing one to replace $\langle\partial\boldsymbol{\sigma}_{\mathrm{op}}/\partial t\rangle_{\mathrm{t}}$ by $\partial\boldsymbol{\sigma}/\partial t$ in (3.5). From (3.6) we arrive at the uncertainty relation [17] for small variations $\Delta\boldsymbol{\sigma}$ and $\Delta\varepsilon$ in classical variables $\boldsymbol{\sigma}(\boldsymbol{r},t)$ and $\varepsilon(\boldsymbol{r},t)$, respectively; that is,

$$\Delta\varepsilon\Delta\boldsymbol{\sigma} \sim \hbar\left\langle\frac{\partial\boldsymbol{\sigma}}{\partial t}\right\rangle_{\mathrm{t}} \sim -\frac{\hbar\boldsymbol{\sigma}}{\tau}. \tag{3.7}$$

It is noted that (3.7) is identical to the conventional uncertainty relation $\tau\Delta\varepsilon \sim \hbar$, provided that $\Delta\boldsymbol{\sigma}/\boldsymbol{\sigma} \approx 1$. Combining these equations, Landau wrote the relation

$$\tau T_{\mathrm{c}} \gg \hbar/\mathrm{k_B} \sim 10^{-11}(\mathrm{secK}), \tag{3.8a}$$

suggesting that such a timescale τ in (3.8a) justifies the classical nature of $\boldsymbol{\sigma}$ at T_{c}, where the fluctuation $\Delta\boldsymbol{\sigma}$ can be infinitesimal, signifying that $\boldsymbol{\sigma}$ is a classical variable. While thermally inaccessible, such a small classical excitation as $\Delta\varepsilon < \mathrm{k_B}T$ should be originated from an internal mechanical *work* in the stressed lattice, which is *non-ergodic* as will be discussed in Section 3.4, and hence in contradiction to Landau's thermal mechanism. According to the lattice dynamical theory, such a small $\Delta\varepsilon$ should be related to an acoustic excitation.

In contrast, for a quantum mechanical $\boldsymbol{\sigma}$, $\Delta\boldsymbol{\sigma}$ should be of a finite magnitude, and so $\boldsymbol{\sigma}/\Delta\boldsymbol{\sigma} < 1$ for a small $\boldsymbol{\sigma}$, for which $\Delta\varepsilon$ should be in excess of thermal energy $\mathrm{k_B}T$. Therefore, for quantum-mechanical systems the conventional criterion

$$\Delta\varepsilon \geq \mathrm{k_B}T_{\mathrm{c}} \quad \text{or} \quad \tau T_{\mathrm{c}} \leq \hbar/\mathrm{k_B} \tag{3.8b}$$

should be applied, being specified by $\Delta\varepsilon$ larger than the thermal energy at T_c.

According to (3.8a), at around $T_c = 200$K fluctuations can be considered as classical, if $\tau \gg 0.5 \times 10^{-13}$sec. For example, in displacive phase transitions in perovskites in the range of 100K $< T_c <$ 200K, the value of τ is estimated to be of the order of 10^{-11}sec., so that the order variables is classical. On the other hand, for proton-tunneling in hydrogen-bonding crystals, with the estimated relaxation time $\tau \sim 10^{-13}$sec. at $T_c \sim 122$K, τT_c gives $\sim 10^{-11}$, which is the limit of the criterion (3.8a). It is well known that phase transitions in hydrogen-bonding crystals are quantum-mechanical and of first-order thermodynamically, due to discontinuous proton rearrangements at T_c, as will be discussed in Chapter 10.

3.3 Quantum-Mechanical Pseudospins and their Correlations

A particle tunneling through a double-well potential between the two minima requires an energy, depending on the height and width of the central barrier, where the motion is by no means classical. For such double-minimum potentials placed at lattice sites, correlations with the nearest neighbors should become significant in the packed crystal structure, shifting minima to off-center positions, as determined by the quartic energy emerging in the lattice at T_c. In this context, pseudospin correlations should be closely related to anharmonic potentials. On the other hand, correlated pseudospins in collective motion can be described as classical, although quantum-mechanical pseudospins are valid in crystals as long as they are uncorrelated. Nevertheless, the quantum nature of a pseudospin is basic, and hence we outline the model in this section, following Blinc and Zeks [12], prior to discussing their correlated motion.

A particle tunneling through the central barrier V in a double-well potential is signified by identical energies ε_o in the two wells, which is then perturbed by V. The unpertubed energy is doubly degenerate, and given as the ground state of the Schrödinger equation

$$\mathbf{H}_m \varphi_m = \varepsilon_o \varphi_m,$$

where $\mathbf{H}_m$ is the Hamiltonian of the unperturbed particle at the site m. In this case, the eigenstate ε_o is invariant by inversion that is expressed by a matrix

$$\sigma_{mz} = \begin{pmatrix} 1 & 0 \\ 0 & -1 \end{pmatrix}, \qquad (3.9)$$

where the index z is referred to the tunneling direction in this case, while we can leave it unspecified for the general argument. The operators $\mathbf{H}_m$ and σ_{mz} are commutable, i.e. $[\mathbf{H}_m, \sigma_{mz}] = 0$, and hence σ_{mz} can be regarded as the variable with eigenvalues ± 1, specifying inversion-related degenerate states of the ground energy ε_o.

With the perturbation potential V turned on, the degeneracy can be removed, separating the ground energy ε_o into two levels $\varepsilon_\pm = \varepsilon_o \pm \frac{1}{2}V$, corresponding to symmetric and antisymmetric combinations of wavefunctions φ_{mL} and φ_{mR}, i.e.

$$\psi_{m+} = (\varphi_{mL} + \varphi_{mR})/2^{1/2} \quad \text{and} \quad \psi_{m-} = (\varphi_{mL} - \varphi_{mR})/2^{1/2}. \tag{3.10a}$$

Here, the indexes L and R refer to the left and right minima, respectively, of the potential $V_<(\sigma_{mz})$. In the second quantization scheme, the functions φ_{mL} and φ_{mR} and their conjugates $\varphi_{mL}{}^*$ and $\varphi_{mR}{}^*$ are annihilation and creation operators of a particle at the left and the right minima of the site m, respectively. Therefore, we can write the normalization relation

$$\psi_{m+}{}^*\psi_{m+} + \psi_{m-}{}^*\psi_{m-} = \varphi_{mL}{}^*\varphi_{mL} + \varphi_{mR}{}^*\varphi_{mR} = 1. \tag{3.10b}$$

Here, the products $\varphi_{mL}{}^*\varphi_{mL}$ and $\varphi_{mR}{}^*\varphi_{mR}$ represent probability densities at the left and right minima, respectively, by which we can define the order variable as

$$\sigma_{mz} = \varphi_{mL}{}^*\varphi_{mL} - \varphi_{mR}{}^*\varphi_{mR} = \psi_{m+}{}^*\psi_{m+} - \psi_{m-}{}^*\psi_{m-}. \tag{3.11a}$$

For an uncorrelated particle, probabilities at the left and right minima are the same, and so $\sigma_{mz} = 0$. Combining φ_{mL}, φ_{mR} and conjugates in the following, we can also define

$$\sigma_{mx} = \varphi_{mL}{}^*\varphi_{mR} + \varphi_{mR}{}^*\varphi_{mL} \tag{3.11b}$$

and

$$\sigma_{my} = \varphi_{mL}{}^*\varphi_{mR} - \varphi_{mR}{}^*\varphi_{mL}, \tag{3.11c}$$

which are found to satisfy the commutation relations

$$[\sigma_{mx}, \sigma_{my}] = i\sigma_{mz}, \quad [\sigma_{my}, \sigma_{mz}] = i\sigma_{mx} \quad \text{and} \quad [\sigma_{mz}, \sigma_{mx}] = i\sigma_{my}. \tag{3.12}$$

Therefore, these three quantities σ_{mx}, σ_{my} and σ_{mz} are considered to be components of a quantum-mechanical pseudospin vector $\boldsymbol{\sigma}_m$. When σ_{mz} is diagonalized as in (3.9), the transversal components are expressed by matrices

$$\sigma_{mx} = \begin{pmatrix} 0 & 1 \\ 1 & 0 \end{pmatrix} \quad \text{and} \quad \sigma_{my} = \begin{pmatrix} 0 & -i \\ i & 0 \end{pmatrix},$$

which are responsible for transitional motion between left and right minima, and the Hamiltonian of a uncorrelated particle can then be expressed a

$$\begin{aligned}\mathbf{H}_m &= \varepsilon_{m+}\psi_{m+}{}^*\psi_{m+} + \varepsilon_{m-}\psi_{m-}{}^*\psi_{m-} \\ &= \varepsilon_o(\psi_{m+}{}^*\psi_{m+} + \psi_{m-}{}^*\psi_{m-}) - \tfrac{1}{2}V(\psi_{m+}{}^*\psi_{m+} - \psi_{m-}{}^*\psi_{m-}) \\ &= \varepsilon_o - \tfrac{1}{2}V\sigma_{mz}.\end{aligned} \tag{3.13}$$

For correlated pseudospins, we further consider interaction potentials v_{mn} between different sites m and n that are expressed as a 4×4 density matrix $(\psi_{\mathrm{m}\alpha}{}^{*}\psi_{\mathrm{n}\beta})$, where α and β represent $\pm$ states of pseudospins. Consequently, the short-range correlations can be determined by

$$\mathbf{H}'_{\mathrm{m}} = \sum_{\mathrm{n}} \mathbf{H}'_{\mathrm{mn}} = \sum_{\mathrm{n}} \sum_{\alpha\beta\gamma\delta} \langle \psi_{\mathrm{m}\alpha}{}^{*}\psi_{\mathrm{m}\beta} | v_{\mathrm{mn}} | \psi_{\mathrm{n}\gamma}{}^{*}\psi_{\mathrm{n}\delta} \rangle, \tag{3.14}$$

consisting of 16 interaction terms, which can be simplified by using pseudospin components, namely

$$\begin{aligned} \psi_{\mathrm{m}+}{}^{*}\psi_{\mathrm{m}+} &= \tfrac{1}{2}(\varphi_{\mathrm{mL}}{}^{*}\varphi_{\mathrm{mL}} + \varphi_{\mathrm{mR}}{}^{*}\varphi_{\mathrm{mR}} + \varphi_{\mathrm{mL}}{}^{*}\varphi_{\mathrm{mR}} + \varphi_{\mathrm{mR}}{}^{*}\varphi_{\mathrm{mL}}) \\ &= \tfrac{1}{2}(1 + \sigma_{\mathrm{mx}}), \\ \psi_{\mathrm{m}+}{}^{*}\psi_{\mathrm{m}-} &= \tfrac{1}{2}(\varphi_{\mathrm{mL}}{}^{*}\varphi_{\mathrm{mL}} - \varphi_{\mathrm{mR}}{}^{*}\varphi_{\mathrm{mR}} - \varphi_{\mathrm{mL}}{}^{*}\varphi_{\mathrm{mR}} + \varphi_{\mathrm{mR}}{}^{*}\varphi_{\mathrm{mL}}) \\ &= \tfrac{1}{2}(\sigma_{\mathrm{mz}} - \sigma_{\mathrm{my}}), \text{ etc.,} \end{aligned}$$

and for the term $\langle \psi_{\mathrm{m}\alpha}{}^{*}\psi_{\mathrm{m}\beta} | v_{\mathrm{mn}} | \psi_{\mathrm{n}\gamma}{}^{*}\psi_{\mathrm{n}\delta} \rangle$ abbreviated by $v_{\alpha\beta\gamma\delta}$, we have symmetric relations

$$v_{++--} = v_{--++}, \quad v_{+-+-} = v_{-+-+} = v_{-++-},$$

but all other asymmetric elements, such as v_{++-+}, vanish. Applying these results to $\boldsymbol{\sigma}_{\mathrm{m}}$ and $\boldsymbol{\sigma}_{\mathrm{m}+1}$, the short-range energy can be expressed as

$$\mathbf{H}'_{\mathrm{m,m+1}} = -J_{\mathrm{m,m+1}}\sigma_{\mathrm{mz}}\sigma_{\mathrm{m+1,z}} - K_{\mathrm{m,m+1}}\sigma_{\mathrm{mx}}\sigma_{\mathrm{m+1,x}}, \tag{3.15a}$$

where

$$J_{\mathrm{m,m+1}} = 4v_{+-+-} \quad \text{and} \quad K_{\mathrm{m,m+1}} = 2v_{+-+-} - v_{++++} - v_{----}. \tag{3.15b}$$

The total Hamiltonian for two interacting pseudospins σ_{m} and $\sigma_{\mathrm{m}+1}$ is therefore written as

$$\begin{aligned} \mathbf{H} &= \mathbf{H}_{\mathrm{m}} + \mathbf{H}_{\mathrm{m}+1} + \mathbf{H}'_{\mathrm{m,m+1}} \\ &= \varepsilon_{\mathrm{o}} - \Omega(\sigma_{\mathrm{mz}} + \sigma_{\mathrm{m+1,z}}) \qquad (3.15c) \\ &\quad - J_{\mathrm{m,m+1}}\sigma_{\mathrm{mz}}\sigma_{\mathrm{m+1,z}} - K_{\mathrm{m,m+1}}\sigma_{\mathrm{mx}}\sigma_{\mathrm{m+1,x}}, \qquad (3.15d) \end{aligned}$$

where

$$\Omega = \tfrac{1}{2}V + v_{++++} - v_{----} \approx \tfrac{1}{2}V.$$

Equations (3.15cd) is the Hamiltonian for two interacting quantum-mechanical particles $\boldsymbol{\sigma}_{\mathrm{m}}$ and $\boldsymbol{\sigma}_{\mathrm{m}+1}$ with their correlation energy. For a proton tunneling system, Blinc and Zeks [12] showed that the terms Ω and $K_{\mathrm{m,m+1}}$ are negligibly small as compared with $J_{\mathrm{m,m+1}}$, and $\mathbf{H}$ is dominated by the correlation term $J_{\mathrm{m,m+1}}\sigma_{\mathrm{mz}}\sigma_{\mathrm{m+1,z}}$ for the Ising model. Nevertheless, the Ising term signifies statistically both quantum and classical correlations, whereas transversal components σ_{mx} and $\sigma_{\mathrm{m+1,x}}$ may play a different role in dynamical correlations.

From the quantum-mechanical treatment of correlations, the variable σ_{mz} defined by (3.11a) represents the density difference between left and right minima of the double-well potential. However, in a linear chain, no asymmetry expressed by nonzero σ_{mz} can occur, unless these σ_{mz} are heavily correlated in the domain structure or under an externally applied field. On the other hand, statistically a nonzero $\langle\sigma_{mz}\rangle$ was attributed to unequal thermal probabilities of binary states, as discussed in Chapter 2, where the collective feature was not taken into consideration.

3.4 The Born-Huang Theory and Structural Ordering in Crystals

Born and Huang [18] discussed the general theory of strained crystals, in which they have shown theoretically that the crystal structure cannot be deformed uniformly, because the lattice stability is maintained with internal and applied stresses, where distributed strains are generally associated with a lattice excitation at a long wavelength. In spite of their prediction, the theory had never been tested experimentally, until such an excitation was detected in recent magnetic resonance experiments in the critical region of structural changes. (See Chapters 7 and 9 for the experimental detail.)

For thermal stability of a crystal, they have proposed two conditions: (1) the constituent atoms and molecules should all be in mechanical equilibrium in the lattice, and (2) their configuration should be determined for vanishing stresses. The first condition (1) can normally be set for the lattice potential to be at the minimum, but more importantly the lattice stability cannot be warranted by condition (1) only. Although condition (1) is sufficient in practice for a stable crystal, condition (2) is necessary for the stability of a stressed crystal. When the potential is expanded into power series of displacements, anharmonic terms of at least the fourth order become significant for dynamic distortion, whereas the harmonic second-order terms are necessary for maintaining lattice stability. Condition (2) is generally required for a crystal as a whole to have a stable structure under stresses and, as the result, a lattice excitation emerges at a long wavelength. As applied to crystals undergoing structural phase transitions, the quartic potential emerging at T_c is considered as the logical consequence of the Born-Huang theory.

Noting that order variables $\boldsymbol{\sigma}_m$ represent parts of the active groups in a crystal, the center-of-mass coordinates should also be displaced by $\mathbf{u}_m$ from the regular site m to offset strains due to displacements $\boldsymbol{\sigma}_m$. Assuming that in the critical region these $\boldsymbol{\sigma}_m$ and $\mathbf{u}_m$ are linearly related for the local equilibrium at each site m, we may write

$$\boldsymbol{\sigma}_m \approx \mathbf{A}.\mathbf{u}_m \tag{3.16}$$

for small amplitudes, where $\mathbf{A}$ is generally a tensor quantity. However, the condition (3.16) may not be exact at all sites, unless their characteristic wavelengths match with each other closely in phase. The displacement $\mathbf{u}_m$ varies from one site to another in the long-wave approximation, according to the Born-Huang theory. These $\boldsymbol{\sigma}_m$ and $\mathbf{u}_m$ are correlated below T_c as given by (3.16), so that the net strain energy can be expressed by

$$W = \sum_{mn} J_{mn}(\boldsymbol{\sigma}_m - \mathbf{A}.\mathbf{u}_m).(\boldsymbol{\sigma}_n - \mathbf{A}.\mathbf{u}_n).$$

This W should be set as equal to zero to fulfill the condition (2) for the distorted structure in equilibrium at a given temperature. We can write W as

$$W = W_\sigma + W_u + w,$$

where

$$W_\sigma = \sum_{mn} J_{mn}\boldsymbol{\sigma}_m.\boldsymbol{\sigma}_n, \tag{3.17a}$$

$$W_u = \sum_{mn} J_{mn}\mathbf{u}_m.\mathbf{A}^{-1}\mathbf{A}.\mathbf{u}_n \tag{3.17b}$$

and

$$w = -\sum_{mn} J_{mn}(\boldsymbol{\sigma}_m.\mathbf{A}.\mathbf{u}_n + \mathbf{u}_m.\mathbf{A}^{-1}.\boldsymbol{\sigma}_n). \tag{3.17c}$$

It is realized that W_u for a modified lattice is a function of space coordinates due to the spatial variation of $\mathbf{u}_m$. In the long-wave approximation, the displacements $\mathbf{u}_m = (u_1, u_2, u_3)$ can be considered as a continuous function of normal lattice points $\boldsymbol{r}_m$, and so we expand $\mathbf{u}_m$ into a power series of lattice distortion $\boldsymbol{r} - \boldsymbol{r}_m = \boldsymbol{\delta r} = (\delta r_1, \delta r_2, \delta r_3)$; that is

$$u_i = \sum_j \left(\frac{\partial u_i}{\partial r_j}\right)\delta r_j + \tfrac{1}{2}\sum_{jk}\delta r_j\left(\frac{\partial^2 u_i}{\partial r_j \partial r_k}\right)\delta r_k + \dots . \tag{3.18}$$

Using (3.18) into (3.17b), we arrive at an expression for the lattice energy

$$W_u = \sum_{ij} D_{ij}\delta r_j + \sum_{ijk} E_{ijk}\delta r_j \delta r_k + \sum_{ijkl} B_{ijkl}\delta r_i \delta r_j \delta r_k \delta r_l + \dots . \tag{3.19}$$

Here, the quartic terms of B_{ijkl} represent strain energy due to lattice fluctuations $\boldsymbol{\delta r}$, whereas the terms of D_{ij} and E_{ijk} express the lattice energy in distorted structure due to $\boldsymbol{\delta r}$ during ordering process. The significance of such anharmonicity is expressed in the expansion (1.13) of the Landau theory, which is in fact consistent with the modified lattice potential by displacements $\boldsymbol{\sigma}_m$ in adiabatic approximation [18] (see Appendix). Cowley considered the quartic lattice potential for softening a lattice mode, as will be discussed in Chapter 4. Later, we also discuss anisotropic quartic potentials for symmetry conversion at T_c.

Applying the Born-Huang condition (2) for a stress-free crystal as given by the condition $W = 0$, the structural phase transition can be signified as follows:

$$W_\sigma = 0, \quad W_u = 0, \quad w = 0 \quad \text{in the normal phase above } T_c \tag{3.20a}$$

$$W_\sigma + W_u + w = 0 \quad \text{in the critical region} \tag{3.20b}$$

$$W_\sigma + W_u \to 0, \quad w \to 0 \quad \text{in the noncritical phase below } T_c. \tag{3.20c}$$

Ordering in crystals can be attributed to a combined property of pseudospins and the lattice, exhibiting temperature-dependent anomalies. Aside from the temperature dependence, $\boldsymbol{\sigma}_{\mathrm{m}}$ and $\mathbf{u}_{\mathrm{m}}$ are primarily independent unless $w \neq 0$, so that it is not surprising to see experimental results that appear to be incompatible with one another. Namely, when observing $\boldsymbol{\sigma}_{\mathrm{m}}$, the phase transition is signified by the outset of pseudospin correlations, whereas in terms of $\mathbf{u}_{\mathrm{m}}$ the presence of soft modes characterizes a structural change. Nevertheless, as inspired by the Born-Huang theory, we consider a coupling w as described by (3.16), which plays an essential role in the critical region.

Accordingly, our experimental task is threefold. First, we have to verify that the threshold of a phase transition is signified by the outset of "minimum pseudospin correlations" in a periodic lattice (Sections 3.5 and 3.6). Second, we wish to obtain evidence that a collective pseudospin mode can interact with a lattice mode in near-phase relation to obtain a significant coupling w (Section 4.1). Third, we investigate the role played by quartic potentials for structural transformations (Section 4.4). These problems should be discussed in light of the fluctuating interaction w given by (3.16) to establish the coherent view of structural transitions.

3.5 Collective Pseudospin Modes in Displacive Systems

In the critical region of a displacive transition, the collective pseudospin mode can be expressed by a Fourier transform of the classical variables $\boldsymbol{\sigma}_{\mathrm{m}}$ with periodic boundaries, i.e. $\boldsymbol{\sigma}_{\mathrm{m}} = \sum_q \boldsymbol{\sigma}_q \exp i(\boldsymbol{q}.\boldsymbol{r}_{\mathrm{m}} - \omega t)$, i.e. a linear combination of $\boldsymbol{\sigma}_q$, where $\boldsymbol{q}$ are wavevectors, expressing spatial forms of sinusoidal correlations. Generally, there are many values of $\boldsymbol{q}$ that satisfy the boundary conditions, but we are only interested in a particular wavevector for the minimum correlation energy among $\boldsymbol{\sigma}_{\mathrm{m}}$ to signify the collective mode at the threshold of ordering.

Being a real quantity in a crystal idealized with periodic boundary conditions, the vector $\boldsymbol{\sigma}_{\mathrm{m}}$ can be written at a given time t as

$$\begin{aligned}\boldsymbol{\sigma}_{\mathrm{m}} &= \exp(-i\varpi t)\{\boldsymbol{\sigma}_q \exp(+i\boldsymbol{q}.\boldsymbol{r}_{\mathrm{m}}) + \boldsymbol{\sigma}_{-q} \exp(-i\boldsymbol{q}.\boldsymbol{r}_{\mathrm{m}})\} \\ &= \exp(-i\varpi t)\{\boldsymbol{\sigma}_q \exp(i\boldsymbol{q}.\boldsymbol{r}_{\mathrm{m}}) + \boldsymbol{\sigma}_q{}^* \exp(i\boldsymbol{q}.\boldsymbol{r}_{\mathrm{m}})^*\},\end{aligned}$$

where the relation

$$\boldsymbol{\sigma}_q{}^* = \boldsymbol{\sigma}_{-q}, \qquad (3.21\text{a})$$

indicates that there are always two running waves at wavevectors $\pm\boldsymbol{q}$ in opposite directions with an equal amplitude. Here ϖ is the characteristic frequency, corresponding to the wavevector $\boldsymbol{q}$ at the minimum correlation energy. If the amplitude σ_{o} is considered as constant in temperature and only the direction is significant, we can write

$$\boldsymbol{e}_q{}^* = \boldsymbol{e}_{-q}, \qquad (3.22\text{a})$$

and

$$\boldsymbol{e}_{\pm q} = \sum_{\mathrm{m}} \boldsymbol{e}_{\mathrm{m}} \exp\{-i(\pm \boldsymbol{q}.\boldsymbol{r}_{\mathrm{m}} - \varpi t_{\mathrm{m}})\}. \tag{3.22b}$$

In this case, the short-range interactions can be calculated from

$$\mathbf{H}_{\mathrm{mn}} = -J_{\mathrm{mn}}\sigma_{\mathrm{o}}^2 \boldsymbol{e}_{\mathrm{m}}.\boldsymbol{e}_{\mathrm{n}}, \quad \text{where} \quad \boldsymbol{e}_{\mathrm{n}} = \boldsymbol{e}_{\pm q} \exp i(\mp \boldsymbol{q}.\boldsymbol{r}_{\mathrm{n}} + \varpi t_{\mathrm{n}}),$$

and its time average can be expressed as

$$\langle \mathbf{H}_{\mathrm{m}} \rangle_{\mathrm{t}} = \sum_{\mathrm{n}} \langle \mathbf{H}_{\mathrm{mn}} \rangle_{\mathrm{t}} = -\sigma_{\mathrm{o}}{}^2 \Gamma_{\mathrm{t}} \sum_{\mathrm{n}} J_{\mathrm{mn}} \exp\{i\boldsymbol{q}.(\boldsymbol{r}_{\mathrm{m}} - \boldsymbol{r}_{\mathrm{n}})\} \boldsymbol{e}_q.\boldsymbol{e}_{-q},$$

where

$$\Gamma_{\mathrm{t}} = (2t_{\mathrm{o}})^{-1} \int_{-t_{\mathrm{o}}}^{t_{\mathrm{o}}} \exp\{-i\varpi(t_{\mathrm{m}} - t_{\mathrm{n}})\} \mathrm{d}(t_{\mathrm{m}} - t_{\mathrm{n}}) = \frac{\sin \varpi t_{\mathrm{o}}}{\varpi t_{\mathrm{o}}} \tag{3.23}$$

is the time correlation function within the timescale t_{o} of observation. The value of Γ_{t} is less than 1 but close to 1 if $\varpi t_{\mathrm{o}} < 1$, providing a condition for spatial fluctuations to be revealed experimentally. Otherwise, the value of $\langle \mathbf{H}_{\mathrm{mn}} \rangle_{\mathrm{t}}$ is averaged out and undetectable for $\varpi t_{\mathrm{o}} > 1$.

Assuming $\Gamma_{\mathrm{t}} \sim 1$, which is valid for a slow variation, the observable short-range energy at a site m can be expressed as

$$\langle \mathbf{H}_{\mathrm{m}} \rangle_{\mathrm{t}} = -\sigma_{\mathrm{o}}{}^2 \boldsymbol{e}_q.\boldsymbol{e}_{-q} J_{\mathrm{m}}(\boldsymbol{q}), \tag{3.24a}$$

where

$$J_{\mathrm{m}}(\boldsymbol{q}) = \sum_{\mathrm{n}} J_{\mathrm{mn}} \exp i\boldsymbol{q}.(\boldsymbol{r}_{\mathrm{m}} - \boldsymbol{r}_{\mathrm{n}}). \tag{3.24b}$$

The correlation energy $\langle \mathbf{H}_{\mathrm{m}} \rangle_{\mathrm{t}}$ should be minimized to obtain the threshold to the critical region, which can be calculated from the equation

$$\mathrm{grad}_{\boldsymbol{q}} J_{\mathrm{m}}(\boldsymbol{q}) = 0. \tag{3.25}$$

The quantity $J_{\mathrm{m}}(\boldsymbol{q})$ can be determined for interactions at short distances $|\boldsymbol{r}_{\mathrm{m}} - \boldsymbol{r}_{\mathrm{n}}|$, extending to the nearest and next-nearest neighbors, regardless of the site m. Therefore omitting m from $J_{\mathrm{m}}(\boldsymbol{q})$, we can write it as $J(\boldsymbol{q})$ for the short-range interaction. In Section 3.6, we obtain expressions for $J(\boldsymbol{q})$ for representative lattices as examples for such calculations.

Of further importance is that these $\boldsymbol{e}_q$ and $\boldsymbol{e}_{-q}$ are unit vectors, and, hence, subjected to the normalization condition in the reciprocal space, we have the relation

$$N = \sum_{q} \boldsymbol{e}_q{}^*.\boldsymbol{e}_q, \tag{3.26a}$$

where N is the number of pseudospins per volume of the system. If specific wavevectors $\pm\boldsymbol{q}$ to minimize the function $J(\boldsymbol{q})$ can be found for a given system, as shown in Section 3.6, we can use these $\pm\boldsymbol{q}$ in (3.26a); that is,

$$N = \boldsymbol{e}_q{}^*.\boldsymbol{e}_q + \boldsymbol{e}_{-q}{}^*.\boldsymbol{e}_{-q} = 2\boldsymbol{e}_q.\boldsymbol{e}_{-q}. \tag{3.26b}$$

On the other hand, the normalization condition in the crystal space is

$$N = \sum_{\mathrm{m}} \boldsymbol{e}_{\mathrm{m}}{}^{*}.\boldsymbol{e}_{\mathrm{m}} = \sum_{q,-q} \boldsymbol{e}_{q}.\boldsymbol{e}_{-q} \exp[i\{\boldsymbol{q} - (-\boldsymbol{q})\}.\boldsymbol{r}_{\mathrm{m}}]$$
$$= 2\boldsymbol{e}_{q}.\boldsymbol{e}_{-q} + \boldsymbol{e}_{q}{}^{2} \exp(2i\boldsymbol{q}.\boldsymbol{r}_{\mathrm{m}}) + \boldsymbol{e}_{-q}{}^{2} \exp(-2i\boldsymbol{q}.\boldsymbol{r}_{\mathrm{m}}),$$

which should be identical to (3.26b). Therefore, we have the identity

$$\boldsymbol{e}_{q}^{2} \exp(2i\boldsymbol{q}.\boldsymbol{r}_{\mathrm{m}}) + \boldsymbol{e}_{-q}^{2} \exp(-2i\boldsymbol{q}.\boldsymbol{r}_{\mathrm{m}}) = 0, \tag{3.27}$$

at all lattice sites m. This can be satisfied by either

$$\boldsymbol{e}_{q}{}^{2} = \boldsymbol{e}_{-q}{}^{2} = 0 \tag{3.28a}$$

or

$$\exp(2i\boldsymbol{q}.\boldsymbol{r}_{\mathrm{m}}) = \exp(-2i\boldsymbol{q}.\boldsymbol{r}_{\mathrm{m}}) = 0, \tag{3.28b}$$

which should be held in addition to (3.27). In order for the relation (3.28b) to be independent of m, the vector $\boldsymbol{q}$ should be either 0 or $\pm\frac{1}{2}\boldsymbol{G}$, where $\boldsymbol{G}$ is a translational vector in the reciprocal lattice, corresponding to *ferrodistortive* or *antiferrodistortive* collective mode, respectively. In these cases, the arrangements are *commensurate* with the lattice period, signifying a stable long-range order. In contrast, (3.28a) signifies an arrangement that varies with a vector $\boldsymbol{q} = (q_{\mathrm{x}}, q_{\mathrm{y}}, q_{\mathrm{z}})$ such that

$$e_{qx}^{2} + e_{qy}^{2} + e_{qz}^{2} = 0$$

in the reciprocal space. Assuming $e_{qy} = 0$, for example, we obtain $e_{q\mathrm{x}} = \pm i e_{q\mathrm{z}}$. Transforming into the crystal space, the corresponding displacement vector $\boldsymbol{e}_{\mathrm{m}}$ can be expressed by components

$$e_{\mathrm{mx}} = \cos\phi_{\mathrm{m}}, \quad e_{\mathrm{my}} = \pm\sin\phi_{\mathrm{m}} \quad \text{and} \quad e_{\mathrm{mz}} = 0,$$

where $\phi_{\mathrm{m}} = \boldsymbol{q}.\boldsymbol{r}_{\mathrm{m}} + \varpi t + \phi_{\mathrm{o}}$ is the phase for propagation with an arbitrary ϕ_{o}. In this case, the wavevector $\boldsymbol{q}$ gives a periodic variation in the crystal space, which can be independent of the lattice period along the z direction. Called *incommensurate* in this case, $|\boldsymbol{q}|$ is irrational with respect to the reciprocal lattice vector $\boldsymbol{G}$. In the long-wave approximation, the phase ϕ_{m} is considered as a continuous variable in the crystal, so that the coordinate may be written as $\boldsymbol{r}$ instead of $\boldsymbol{r}_{\mathrm{m}}$. Although arbitrary at each site m, the phase constant ϕ_{o} is insignificant in this approximation if the phase variable $\phi = \phi(\boldsymbol{r})$ is defined for the continuous $\boldsymbol{r}$, varying in the range $0 \leq \phi \leq 2\pi$ in repetition. We realize that such a sinusoidal variation is only detectable in the timescale t_{o} for $\varpi t_{\mathrm{o}} \leq 1$, otherwise averaged out and undetectable. In the critical region, the condition $\varpi t_{\mathrm{o}} \approx 1$ is fulfilled for most microscopic observations.

We have remarked that the sign of the correlation parameter J_{mn} is significant for the stable pseudospin arrangement, for which the correlation energy $\mathbf{H}_{\mathrm{mn}}$ must take negative values. In the reciprocal lattice, the same principle applies to (3.24a), whose sign is determined by $J(\boldsymbol{q})\boldsymbol{e}_q.\boldsymbol{e}_{-q}$. In a binary system, the order variable $\boldsymbol{\sigma}_q = \sigma_o \boldsymbol{e}_q$ must be subjected to inversion symmetry, so that $\boldsymbol{e}_q = -\boldsymbol{e}_{-q}$ under a spatial inversion $\boldsymbol{r} \to -\boldsymbol{r}$. Therefore, noting that $\boldsymbol{e}_q.\boldsymbol{e}_{-q} = -1$, the sign of $J(\boldsymbol{q})$ in (3.24a) must be negative to obtain stable arrangements for the $\pm\boldsymbol{q}$ modes.

3.6 Examples of Collective Pseudospin Modes

In the theory of magnetism, "the method of minimum correlations" has been used for classifying magnetic ordering, such as ferromagnetic, antiferromagnetic, spiral spin arrangements and others [22]. Here, we apply the same method to representative displacive systems to see if such a collective pseudospin mode for minimum correlations can really be derived from (3.25). In fact, in recent magnetic resonance experiments, modulated structures were observed in practical systems as calculated with this principle, in spite of unknown correlation parameters in $J(\boldsymbol{q})$. We may therefore postulate that the structural transition can be initiated spontaneously with such a collective pseudospin mode. In this section, phase transitions in perovskites and in organic calcium chloride crystals, particularly *tris-sarcosine calcium chloride* (TSCC), are selected as model systems for the present argument, because rather comprehensive experimental data are readily available in the literature, which allow us to interpret critical anomalies in light of calculated correlations.

3.6.1 Strontium Titanate and Related Perovskites

Structural phase transitions in perovskite crystals provide typical examples of displacive systems, where the model of classical fluctuations can explain their collective motion of pseudospins. In the ferroelectric phase transition of $BaTiO_3$, the lattice symmetry changes from cubic to tetragonal, when the constituent Ti^{4+} and O^- ions displace along one of the cubic axes. Naturally grown perovskite crystals are thus "twinned" and hence considered as ferroelastic below T_c, being characterized by three differently oriented tetragonal domains, whereas single-domain samples can be cut from twinned crystals for studying the phase transition. Denoting displacements of the central Ti^{4+}, two O^- on the tetragonal z axis and four O^- in the xy plane of an octahedral TiO_3 complex by u_+, u_- and u'_-, respectively, the order variable can be expressed in a long timescale as related to

$$\sigma_{mz} \propto (u_+ - 2u_- - 4u'_-)_m,$$

at a site m, with respect to the z axis that is taken along the cubic a axis. Noting that σ_{mz} and σ_{-mz} are related by inversion, the order variable can be defined statistically as the z component of a vector $\boldsymbol{\sigma}_m$, i.e.

$$\sigma_{mz} = p_m(+z) - p_m(-z),$$

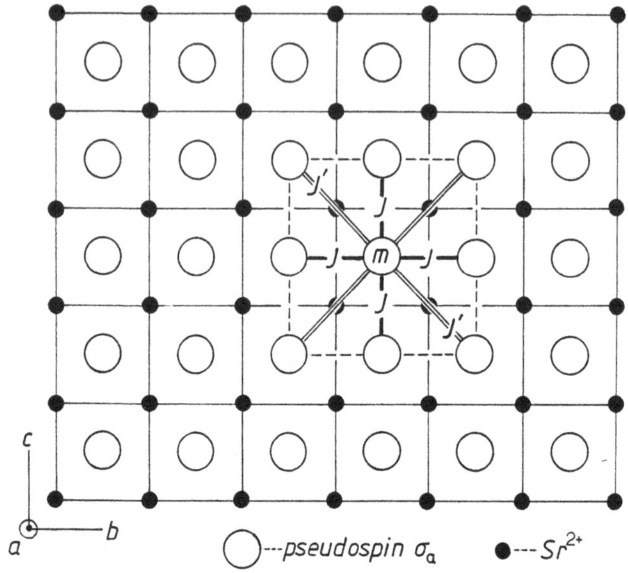

Fig. 3.3. The pseudospin arrangement in a perovskite lattice in the *bc* plane. The square of broken lines indicates the range of the short-range interactions proposed for collective motion at the threshold of phase transitions.

where $p_m(\pm z)$ are probabilities for binary states of σ_{mz}.

On the other hand, the structural change in $SrTiO_3$ crystals is considered to arise from axial rotation of ${TiO_6}^{2-}$ octahedra, like rigid bodies. Letting $\delta\theta_\pm$ be small rotational angles in opposite sense around the z axis, the order variable can also be defined statistically as a vector with the z component:

$$\sigma_{mz} \propto (\delta\theta_+ - \delta\theta_-)_m,$$

where $\delta\theta_\pm$ represent probabilities for angular displacements $\pm\delta\theta$.

In Fig. 3.3 shown is the arrangement of such pseudospins in the *bc* plane of a perovskite crystal, where significant short-range correlations are conceivable between $\boldsymbol{\sigma}_m$ and six $\boldsymbol{\sigma}_n$ in the nearest-neighbor sites and between $\boldsymbol{\sigma}_m$ and eight $\boldsymbol{\sigma}_n$ in the next nearest-neighbor sites, for which the interaction parameters are denoted by J and J', respectively. Considering such a cluster of pseudospins at each site m, the short-range correlations can be expressed for $+\boldsymbol{q}$ and $-\boldsymbol{q}$ modes, for which the interaction parameter $J(\boldsymbol{q})$ should be minimized to determine a specific $\boldsymbol{q}$ at the threshold of the critical region in this system. The function $J(\boldsymbol{q})$ in (3.24b) can be expressed explicitly for the nearest and next-nearest interactions as

$$J(\boldsymbol{q}) = 2J\{\cos(q_a a) + \cos(q_b b) + \cos(q_c c)\}$$
$$+J'\{\cos(q_b b)\cos(q_c c) + \cos(q_c c)\cos(q_a a) + \cos(q_a a)\cos(q_b b)\}.$$

Differentiating with respect to q_a, q_b and q_c, the minimum of $J(\boldsymbol{q})$ can be determined by solving the equations

$$\sin(q_a a)\{J + 2J' \cos(q_b b) + 2J' \cos(q_c c)\} = 0,$$
$$\sin(q_b b)\{J + 2J' \cos(q_c c) + 2J' \cos(q_a a)\} = 0$$

and

$$\sin(q_c c)\{J + 2J' \cos(q_a a) + 2J' \cos(q_b b)\} = 0,$$

which give solutions as follows:

$$\begin{array}{ll}
\text{(i)} & \sin(q_a a) = \sin(q_b b) = \sin(q_c c) = 0, \\
\text{(ii.1)} & \sin(q_a a) = 0, \quad \cos(q_b b) = \cos(q_c c) = -(1 - J/2J') \\
\text{(ii.2)} & \sin(q_b b) = 0, \quad \cos(q_c c) = \cos(q_a a) = -(1 - J/2J') \\
\text{(ii.3)} & \sin(q_c c) = 0, \quad \cos(q_b b) = \cos(q_a a) = -(1 - J/2J') \\
\text{(iii.1)} & \cos(q_a a) = 0, \quad \cos(q_b b) = \cos(q_c c) = -J/2J' \\
\text{(iii.2)} & \cos(q_b b) = 0, \quad \cos(q_c c) = \cos(q_a a) = -J/2J' \\
\text{(iii.3)} & \cos(q_c c) = 0, \quad \cos(q_a a) = \cos(q_b b) = -J/2J'.
\end{array}$$

Solution (i) gives the wavevector $\boldsymbol{q}_1$ with components

$$q_{1a} = (\pi/a)l, \quad q_{1b} = (\pi/b)m, \quad q_{1c} = (\pi/c)n,$$

where l, m and n take 0 or plus integers. In these calculations, only the magnitude $|\boldsymbol{q}|$ is significant for the function $J(\boldsymbol{q})$ in (3.24b). Hence, the calculated wavevector $\boldsymbol{q}_1$ for positive angles gives a commensurate arrangement with the lattice vector $\boldsymbol{G}$, and $J(\boldsymbol{q}_1) = 6J + 12J'$, where pseudospins are all parallel.

Solution (ii.1) indicates that the q_{2a} is rational in the reciprocal unit $\boldsymbol{a}^*$, whereas q_{2b} and q_{2c} are irrational in the units of $\boldsymbol{b}^*$ and $\boldsymbol{c}^*$, provided that $|1 - J'/2J'| < 1$. Thus, the wavevector $\boldsymbol{q}_2$ for Solution (ii.1) gives an incommensurate arrangement of pseudospins in the bc plane, while commensurate in the a direction. For convenience, we can rewrite these components as

$$q_{2a} = la^* \quad (l = 0, \text{ or half integers}),$$
$$q_{2b} = \left(\tfrac{1}{2} - \delta_b\right) b^* \quad \text{and} \quad q_{2c} = \left(\tfrac{1}{2} - \delta_c\right) c^*,$$

where δ_b and δ_c are so-called *incommensurate parameters* in the b and c directions, respectively. Corresponding to $\boldsymbol{q}_2$, we have $J(\boldsymbol{q}_2) = 2J - 4J' - J^2/J'$, which should be negative for the mode to give a lower correlation energy below T_c.

Solution (iii.1) also give a two-dimensional incommensurate arrangement in the bc plane, if $|J/2J'| < 1$, while commensurate in the a direction, and $J(\boldsymbol{q}_3) = -J^2/2J' < 0$ for $J' > 0$. Solutions (ii) and (iii) are very similar, although giving different correlation energies, both being incommensurate in two dimensions while commensurate in the perpendicular direction. Experimentally, observed anomalies in perovskite crystals exhibited such features consistent with these calculated results, as discussed in Chapter 10.

In fact, the structural change in perovskites is known as from cubic to tetragonal, whereas we calculated for an orthorhombic case in the above. Considering for example the displacement along the c axis, the wavevectors of types (ii) and (iii) are characterized by q_a and q_b, which are identical in the tetragonal structure. Therefore, a binary domain structure may logically be expected in the low-temperature phase, for which some evidence was exhibited in magnetic resonance anomalies from $SrTiO_3$ crystals, as discussed in Section 10.1.

3.6.2 Tris-Sacosine Calcium Chloride and Related Crystals

Tris-sarcosine calcium chloride (TSCC) with the formula unit (sarcosine)$_3$ $CaCl_2$, where sarcosine is H_3C-NH_2-CH_2COOH, crystallizes in an orthorhombic structure. At room-temperature, TSCC crystals exhibit twinning due to the pseudostructure that strains the structure. *Ferroelastic domains* in twinned crystals can be easily identified by viewing through a pair of crossed polarizers [23]. A single-domain sample can then be cut out of a naturally grown crystal for studies of the ferroelectric phase transition at 120K, leading to the ordered phase polarized along the b axis. TSCC crystals offer a prototype example of uniaxial structural phase transitions, characterized by the loss of mirror symmetry on the b plane below T_c.

Figure 3.4a illustrates the molecular arrangement in TSCC at room temperature, which was determined by Ashida et al. [24] from their X-ray studies. In this figure, quasi-mirror symmetry on the b plane is clearly visible in the quasi-triginal structure. Nakamura et al. [25] confirmed from their diffuse X-ray results near T_c that the active group is the Ca(sarcosine)$_6$ complex, where the Ca^{2+} ion is surrounded near-octahedrally by six carbonyl oxygens, as shown in Fig. 3.4b.

Figure 3.5 shows the pseudospin lattice in the bc plane of TSCC abstracted from Fig. 3.4a, where we consider correlations with the nearest neighbors along the a, d and c directions and with the next-nearest neighbors along the b axis, although the perpendicular interaction J_a to the plane is not shown. For such a cluster of pseudospins, the function $J(\boldsymbol{q})$ can be written as

$$J(\boldsymbol{q}) = 2J_\mathrm{a}\cos(q_\mathrm{a}a) + 2J_b\cos(q_b b) + 2J_\mathrm{c}\cos(q_\mathrm{c}c) + 8J_\mathrm{d}\cos\left(\tfrac{1}{2}q_\mathrm{b}b\right)\cos\left(\tfrac{1}{2}q_\mathrm{c}c\right), \tag{3.29}$$

where the lattice constant b is larger than $a \approx c$, according to the X-ray data [24]. Solutions of the equation (3.25) for this $J(\boldsymbol{q})$ can be specified by the following wavevector components

$$\begin{aligned}
&\text{(i)} \quad q_{1\mathrm{a}} = (\pi/a)l, \quad q_{1\mathrm{b}} = (\pi/b)m, \quad q_{1\mathrm{c}} = (\pi/c)n;\\
&\text{(ii)} \quad q_{2\mathrm{a}} = (\pi/a)l, \quad \cos\left(\tfrac{1}{2}q_{2\mathrm{b}}b\right) = -J_\mathrm{d}/2J_\mathrm{b}, \quad q_{2\mathrm{c}} = (2\pi/c)n;\\
&\text{(iii)} \quad q_{3\mathrm{a}} = (\pi/a)l, \quad q_{3\mathrm{b}} = (\pi/b)m, \quad \cos\left(\tfrac{1}{2}q_{3\mathrm{c}}c\right) = -J_\mathrm{d}/2J_\mathrm{b},
\end{aligned}$$

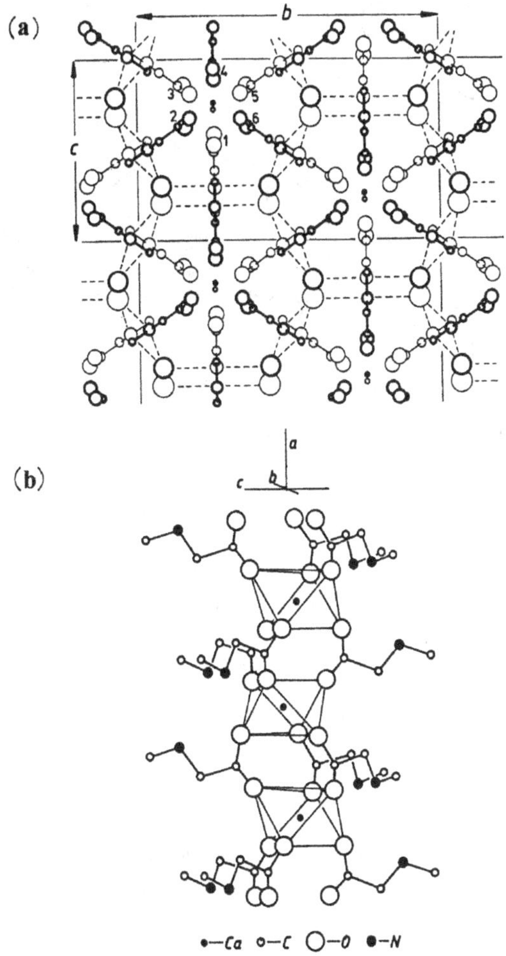

Fig. 3.4. (a) The molecular arrangement in the normal (ferroelastic) phase in TSCC crystals; (b) the ligand structure in Ca^{2+}(sarcosine)$_6$ complexes in TSCC.

where l, m and n are zero or positive integers.

Solution (i) represents a commensurate pseudospin arrangement, for which

$$J(\boldsymbol{q}_1) = 2J_\mathrm{a} + 2J_\mathrm{b} + 2J_\mathrm{c}.$$

Assuming that it is negative, the commensurate pseudospin mode $\boldsymbol{q}_1$ can be stable, which, however, cannot be verified unless such a state is identified.

On the other hand, Solutions (ii) and (iii) give incommensurate arrangements in the b and c directions if $|-J_\mathrm{d}/2J_\mathrm{b}| < 1$. The ferroelectric phase transition in TSCC exhibits the unique axis of polarization along the b axis, for which Solution (ii) should be an obvious choice. In this case, the function

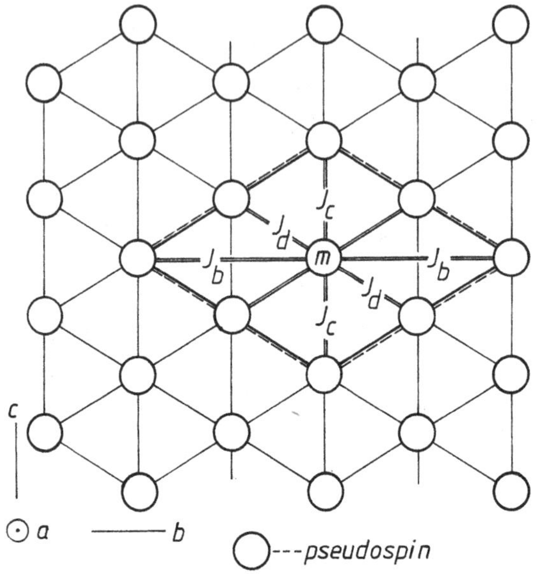

Fig. 3.5. The pseudospin lattice proposed for TSCC crystals, as viewed along the *a* axis. The short-range correlations at the transition threshold are indicated by the range shown by broken lines.

$J(\boldsymbol{q})$ can be written as

$$J(\boldsymbol{q}_2) = 2J_\mathrm{a} + 2J_\mathrm{b} + 2J_\mathrm{c} - 8{J_\mathrm{d}}^2/J_\mathrm{b},$$

which should be negative for stability of the mode $\boldsymbol{q}_2$. For one-dimensional correlations along the b direction, we can assume that $J_\mathrm{a} = J_\mathrm{c} = 0$ in $J(\boldsymbol{q}_2)$, which can then be written as

$$J(\boldsymbol{q}_2) = 2J_\mathrm{b}\{1 - (J_\mathrm{d}/2J_\mathrm{b})^2\},$$

depending on the value of J_d which can vary in the range $0 \le J_\mathrm{d} \le J_\mathrm{b}$. We can consider that a change $J_\mathrm{d} \to 2J_\mathrm{b}$ describes the approach to complete order. The complete ferroelectric order is established by lowering the temperature in TSCC, where the ordering mechanism can logically be attributed to the temperature-dependent J_d.

Aside from the origin for variable J_d, the vector $\boldsymbol{e}_\mathrm{m}$ for $\boldsymbol{q}_2$ can be expressed as

$$e_\mathrm{ma} = 0, \quad e_\mathrm{mb} = \cos\phi_\mathrm{m} \quad \text{and} \quad e_\mathrm{mc} = \pm\sin\phi_\mathrm{m},$$

where $\phi_\mathrm{m} = q_{2b}y_\mathrm{m} + \phi_\mathrm{o}$ is the phase of sinusoidal propagation along the b axis, as verified by Fujimoto et al. [21] from observed electron paramagnetic resonance (EPR) anomalies in Mn^{2+} spectra in the critical region of TSCC.

Although only short-range correlations were considered in the above, the parameter J_d can also depend on the long-range dipolar interactions that are temperature dependent. Judging from the EPR results, pseudospin correlations in TSCC are primarily due to J_b in the one-dimensional chain in the b direction, but are modified by the interchain interaction J_d. For one-dimensional charge-density waves in a uniaxial conductor, Lee et al. [27] proposed a model similar to J_d in TSCC. Considering that the phase ϕ represents an angle between the vector $\boldsymbol{\sigma}$ and the chain direction geometrically, they wrote a dipolar interaction between such chains $\sigma_o \exp i\phi_1$ and $\sigma_o \exp i\phi_2$ as

$$-J_{12}\boldsymbol{\sigma}_1.\boldsymbol{\sigma}_2 = -J_{12}{\sigma_o}^2 \cos(\phi_1 - \phi_2),$$

where J_{12} is a scalar parameter that depends on the interchain distance. It is noted that the interaction is maximum if $\phi_1 - \phi_2 = 0$, although it is zero when $\phi_1 - \phi_2 = \frac{1}{2}\pi$, suggesting a phase shift in neighboring-chain interactions that can.be vary with temperature. Applying this model to the TSCC structure illustrated by Fig. 3.5, we may consider that the parameter J_d is temperature dependent, and that two adjacent chains $\boldsymbol{\sigma}_m(1)$ and $\boldsymbol{\sigma}_{m+1/2}(2)$ are signified by a phase difference $\frac{1}{2}\pi$ at the threshold of the transition. In this context, we can express the interchain phase difference as $\phi_1 - \phi_2 = (\frac{1}{2} - \delta_b)\pi$, where δ_b is the temperature-dependent incommensurate parameter.

Solution (ii) represents generally a collective mode of pseudospins (or spins) in one-dimensional chain, in which the nearest-neighbor ferrodistortive interaction J_b is counteracted by eight interchain interactions J_d (or two next-nearest antiferrodistortive interaction). In (3.29), we can write $\frac{1}{2}q_b b = \varphi$, and $\cos(\frac{1}{2}q_c c) = 1$,

$$J(\varphi) = 2J_a + 2J_c + 2J_b \cos(2\varphi) + 8J_d \cos\varphi, \tag{3.30}$$

which is a well-known formula for a linear chain model of collective pseudospin vectors. If the vectors are transversally rotatable around the b axis, the collecive mode can be described as a spiral distribution of the spin vectors [28, 29].

3.7 The Variation Principle and the Weiss Singularity

We have considered order variables emerging at T_c with infinitesimal amplitude, which are in collective motion as initiated by minimum correlations. With lowering the temperature, their amplitudes increase due to increasing long-range order. Although weak, such long-range interactions exist in the critical region, but are not easily evaluated. Nevertheless, responding to an applied field, the susceptibility shows a singular behavior near T_c in the mean-field approximation, for which the averaged local field, called the Weiss field, is considered responsible. Although introduced as a theoretical concept, the Weiss field behaves as if applied externally, as the Curie-Weiss law of the susceptibility can be derived as a response to the combined internal and applied fields. In this section, we show that such local fields can also be derived by the variation principle applied to the correlation energy.

In a pseudospin system in equilibrium at a given temperature, the correlation energy should take a minimum value, around which the pseudospin mode can fluctuate. We consider that the correlation energy

$$\mathbf{H} = -\sum_{\mathrm{m}}\sum_{\mathrm{n}} J_{\mathrm{mn}}\boldsymbol{\sigma}_{\mathrm{m}}.\boldsymbol{\sigma}_{\mathrm{n}}, \tag{i}$$

where correlations beyond the short-range can also be included in the summation. The energy in (i) can include all the correlations of this type in the system, and the effect of long-range correlations can be discussed with this expression **H**. Applying the variation principle to (i) for all correlations of the same type, equilibrium can be obtained by minimizing **H** for arbitrary variations $\delta\boldsymbol{\sigma}_{\mathrm{m}}$ and $\delta\boldsymbol{\sigma}_{\mathrm{n}}$, for which the magnitude σ_{o} is assumed as constant of a given temperature; that is,

$$\boldsymbol{\sigma}_{\mathrm{m}}{}^2 = \boldsymbol{\sigma}_{\mathrm{n}}{}^2 = \sigma_{\mathrm{o}}{}^2 = \mathrm{const.} \tag{ii}$$

For the stationary state, the variation of (i) is

$$\delta\mathbf{H} = -\sum_{\mathrm{m}} \delta\boldsymbol{\sigma}_{\mathrm{m}}.\left(\sum_{\mathrm{n}} J_{\mathrm{mn}}\boldsymbol{\sigma}_{\mathrm{n}}\right) - \sum_{\mathrm{n}}\left(\sum_{\mathrm{m}} J_{\mathrm{mn}}\boldsymbol{\sigma}_{\mathrm{m}}\right).\delta\sigma_{\mathrm{n}} = 0,$$

and from (ii)

$$\boldsymbol{\sigma}_{\mathrm{m}}.\delta\boldsymbol{\sigma}_{\mathrm{m}} = \boldsymbol{\sigma}_{\mathrm{n}}.\delta\boldsymbol{\sigma}_{\mathrm{n}} = 0.$$

For arbitrary variations $\delta\boldsymbol{\sigma}_{\mathrm{m}}$ and $\delta\boldsymbol{\sigma}_{\mathrm{n}}$, we can write relations

$$-\sum_{\mathrm{n}} J_{\mathrm{mn}}\boldsymbol{\sigma}_{\mathrm{n}} + \lambda_{\mathrm{m}}\boldsymbol{\sigma}_{\mathrm{m}} = 0 \quad \text{and} \quad -\sum_{\mathrm{m}} J_{\mathrm{mn}}\boldsymbol{\sigma}_{\mathrm{m}} + \lambda_{\mathrm{n}}\boldsymbol{\sigma}_{\mathrm{n}} = 0,$$

where λ_{m} and λ_{n} are constants called the Lagrange multipliers. Here, we note that the quantity

$$\boldsymbol{F}_{\mathrm{m}} = \sum_{\mathrm{n}} J_{\mathrm{mn}}\boldsymbol{\sigma}_{\mathrm{n}} = \lambda_{\mathrm{m}}\boldsymbol{\sigma}_{\mathrm{m}} \tag{3.31}$$

represents the effective local field at site m due to all distant $\boldsymbol{\sigma}_{\mathrm{n}}$ interacting with $\boldsymbol{\sigma}_{\mathrm{m}}$, regardless of distances. Equation (3.31) indicates that $\boldsymbol{\sigma}_{\mathrm{m}}$ is proportional to $\boldsymbol{F}_{\mathrm{m}}$ which increases with increasing order. If $\boldsymbol{\sigma}_{\mathrm{m}}$ is considered like an electric dipole moment, the correlation energy $\mathbf{H}_{\mathrm{m}}$ expresses an energy for $\boldsymbol{\sigma}_{\mathrm{m}}$ to be in the field $\boldsymbol{F}_{\mathrm{m}}$; that is,

$$\mathbf{H}_{\mathrm{m}} = -\boldsymbol{\sigma}_{\mathrm{m}}.\boldsymbol{F}_{\mathrm{m}}. \tag{3.32}$$

The mean-field average of $\boldsymbol{F}_{\mathrm{m}}$, called the Weiss field, is defined as

$$\boldsymbol{F} = \langle \boldsymbol{F}_{\mathrm{m}}\rangle_{\mathrm{s}} = \left(\sum_{\mathrm{m}} \boldsymbol{F}_{\mathrm{m}}\right)/N,$$

where N is the number of pseudospins in the crystal. The above variational argument applies in principle to any stage of collective motion, where the amplitude σ_{o} is finite, arising from the long-range field $\boldsymbol{F}_{\mathrm{m}}$. Nonetheless, the amplitude σ_{o} is infinitesimal at the transition threshold T_{c}, becoming finite due to the mean-field average $\langle \boldsymbol{F}_\mathrm{m}\rangle_\mathrm{s}$ of long-range interactions. In the mean-field approximation, the susceptibility shows a singularity with vanishing $\langle \boldsymbol{F}_\mathrm{m}\rangle_\mathrm{s}$ at T_o, which is experimentally found higher than T_c.

For a collective mode of $\boldsymbol{\sigma}_\mathrm{m}$ with finite amplitudes, the internal Weiss field $\boldsymbol{F}_\mathrm{m}$ is present in the crystal as if a real field, although not directly measurable under normal circumstances. On the other hand, in susceptibility measurements, we can obtain the response of such a collective mode to the effective field composed of the internal Weiss field plus an applied field.

The collective mode can be conveniently expressed by the Fourier tramsforms $\boldsymbol{\sigma}_{\pm q}$, i.e.

$$\boldsymbol{\sigma}_\mathrm{m} = N^{-1/2}\{\boldsymbol{\sigma}_{-q}\exp(i\boldsymbol{q}.\boldsymbol{r}_\mathrm{m}) + \boldsymbol{\sigma}_q\exp(-i\boldsymbol{q}.\boldsymbol{r}_\mathrm{m})\},$$

where the time factor $\exp(-i\varpi t)$ is omitted, because the spatial variation expressed by $\boldsymbol{\sigma}_{\pm q}$ is significant and explicit under the condition $\varpi t_\mathrm{o} \le 1$. Corresponding to $\boldsymbol{\sigma}_\mathrm{m}$, we can write

$$\boldsymbol{F}_\mathrm{m} = N^{-1/2}\{\boldsymbol{F}_{-q}\exp(i\boldsymbol{q}.\boldsymbol{r}_\mathrm{m}) + \boldsymbol{F}_q\exp(-i\boldsymbol{q}.\boldsymbol{r}_\mathrm{m})\},$$

where

$$\begin{aligned}\boldsymbol{F}_{\pm q} &= N^{-1/2}\sum\nolimits_\mathrm{m}\boldsymbol{F}_\mathrm{m}\exp(\pm i\boldsymbol{q}.\boldsymbol{r}_\mathrm{m}) = N^{-1/2}\sum\nolimits_\mathrm{m}\left(\sum\nolimits_\mathrm{n} J_\mathrm{mn}\boldsymbol{\sigma}_\mathrm{n}\right)\exp(\pm i\boldsymbol{q}.\boldsymbol{r}_\mathrm{m}\Big). \\ &= N^{-1}\sum\nolimits_\mathrm{mn} J_\mathrm{mn}\exp\{\pm i\boldsymbol{q}.(\boldsymbol{r}_\mathrm{m}-\boldsymbol{r}_\mathrm{n})\}\boldsymbol{\sigma}_{\pm q} = J_\mathrm{long}(\boldsymbol{q})\boldsymbol{\sigma}_{\pm q},\end{aligned}$$

which is considered as the long-range field acting on the collective mode $\boldsymbol{\sigma}_{\pm q}$, prevailing over the short-range correlations at temperatures below T_c. Therefore, in the presence of an external field $\boldsymbol{E}$, the internal field is effectively the resultant $\boldsymbol{E} + \boldsymbol{F}_{\pm q}$, so that we can write

$$\boldsymbol{\sigma}_{\pm q} = \chi_\mathrm{o}(\boldsymbol{E} + \boldsymbol{F}_{\pm q}) = \frac{C}{T}\{\boldsymbol{E} + J_\mathrm{long}(\boldsymbol{q})\boldsymbol{\sigma}_{\pm q}\},$$

where $\chi_\mathrm{o} = C/T$ is the static susceptibility of uncorrelated pseudospins. From this, we can derive the Curie-Weiss law for the collective modes $\boldsymbol{\sigma}_{\pm q}$.:

$$\chi_q = \sigma_{\pm q}/E = C/\{T - T_\mathrm{o}(\boldsymbol{q})\}, \quad T_\mathrm{o}(\boldsymbol{q}) = CJ_\mathrm{long}(\boldsymbol{q}),$$

which is independent of the sign of $\boldsymbol{q}$. It is noted that this expression of χ_q is exactly the same as the static Curie-Weiss formula, so that it does not distinguish the collective mode $\boldsymbol{\sigma}_q$ from uncorrelated pseudospins $\boldsymbol{\sigma}_\mathrm{m}$.

Although by the mean-field approximation, the Weiss field is significant in the noncritical region below T_o, where the long-range correlations dominates. The collective pseudospins below T_c are in a propagating mode at a long wavelength, as signified by the phase variable $\phi = \boldsymbol{q}.\boldsymbol{r} - \varpi t + \phi_\mathrm{o}$ for a continuum crystal, in which $\boldsymbol{r}$ and t are space-time coordinates. If the amplitude σ_o is infinitesimal, the speed of propagation $v = \varpi/q$ is constant, whereas the speed for a finite σ_o is no longer constant of time, because the propagation is forced to be modified by the internal field $\boldsymbol{F}_q(\phi)$. By hydrodynamical analogy, we can consider $\boldsymbol{\sigma}(\phi)$ as if associated with a one-dimensional flow of liquid-like material that can be represented by the density $\boldsymbol{\sigma}^*.\boldsymbol{\sigma}$, for which we can write the equation

$$\partial v/\partial t + v\partial v/\partial x = -\alpha \partial F_q/\partial x$$

for the speed $v = v(x,t)$ that is not constant in the presence of $F_q \neq 0$, where α is constant. In this case, the term $v\partial v/\partial x$ on the left makes the flow typically nonlinear. The mathematical detail are discussed in Chapter 5, where we can show that the amplitude increases with increasing internal field F_q of long-range correlations.

4

Soft Modes, Lattice Anharmonicity and Pseudospin Condensates in the Critical Region

4.1 The Critical Modulation

At the threshold of the critical region, pseudospins are in collective motion at a specific wavevector for minimum correlations, as described in Chapter 3. On the other hand, Cochran [30] and Anderson [31] proposed considering soft modes with characteristic frequencies, that diminish toward the transition temperature T_c, causing lattice instability. Experimentally, pseudospin modes were observed in magnetic resonance anomalies, and soft modes were recognized in light- and neutron-scattering measurements.

Pseudospins and soft phonons are primarily independent in the harmonic approximation. Therefore, if excitations in these constituents are exclusively observed, the experimental results from these variables are seemingly incompatible with one another. Whereas the critical anomalies can be attributed to lattice anharmonicity, the phase transition is recognized from the singular behavior of pseudospin correlations. Nevertheless, representing the same phase transition, pseudospins and soft phonons should interact in higher order, and the resulting anomalies should be interpreted in light of the Born-Huang theory as related to the coupling w for vanishing strains.

When a collective pseudospin mode $\boldsymbol{\sigma}_{\pm q}$ occurs at a specific $\boldsymbol{q}$ of a long wavelength, the hosting crystal should be deformed in principle, where a lattice excitation $\mathbf{u}_{\pm q'}$ is induced by the coupling $\boldsymbol{\sigma}_\mathrm{m} = \mathbf{A}.\mathbf{u}_\mathrm{m}$ at each site m, and $\boldsymbol{\sigma}_\mathrm{m}$ and $\mathbf{u}_\mathrm{m}$ can be effectively in near-phase in the critical region. Such a phase-matching condition given by (3.16) can be expressed in the reciprocal space with the Fourier transforms between $\boldsymbol{\sigma}_{\pm q}$ and $\mathbf{u}_{\pm q}$ at T_c. Writing

$$\boldsymbol{\sigma}_\mathrm{m} = \{\boldsymbol{\sigma}_{-q}\exp(i\boldsymbol{q}.\boldsymbol{r}_\mathrm{m}) + \boldsymbol{\sigma}_{+q}\exp(-i\boldsymbol{q}.\boldsymbol{r}_\mathrm{m})\}\exp(-i\varpi t)$$

and

$$\mathbf{u}_\mathrm{m} = \{\mathbf{u}_{-q'}\exp(i\boldsymbol{q}'.\boldsymbol{r}_\mathrm{m}) + \mathbf{u}_{+q'}\exp(-i\boldsymbol{q}'.\boldsymbol{r}_\mathrm{m})\}\exp(-i\omega t),$$

for (3.16), these Fourier transforms are related by

$$\boldsymbol{\sigma}_{\pm q} = \mathbf{A}.\mathbf{u}_{\pm q'}\exp i\{\pm(\boldsymbol{q}'-\boldsymbol{q}).\boldsymbol{r}_\mathrm{m} - (\omega-\varpi)t\},$$

indicating that both $\boldsymbol{\sigma}_{\mathrm{m}}$ and $\mathbf{u}_{\mathrm{m}}$ are *amplitude modulated* in the critical region, as described by

$$\pm\Delta \boldsymbol{q} = \boldsymbol{q}' - \boldsymbol{q} \quad \text{and} \quad \mp\Delta\omega = \varpi - \omega. \tag{4.1}$$

Here, attached double signs are chosen for convenience in neutron inelastic scattering experiments, where the conservation laws for inelastic scattering demand that the wavevector difference ($\pm\Delta\boldsymbol{q}$) are related to the loss and gain of lattice energies ($\mp\hbar\Delta\omega$). As will be explained in Chapter 6, anomalous scattering intensities of neutrons can be expressed in terms of distributed phases $\Delta\phi_{\mathrm{m}} = \pm(\Delta\boldsymbol{q}.\boldsymbol{r}_{\mathrm{m}} - \Delta\omega.t)$, as allowed by (4.1), at each site.

It is noted that $\Delta\boldsymbol{q}$ may not necessarily be small, whereas $\Delta\omega$ is very small in practice. For the present problem in crystals, such wavevector exchanges between pseudospins and phonons can be expressed as

$$\pm\Delta\boldsymbol{q} + \boldsymbol{G} = \boldsymbol{q}' - \boldsymbol{q},$$

where $\boldsymbol{G}$ is the reciprocal lattice vector, corresponding to the repeat period of the lattice. Therefore, the problem can always be reduced to the center of the Brillouin zone, i.e. $\boldsymbol{G} = 0$, at which $\pm\Delta\boldsymbol{q} = \boldsymbol{q}' - \boldsymbol{q}$ represent small fluctuations. In practical crystals however, obstacles may exist against propagation in the crystal at nonlattice points $\boldsymbol{G}_{\mathrm{i}}$, where the translational symmetry is disrupted [32]. Such obstacles can be lattice imperfections, aperiodic structures or any other defects, at which the vector $\boldsymbol{G}_{\mathrm{i}}$ is irrational in the reciprocal lattice. In such a practical crystal, the relation (4.1) can be modified as

$$\boldsymbol{q}' - (\boldsymbol{q} + \boldsymbol{G}_{\mathrm{i}}) = \pm\Delta\boldsymbol{q}, \tag{4.2}$$

suggesting the presence of small fluctuations $\pm\Delta\boldsymbol{q}$ in the vicinity of the irrational point $\boldsymbol{G}_{\mathrm{i}}$.

Nevertheless, considering an imperfection-free crystal expressed by $\boldsymbol{G}_{\mathrm{i}} = 0$, for simplicity, illustrated in Fig. 4.1 are one-dimensional chains of active groups that are indicated by squares, in which hypothetical mass particles are located to represent displacive pseudospins $\boldsymbol{\sigma}_{\mathrm{m}}$. Such displacements $\boldsymbol{\sigma}_{\mathrm{m}}$ do not generally occur in the same way as $\mathbf{u}_{\mathrm{m}}$, because $\boldsymbol{\sigma}_{\mathrm{m}}$ can represent only a part of the active group at site m, whereas $\mathbf{u}_{\mathrm{m}}$ occurs associated with the squared group of a heavier mass, being distinguishable mechanically with regard to different masses. Such sketches are helpful for interpreting more complicated cases as well. For the purpose of illustration, a transversal vibrational mode of $\mathbf{u}_{\mathrm{m}}$ of a long wavelength near T_{c} is shown in Fig. 4.1a, assuming that $\boldsymbol{\sigma}_{\mathrm{m}} = 0$, whereas in Fig. 4.1b shown is the displacement mode $\boldsymbol{\sigma}_{\mathrm{m}}$, assuming $\mathbf{u}_{\mathrm{m}} = 0$. Figure 4.1c shows two waves $\boldsymbol{\sigma}_{\mathrm{m}}$ and $\mathbf{u}_{\mathrm{m}}$, when occurring out of phase, and $\mathbf{A}$ is assumed as a scalar.

It is important to realize that such sinusoidal fluctuations as specified by (4.1) or (4.2) are detectable in such a short timescale t_{o} that $\Delta\omega.t_{\mathrm{o}} \leq 1$, otherwise averaged out in a long timescale t_{o}. In fact, such a condition is fulfilled in most microscopic experiments of diffraction, scattering and magnetic

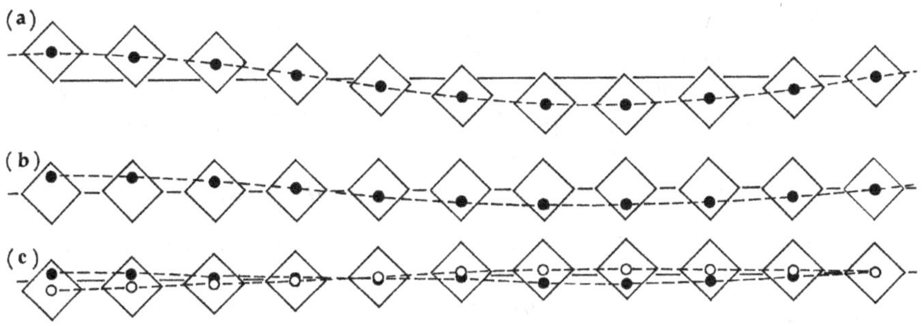

Fig. 4.1. A one-dimensional model for collective displacements. (a) normal displacements at a long wavelength; (b) a transversal pseudospin mode in a rigid lattice, (c) the same pseudospin mode in a deformed lattice.

resonance under the critical condition characterized by a very low frequency $\Delta\omega$.

The coupling w in (3.16) for phase matching between $\boldsymbol{\sigma}_{\mathrm{m}}$ and $\mathbf{u}_{\mathrm{m}}$ is analogous to the coupling between a charge-density wave (CDW) and a periodic lattice distortion (PLD) proposed by Peierls [33] for one-dimensional conductors. Such a coupled object as CDW-PLD is generally referred to as a *condensate*, and we apply a similar concept to the critical region. Related to the wavevector fluctuations $\pm\Delta\boldsymbol{q}$, we can consider the kinetic energy of fluctuating pseudospins $\hbar^2\Delta\boldsymbol{q}^2/2m = \hbar\Delta\omega$ for energy exchanges with the soft lattice mode, where m is the effective mass of a condensate.

The soft mode was a hypothetical concept, when introduced originally for lattice instability. In dielectric spectra observed in the critical region, absorption peaks are considered to represent soft modes, but such an interpretation is based on the assumption that $\boldsymbol{\sigma}_{\mathrm{m}}$ represents the phonon mode. Although signifying the displacive mechanism, pseudospin waves per se do not fully represent the distorted structure. Nevertheless, in traditional theories, pseudospins and phonons were assumed as identical for simplifying arguments.

4.2 The Lyddane-Sachs-Teller Relation

Critical fluctuations are a major objective for the investigation on structural phase transitions, for which susceptibility measurements at various frequencies yield significant information on the dynamical nature of order variables. The Lyddane-Sachs-Teller (LST) formula describes a general relation between the dielectric response and measuring frequencies, although derived from a simplified model for a uniaxial ferroelectric crystal. Although the polarization arises from ionic displacements, they assumed no distinction between $\boldsymbol{\sigma}_{\mathrm{m}}$ and $\mathbf{u}_{\mathrm{m}}$ in their theory, namely $\boldsymbol{\sigma}_{\mathrm{m}} = \mathbf{u}_{\mathrm{m}}$ in (3.16). Accordingly, such fluctuations as described by $\Delta\boldsymbol{q}$ and $\Delta\omega$ do not emerge in the LST theory. On the other hand, a uniaxial crystal, characterized by a specific direction of polarization z, is considered, in which longitudinal vibrational displacements $\mathbf{u}_\mathrm{L}$ play a role distinct from transversal displacements $\mathbf{u}_\mathrm{T}$.

Assuming that the order variable $\mathbf{u}_\mathrm{m} = \boldsymbol{\sigma}_\mathrm{m}$ represents local polarization $\boldsymbol{p}(\boldsymbol{r}, t)$ in the long-wave approximation, the electric field at $\boldsymbol{r}$ and t is expressed as

$$\boldsymbol{E}(\boldsymbol{r}, t) = \{\boldsymbol{E}_{-q} \exp(i\boldsymbol{q}.\boldsymbol{r}) + \boldsymbol{E}_q \exp(-i\boldsymbol{q}.\boldsymbol{r})\} \exp(-i\omega t), \tag{4.3}$$

which is considered as synchronizing with the harmonic displacement

$$\mathbf{u}(\boldsymbol{r}, t) = \{\mathbf{u}_{-q} \exp(i\boldsymbol{q}.\boldsymbol{r}) + \mathbf{u}_q \exp(-i\boldsymbol{q}.\boldsymbol{r})\} \exp(-i\omega t), \tag{4.4}$$

as consistent with (3.31), whereas the related lattice deformation is implicit with this assumption.

The electric displacement $\mathbf{u}(\boldsymbol{r}, t)$ occurs along the unique direction z to cause polarization, which is also induced by the longitudinal component $\boldsymbol{E}_\mathrm{L}$ of the applied field, namely

$$\boldsymbol{P}_\mathrm{L} = \varepsilon_\mathrm{o}(b'\mathbf{u}_\mathrm{L} + \alpha \boldsymbol{E}_\mathrm{L}), \tag{4.5a}$$

where b' is a proportionality factor and α is the ionic polarizability. In transversal directions, on the other hand, a polarization occurs only by an applied transversal field $\boldsymbol{E}_\mathrm{T}$, and so we can write

$$\boldsymbol{P}_\mathrm{T} = \varepsilon_\mathrm{o} \alpha \boldsymbol{E}_\mathrm{T}. \tag{4.5b}$$

We derive a dynamic response function of such a displacive system against an applied time-dependent electric field $\boldsymbol{E}(t) = \boldsymbol{E}_\mathrm{L} + \boldsymbol{E}_\mathrm{T}$. The lattice exhibits a sinusoidal excitation to meet the stress-free condition, as described by the linear equation

$$\partial^2 \mathbf{u}/\partial t^2 - v^2 \nabla^2 \mathbf{u} = b\boldsymbol{E}(t), \tag{4.6}$$

where v is the speed of propagation in the crystal, $b = e/m$ is a constant of the mode $\mathbf{u}$, and the damping is ignored for simplicity. In (4.6) for a forced oscillation, $b\boldsymbol{E}(t)$ is an effective electrical force, consisting of those originating from the interaction between the lattice and ionic charges in the crystal, and with the applied field. In this case, the problem is essentially nonlinear, but the linear equation (4.6) provides only solutions accurate at small amplitudes of these electric fields. Nevertheless, in (4.5a) the effect of long-range order is expressed in the mean-field accuracy by the constant b'. Further, it is important that in such a uniaxial system, the response function can be calculated for a longitudinal field $\boldsymbol{E}_\mathrm{L}$ as well as for a transversal field $\boldsymbol{E}_\mathrm{T}$ separately, and we discuss it following the textbook by Elliott and Gibson [34].

For dielectric studies, we normally use low frequencies, at which the rate $\partial \boldsymbol{E}_\mathrm{T}/\partial t$ is negligible, so that curl $\boldsymbol{E}_\mathrm{T} \approx 0$. Therefore, applying "curl" to (4.6), we obtain the relation for the transversal displacement $\mathbf{u}_\mathrm{T}$,

$$\mathrm{curl}(\partial^2 \mathbf{u}_\mathrm{T}/\partial t^2 - v^2 \nabla^2 \mathbf{u}_\mathrm{T}) = 0. \tag{4.7}$$

For a sinusoidal displacement along the z direction (unit vector $\mathbf{k}$), we can write

$$\mathbf{u}_\mathrm{T}(z,t) = (\mathbf{u}_{\mathrm{T}+} + \mathbf{u}_{\mathrm{T}-})\exp(-i\omega_\mathrm{T} t), \quad \text{where} \quad \mathbf{u}_{\mathrm{T}\pm} = u_{\mathrm{To}}\exp(\pm iqz)\mathbf{k},$$

and

$$\operatorname{curl}\mathbf{u}_{\mathrm{T}\pm} = i\boldsymbol{q}\times\mathbf{u}_{\mathrm{T}\pm}.$$

Therefore, from (4.7) we obtain

$$(\pm i\boldsymbol{q}\times\mathbf{k})u_{\mathrm{T}\pm}(-\omega_\mathrm{T}{}^2 + \varpi^2) = 0, \quad \text{where } \varpi = vq.$$

Because $\boldsymbol{q}\times\mathbf{u}_{\mathrm{T}\pm} \neq 0$, we obtain $\omega_\mathrm{T} = \varpi$ for the transversal mode.

On the other hand, in a longitudinal field $\boldsymbol{E}_\mathrm{L}$ the spatial variation of displacements should be significant, and so we apply "div" to (4.6) for $\mathbf{u}_{\mathrm{L}\pm}$,

$$\operatorname{div}(\partial^2\mathbf{u}_{\mathrm{L}\pm}/\partial t^2 - \varpi^2\mathbf{u}_{\mathrm{L}\pm}) = b\operatorname{div}\boldsymbol{E}_\mathrm{L}.$$

Here, from the Maxwell equations we note that $\operatorname{div}\boldsymbol{D}_\mathrm{L} = 0$, and so $\boldsymbol{E}_\mathrm{L} = -\boldsymbol{P}/\varepsilon_\mathrm{o}$. Using (4.5a), $\operatorname{div}\boldsymbol{E}_\mathrm{L} = -\operatorname{div}(b'\mathbf{u}_{\mathrm{L}\pm} + \alpha\boldsymbol{E}_\mathrm{L})$ and, therefore,

$$\operatorname{div}\boldsymbol{E}_\mathrm{L} = -b'\operatorname{div}\mathbf{u}_{\mathrm{L}\pm}/(1+\alpha),$$

with which the wave equation for $\mathbf{u}_{\mathrm{L}\pm}$ is written as

$$\partial^2\mathbf{u}_{\mathrm{L}\pm}/\partial t^2 + \{\varpi^2 + bb'/(1+\alpha)\}\mathbf{u}_{\mathrm{L}\pm} = 0.$$

The characteristic frequency ω_L for the longitudinal mode can thus be obtained as

$$\omega_\mathrm{L}{}^2 = \varpi^2 + bb'/(1+\alpha). \tag{4.8}$$

We notice that the difference between ω_L and $\omega_\mathrm{T} = \varpi$ arises clearly from the nonvanishing factor bb', signifying a nonzero coupling between $\boldsymbol{p}$ and $\mathbf{u}$ along the unique z axis.

The dielectric behavior of polar pseudospins as observed at an arbitrary frequency ω can be described in terms of the response function $\varepsilon(\omega)$ that is defined as follows. From (4.6) and (4.5a), we have

$$(\varpi^2 - \omega^2)\mathbf{u}_{\mathrm{L}\pm} = b\boldsymbol{E}_\mathrm{L}, \quad \text{and} \quad \boldsymbol{P} = \varepsilon_\mathrm{o}\{\alpha + bb'/(\varpi^2-\omega^2)\}\boldsymbol{E}_\mathrm{L}.$$

Hence,

$$\boldsymbol{D}_\mathrm{L} = \varepsilon(\omega)\boldsymbol{E}_\mathrm{L} = \varepsilon_\mathrm{o}\boldsymbol{E}_\mathrm{L} + \boldsymbol{P} = \varepsilon_\mathrm{o}\{\alpha + bb'/(\varpi^2-\omega^2)\}\boldsymbol{E}_\mathrm{L},$$

where

$$\varepsilon(\omega) = \varepsilon_\mathrm{o}\{1 + \alpha + bb'/(\varpi^2-\omega^2)\}.$$

Writing

$$\varepsilon(\infty) = \varepsilon_\mathrm{o}(1+\alpha) \quad \text{and} \quad \varepsilon(0) = \varepsilon(\infty) + bb'/\varpi^2, \tag{4.9a}$$

we obtain the dielectric function in general form

$$\varepsilon(\omega) = \varepsilon(\infty) + \{\varepsilon(0) - \varepsilon(\infty)\}/\{1 - (\omega/\varpi)^2\}. \tag{4.9b}$$

From (4.9b) and (4.8), it is noted that the function $\varepsilon(\omega)$ shows specific behaviors

$$\varepsilon(\omega_T) = \pm\infty \quad \text{and} \quad \varepsilon(\omega_L) = 0 \tag{4.10}$$

at $\omega = \omega_T$ and ω_L, respectively. Fig. 4.2 shows a sketch of $\varepsilon(\omega)$, where there is a forbidden gap between $\varepsilon(0)$ and $\varepsilon(\infty)$. The frequencies ω_L and ω_T are characteristic for the pseudospin mode that is coupled with an optic lattice mode. From (4.8), (4.9a) and (4.9b) in the above argument, we can derive the equation

$$\omega_L{}^2/\omega_T{}^2 = \varepsilon(0)/\varepsilon(\infty), \tag{4.11}$$

which is known as the Lyddane–Sachs–Teller (LST) relation.

The ferroelectric phase transition is characterized by the static Curie-Weiss law at zero frequency:

$$\varepsilon(0) = C/(T - T_o),$$

therefore from the LST relation, we can state that

$$\omega_T{}^2 = \varpi^2 \propto (T - T_o),$$

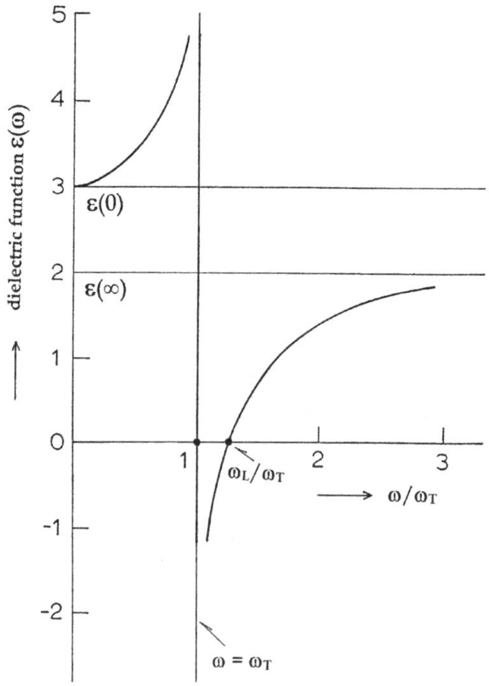

Fig. 4.2. A sketch of a dielectric function $\epsilon(\omega)$ in a uniaxial polar crystal, where $\epsilon(\omega_L) = 0$ and $\epsilon(\omega_T) = \pm\infty$.

indicating that the transversal frequency is softened in the mean-field approximation, as T_o is approached from above. According to the LST relation, the mode softening occurs toward the Curie-Weiss singularity at T_o. It is also noted that in the Landau theory, the coefficient A in the Gibbs potential is given by $A'(T - T_o)$, and, hence, $G(\eta) \propto \frac{1}{2}\varpi^2\eta^2$, thereby allowing one to consider that η in equilibrium is characterized by zero frequency, i.e. $\varpi = 0$ at $T = T_o$.

The LST formula is widely used for analyzing dielectric results. It is realized, however, that the distinction between two displacements $\mathbf{u}_q$ and $\boldsymbol{p}_q$ is clearly made by $b' \neq 0$, although their proportionality relation yields no such coupling w as proposed by (3.16). Signifying the nature of order variables $\boldsymbol{p}$, the Curie-Weiss law is valid in the limit of $b' \to 0$, and the amplitude $|\boldsymbol{p}|$ is infinitesimal in the harmonic approximation.

4.3 Long-Range Interactions and the Cochran Theory

Although inadequate for the critical region, the mean-field theory of distant interactions is capable of predicting a phase transition with respect to their average, playing the decisive role in determining the transition temperature T_o. The Weiss field signifies emerging long-range correlations at T_o, which is responsible for the singular behavior of the susceptibility in the mean-field approximation. On the other hand, Cochran [30] introduced the concept of soft modes whose characteristic frequency diminishes to zero as the transition temperature is approached. Although such a soft mode is predictable from the LST relation, the softening mechanism should be found with the effective field in conjunction with lattice anharmonicity, as will be discussed later. Using a simplified model of ionic crystals, Cochran showed that the singularity can arise from the short-range correlations as competitive with long-range interactions in the critical region.

In his model of an ionic crystal, an electric polarization appears when constituent ions are displaced from normal positions. A pair of charges $(+e, -e)$ can effectively be generated, if an ion $(+e)$ is displaced from its regular lattice site, and the crystal structure remains unchanged. If a postive ion $+e$ at a site m and a negative charge $-e$ at an adjacent site m′ are displaced by $\mathbf{u}_m$ and $-\mathbf{u}_{m'}$ by an applied field $\boldsymbol{E}$, respectively, an electric dipole moment $\boldsymbol{p}_m = e(\mathbf{u}_m - \mathbf{u}_{m'})$ is generated in the neutral crystal. If the wavelength of an applied field $\boldsymbol{E}$ is sufficiently long compared with the lattice constant, such a dipole moment formed at m and m′ sites in a short ionic distance may be regarded as located at the same site m. Cochran assumed that the field of these dipoles is uniform over all ionic sites and wrote his equation for a uniformly polarized crystal. In his theory for a one-dimensional ferroelectric, further significant is the local Weiss field that is considered as proportional to the macroscopic polarization P, when ions are displaced longitudinally along the unique axis. In addition, he considered the internal Lorentz field $\mathrm{P}/3\varepsilon_o$

due to classical dipole-dipole interactions in the direction of an applied field E.

For positive and negative ions, the transversal displacements u_{T+} and u_{T-} are driven by the transversally applied field E_T combined with the Lorentz field, and so the equations of motion can be written as

$$m_+\partial^2 u_{T+}/\partial t^2 + C(u_{T+} - u_{T-}) = e(E_T + P_T/3\varepsilon_o)$$

and

$$m_-\partial^2 u_{T-}/\partial t^2 + C(u_{T-} - u_{T+}) = -e(E_T + P_T/3\varepsilon_o),$$

respectively, where C is the elastic constant in the binding force between ionic masses m_+ and m_-. These equations can be combined for the transversal dipole moment $p_T = e(u_{T+} - u_{T-})$, and we have

$$m\partial^2 p_T/\partial t^2 + Cp_T = e^2(E_T + P_T/3\varepsilon_o),$$

where $m = m_+m_-/(m_+ + m_-)$ is the *reduced mass* of the ion pair. This equation can also be expressed in terms of the macroscopic polarization $P_T = vp_T$, where v is the number of these ion pairs per unit volume.

$$m\partial^2 P_T/\partial t^2 + CP_T = (e^2/v)(E_T + P_T/3\varepsilon_o). \tag{4.12a}$$

Therefore, we can express the transversal electric susceptibility as

$$\chi_T(\omega) = P_T/E_T = (e^2/v)/\{C - (e^2/3\varepsilon_o v) - m\omega^2\},$$

where a singularity occurs at a frequency

$$m\omega^2 = m\omega_T{}^2 = C - (e^2/3\varepsilon_o v). \tag{4.12b}$$

On the other hand for the longitudinal mode, the motion is driven by the Weiss field plus the Lorentz field as well as an applied field E_L, so that the equation of motion is given by

$$m\partial^2 P_L/\partial t^2 + CP_L = (e^2/v)(E_L + P_L/\varepsilon_o + P_L/3\varepsilon_o), \tag{4.13a}$$

from which the singularity in the longitudinal susceptibility could be determined by

$$m\omega_L{}^2 = C + 2e^2/3\varepsilon_o v. \tag{4.13b}$$

It is noted that from (4.12b) ω_T can be equal to 0, if $C = e^2/3\varepsilon_o v$, whereas from (4.13b) ω_L cannot vanish. Cochran called such ω_T the soft frequency, because one can derive the expression $\omega_T{}^2 \propto (T - T_o)$ from the LST theorem combined with the Curie-Weiss law for $\varepsilon_o \propto (T - T_o)^{-1}$. The elastic constant C can be correctly interpreted as related to the short-range correlations of order variables, as long as the displacement $\mathbf{u}_m$ is considered to represent the order variable $\boldsymbol{p}_m$. In this case, owing to (4.12b) in the Cochran theory, the transversal mode softening can be attributed to short-range correlations competing with the long-range interaction.

The Cochran theory is abstract, where the short-range interaction is considered only in the mean-field accuracy, whereas the Lorentz field represents classical dipolar interactions in polar crystals. In his theory, the temporal fluctuations in dielectric polarization exhibits a singularity when these interactions are competing.

4.4 The Quartic Anharmonic Potential in the Critical Region

Inspired by the Born-Huang theory, we discussed in Section 3.4 the strain-strain coupling between pseudospins and the hosting lattice, which we consider is responsible for fluctuations in the lattice structure. If we paid attention only to pseudospins in the critical region, ordering inevitably strains the lattice, resulting in a structural change. In Cochran's theory, a structural instability is attributed to a softening vibrational mode, which was nevertheless derived from dipolar forces counteracting ionic correlations. On the other hand, Cowley showed that the phonon frequency may shift with temperature when scattered by anharmonic lattice potentials. In his theory, such anharmonicity was not particularly regarded as related to critical fluctuations. Nevertheless, he showed that anharmonic potentials can play a significant role in scattering phonons in a crystal.

In fact, in the Landau theory, the Gibbs potential below T_o is expressed as a power series of the order parameter, where the quartic term $\frac{1}{4}B\eta^4$ emerging at the transition temperature is essentially due to correlations expressed in mean-field accuracy. Using the relation (3.16) between $\boldsymbol{\sigma}_\mathrm{m}$ and $\mathbf{u}_\mathrm{m}$ and between their Fourier transforms, the quartic term in the Landau expansion can be written as

$$\tfrac{1}{4}B\eta^4 = \tfrac{1}{4}B\left(\sum\nolimits_\mathrm{m}\boldsymbol{\sigma}_\mathrm{m}\right)^4\Big/N = \tfrac{1}{4}(B/N)\sum\nolimits_{qq'q''q''}\mathbf{u}_q\mathbf{u}_{q'}\mathbf{u}_{q''}\mathbf{u}_{q'''}, \qquad (4.14\mathrm{a})$$

to which the conservation rule

$$\boldsymbol{q} + \boldsymbol{q}' + \boldsymbol{q}'' + \boldsymbol{q}''' = 0 \qquad (4.14\mathrm{b})$$

must be applied to obtain a secular average of fluctuating potentials. We have written (4.14a) in terms of displacements $\mathbf{u}_q$, which are, however, rather indirect variables to specify strains in the deformed lattice. Although the deviations $\delta\boldsymbol{r}_\mathrm{m}$ of normal lattice points $\boldsymbol{r}_\mathrm{m}$ should be used as expressed in (3.18), we use $\mathbf{u}_\mathrm{m}$ instead of $\delta\boldsymbol{r}_\mathrm{m}$; nevertheless, by doing so, we can reduce the number of necessary notations without losing generality. In the following discussions, we consider that the displacement $\mathbf{u}_\mathrm{m}$ in the active group at site m represents the deformed structure, unless we encounter with a serious conflict.

4.4.1 The Cowley Theory of Mode Softening

At the threshold of a binary transition, a lattice displacement wave $\mathbf{u}(\boldsymbol{r},t)$ should occur spontaneously at specific wavevectors $\boldsymbol{q}' = \boldsymbol{q} \pm \Delta\boldsymbol{q}$ and at frequencies $\omega = \varpi \mp \Delta\omega$, as expressed by

$$\begin{aligned}\mathbf{u}(\boldsymbol{r},t) &= \mathbf{u}_{-q'} \exp i(\boldsymbol{q}'.\boldsymbol{r} - \varpi t) + \mathbf{u}_{q'} \exp i(-\boldsymbol{q}'.\boldsymbol{r} + \varpi t)\\ &= \mathbf{u}_{-q'} \exp i(\boldsymbol{q}'.\boldsymbol{r} - \Delta\omega.t)\exp(-i\omega t) + \mathbf{u}_{q'} \exp i(-\boldsymbol{q}'.\boldsymbol{r} + \Delta\omega.t)\exp i\omega t.\end{aligned}$$

Writing $\mathbf{u}_{\mp q'} \exp \pm i(\boldsymbol{q}'.\boldsymbol{r} - \Delta\omega.t) = \mathbf{u}_{\pm q'}(\boldsymbol{r};\boldsymbol{q}',\Delta\omega) = \mathbf{u}_{\pm q'}(t)$ for brevity, the wave $\mathbf{u}(\boldsymbol{r},t)$ can be expressed conveniently to deal with the temporal variation as

$$\mathbf{u}(\boldsymbol{r},t) = \mathbf{u}_{-q'}(t)\exp(-i\omega t) + \mathbf{u}_{+q'}(t)\exp(i\omega t),$$

implying that at a given $\boldsymbol{r}$, the amplitudes $\mathbf{u}_{\pm q'}(t)$ are modulated at $\Delta\boldsymbol{q}$ and $\Delta\omega$. In an ionic crystal, the driving force can be considered effectively as an electric field $\boldsymbol{E}_{\pm q'}$ related to $\boldsymbol{\sigma}_{\pm q'}(t)$ as in (3.31). Therefore, for such $\mathbf{u}_{\pm q'}(t)$ we can write the dynamic equation

$$\frac{\mathrm{d}^2\mathbf{u}_{\pm q'}}{\mathrm{d}t^2} + \gamma\frac{\mathrm{d}\mathbf{u}_{\pm q'}}{\mathrm{d}t} + \varpi(\boldsymbol{q}')^2\mathbf{u}_{\pm q'} = \frac{e}{m}\boldsymbol{E}_{\pm q'}\exp(\pm i\omega t), \tag{4.15}$$

where γ is the damping constant, $\varpi(\boldsymbol{q}')$ is the characteristic frequency at $\boldsymbol{q}'$, and e/m is the effective charge/mass ratio for the lattice modes $\mathbf{u}_{\pm q'}$.

Cowley [35] considered anharmonic potentials as scatterers of lattice waves $\mathbf{u}_{\pm q'}(t)$, for which cubic and quartic potentials were calculated as significant perturbations. He derived the expression for the characteristic frequency of the perturbed oscillator. The calculation was standard but lengthy, so that we only quote his theoretical results. Those readers who are interested in the derivation are referred to Ref. [35].

In the foregoing, the unprimed $\boldsymbol{q}$ was used for pseudospins, however, here we use the same $\boldsymbol{q}$ for the lattice mode that constitutes the subject for discussion in this section. Normally the equation (4.15) is written for the lattice response to $\boldsymbol{E}_{\pm q}$, for which Cowley showed that the damping constant can be effectively replaced by a complex factor $\gamma = \Gamma - i\Phi$, as the consequence of anharmonic perturbations, namely

$$\begin{aligned}\mathbf{u}_{\pm}(\boldsymbol{r};\boldsymbol{q},\Delta\omega) &= (e/m)\boldsymbol{E}_{\pm q}/\{-\omega^2 + (\Phi - i\Gamma)\omega(\boldsymbol{q}) + \varpi(\boldsymbol{q})^2\}\\ &= (e/m)\boldsymbol{E}_{\pm q}/\{-\omega^2 + i\Gamma\omega(\boldsymbol{q}) + \varpi(\boldsymbol{q},\Delta\omega)^2\},\end{aligned}$$

where

$$\varpi(q,\Delta\omega)^2 = \varpi(\boldsymbol{q})^2 + \omega\{\Phi(\boldsymbol{q},\Delta\omega) - 2i\Gamma(\boldsymbol{q},\Delta\omega)\}, \tag{4.16}$$

expresses the square of the characteristic frequency of the perturbed lattice mode. Here, $\Phi(\boldsymbol{q},\Delta\omega)$ and $\Gamma(\boldsymbol{q},\Delta\omega)$ were evaluated by the perturbation calculation of the anharmonic potentials (4.14a), for which Cowley has derived the following expressions:

$$\Phi(\boldsymbol{q},\Delta\omega) = \Phi_{\mathrm{o}}(\boldsymbol{q}) + \Phi_1(\boldsymbol{q}) + \Phi_2(\boldsymbol{q},\Delta\omega),$$

where

$$\Phi_o(\boldsymbol{q}) = \frac{\partial \varpi(\boldsymbol{q})}{\partial V}\Delta V = -\varpi(\boldsymbol{q})k_B T \left\{\frac{\phi'''(r_o)^2}{\phi''(r_o)^3}\right\} < 0,$$

$$\Phi_1(\boldsymbol{q}) = \varpi(\boldsymbol{q})k_B T \frac{\phi''''(r_o)}{8\phi''(r_o)}$$
$$= \frac{\hbar}{N\varpi(\boldsymbol{q})}\sum\nolimits_{q'} \frac{(2n'+1)}{2\omega'} V_4(-\boldsymbol{q}, \boldsymbol{q} : \boldsymbol{q}', -\boldsymbol{q}'). \tag{4.17a}$$

and

$$2\Gamma(\boldsymbol{q}, \Delta\omega) = \Phi_2(\boldsymbol{q}, \Delta\omega) = \frac{\pi\hbar}{16N\varpi(\boldsymbol{q})}\sum\nolimits_{q',q''} \frac{|V_3(\boldsymbol{q}; \boldsymbol{q}', \boldsymbol{q}'')|^2}{\omega'\omega''}$$
$$\times[(n'+n''+1)]\{-\delta(\omega+\omega'+\omega'')+\delta(\omega-\omega'-\omega'')\}$$
$$-(n'-n'')\{-\delta(\omega-\omega'+\omega'')+\delta(\omega+\omega'-\omega'')\}]. \tag{4.18a}$$

Here, Φ_o and Φ_1 are expressed in terms of derivatives of the interatomic potential $\phi(r_o)$ in the high-temperature approximation. Corresponding to the wavevectors $\boldsymbol{q}'$ and $\boldsymbol{q}''$, the frequencies are written in the above expressions as ω' and ω'', respectively. Notice that $\Phi_o = 0$ under a constant volume condition and insignificant, whereas Φ_1 is determined by the quartic potential V_4 as related to a phonon scattering

$$\boldsymbol{q} + (-\boldsymbol{q}) \leftrightarrow \boldsymbol{q}' + (-\boldsymbol{q}'), \tag{4.17b}$$

and $\Phi_2 = \Gamma$ is due to the cubic potential V_3 for a dissipative scattering

$$\boldsymbol{q} \rightarrow \boldsymbol{q}' + \boldsymbol{q}''. \tag{4.18b}$$

Therefore, for the response function of $\mathbf{u}_\pm(\boldsymbol{r}; \boldsymbol{q}, \Delta\omega)$, the temperature-dependent frequency shift should be involved in Φ_1 for nondissipative phonon-scattering processes (4.17b), whereas the damping is due to $\Phi_2 = 2\Gamma$ for dissipative processes (4.18b). It is noted that in the expression (4.17a) the sum over the states $\boldsymbol{q}'$ is proportional to the temperature T, if evaluated in the high-temperature approximation. For the process (4.17b), the perturbed characteristic frequency by Φ is given by

$$\varpi(\boldsymbol{q}, \Delta\omega)^2 = \varpi(\boldsymbol{q})^2 + \varpi(\boldsymbol{q})\Phi_1(\boldsymbol{q}) \mp \Delta\omega\Phi_1(\boldsymbol{q}).$$

For $\boldsymbol{q} = 0$, this gives, in particular,

$$\varpi(0, \Delta\omega)^2 = \varpi(0)^2 + \varpi\Phi_1(0) \mp \Delta\omega\Phi_1(0)$$
$$= \{\varpi(0) + \tfrac{1}{2}\Phi_1(0)\}^2 + \{\mp\Delta\omega\Phi_1(0) - \tfrac{1}{4}\Phi_1(0)^2\},$$

which can be equal to zero if $\varpi(0) = -\frac{1}{2}\Phi_1(0)$ and $\Delta\omega = \pm\frac{1}{4}\Phi_1(0)$, i.e. $\varpi(0) = \pm\frac{1}{2}\Delta\omega$. This means that the terminal frequency of the soft mode at

$q = 0$ is determined by $\Delta\omega$, which is an acceptable conjecture in the mean-field approximation. We can consider that these conditions for $\varpi(0, \Delta\omega) = 0$ are met at the transition temperature T_o in the mean-field accuracy. Furthermore, in the first expression for $\varpi(0, \Delta\omega)^2$, the second and third terms on the right, i.e.,

$$\varpi\Phi_1(0) \mp \Delta\omega\Phi_1(0) = \omega_\pm\Phi_1(0) \approx \mp\Delta\omega\Phi_1(0)$$

is proportional to T in the vicinity of $\omega \approx 0$. It is interesting that the phases above and below T_o can be distinguished by $-\Delta\omega$ and $+\Delta\omega$, respectively, if these signs are chosen as consistent to the wavevector transfer, referring to the energy flow direction into and out of the soft mode. In this context, it is logical to write $\omega_\pm\Phi_1(0) = \mp A'T$ near $\omega = 0$, where A' changes signs from positive and negative.

Although $\varpi(0, \Delta\omega)^2$ is zero at T_o, the small discontinuity of $\varpi(0) = \pm\frac{1}{2}\Delta\omega$ signifies switching a stable thermal mode above T_o to a unstable mode below T_o, so that we can write $\varpi(0)^2 = \pm A'T_o$ in the limit of $T \rightarrow T_o$. Summarizing these arguments, the soft mode frequencies are expressed as

$$\varpi(0, \Delta\omega)^2 = A'(T - T_o) \quad \text{for} \quad T > T_o, \tag{4.19a}$$

and

$$\varpi(0, \Delta\omega)^2 = A'(T_o - T) \quad \text{for} \quad T < T_o, \tag{4.19b}$$

representing the temperature dependences of the frequency $\varpi(0, \Delta\omega)$ in the mean-field approximations.

In the Cowley theory, anharmonic potentials were considered as perturbations, giving rise to the temperature-dependent frequency. However, the critical region is not sufficiently described in this approximation, where T_o is only a mean-field parameter for the phase transition.

4.4.2 Symmetry Change at a Continuous Phase Transition

In the foregoing, we considered a quartic potential in one dimension along the unique axis, and obtained expressions for softening frequencies (4.19a) and (4.19b). However, the structural change is signified by a change of lattice symmetry, for which a quartic potential with transversal coordinates should also be considered as responsible in three-dimensional crystals. Experimentally, symmetry changes were clearly noticed in the soft-mode spectra observed in the cell-doubling phase transition in $SrTiO_3$ at 105K and in the ferroelectric phase transition in *tris-sarcosine calcium chloride* (TSCC) at 130K, as shown in Fig. 4.3 and 4.4, respectively. In the former transition, neutron inelastic scattering experiments were performed at various points on the Brillouin-zone boundaries, whereas in the latter infrared and Raman studies at the zone center showed scattering anomalies as related to the soft modes. In TSCC, the loss of mirror symmetry at the ferroelectric phase transition was evident from the soft modes observed above and below T_c. In this subsection, we discuss

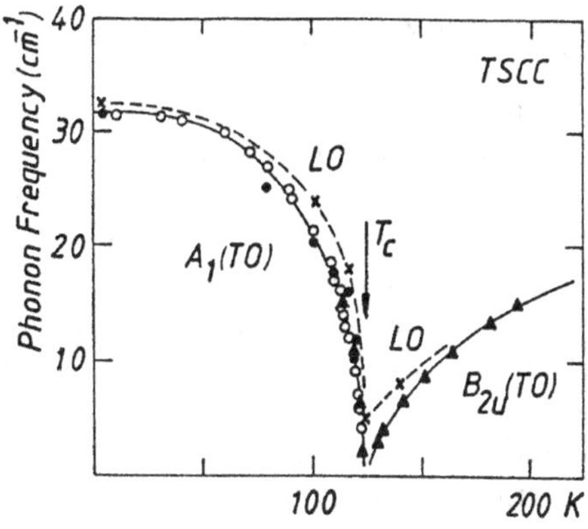

Fig. 4.3. Phonon frequencies as a function of temperature near T_c = 130K in the ferroelectric phase transition of TSCC crystals. Data by Raman and infra-red experiments are marked by (∘, •) and (▲, ×), respectively. (From J. F. Scott, Raman Spectroscopy of Structural Phase Transitions, in *Structural Phase Transitions I*, edited by K. A. Müller and H. Thomas, Springer Verlag, Heidelberg (1981).)

the role of quartic potentials that can be responsible for observed changes in soft-mode spectra at T_c in these typical examples.

As a consequence of displacements $\mathbf{u}_q$, additional lattice potentials appear that can be expressed by

$$V = \tfrac{1}{2}A\sum\nolimits_{q,q'} \mathbf{u}_q.\mathbf{u}_{q'} + \tfrac{1}{4}B\sum\nolimits_{q,q',q'',q'''} \mathbf{u}_q\mathbf{u}_{q'}\mathbf{u}_{q''}\mathbf{u}_{q'''},$$

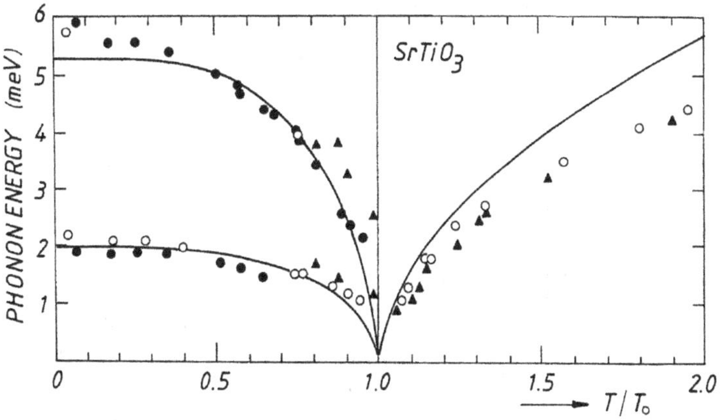

Fig. 4.4. Phonon energy vs. temperature near T_c = 105K in $SrTiO_3$. Data: ▲ Cowley et al.; • Fleury et al.; ∘ and Shirane et al. (From J. Feder and E. Pytte, Phys. Rev. B1, 4805 (1970).)

where conservation relations $\boldsymbol{q} + \boldsymbol{q}' = 0$ and $\boldsymbol{q} + \boldsymbol{q}' + \boldsymbol{q}'' + \boldsymbol{q}''' = 0$ hold for the first and second terms, respectively, so that V can be regarded as secular perturbations. For simplicity, damping and higher-order terms are excluded from V. The first harmonic term in V keeps the distorted lattice structure in a stable condition, whereas the second quartic term can be considered as responsible for a possible transformation of lattice symmetry. At room temperature, TSCC crystals are monoclinic and in a ferroelastic phase, where the mirror b plane can be easily identified in each domain crystal by viewing through crossed optical polarizers. For convenience, we take the z axis perpendicular to the b plane, and the x axis along the a direction in the mirror plane. With respect to the rectangular xyz reference system, the potentials V of B_{2u} and A_1 symmetries can be written as

$$V(\mathrm{B_{2u}}) = \tfrac{1}{2}A\mathrm{u_z}^2 + \tfrac{1}{4}B\mathrm{u_z}^4 \quad \text{and} \quad V(\mathrm{A_1}) = \tfrac{1}{2}A\mathrm{u_x}^2 + \tfrac{1}{4}B\mathrm{u_x}^4;$$

However, these are not quite independent, when V consists of additional terms mixing these modes under critical conditions.

For the symmetry change $B_{2u} \rightarrow A_1$ in the critical region, we consider a phonon-scattering process

$$\{q_\mathrm{z} + (-q_{-\mathrm{z}})\}\mathbf{k} \rightarrow \{q'_\mathrm{x} + (-q'_{-\mathrm{x}})\}\mathbf{i} + \{(-q'_\mathrm{z}) + q'_\mathrm{z}\}\mathbf{k},$$

to which the corresponding change in displacements is expressed as

$$\mathbf{u} = \pm\mathrm{u_z}\mathbf{k} \quad \text{for} \quad T > T_\mathrm{c} \quad \rightarrow \quad \mathbf{u}' = \pm\mathrm{u}'_\mathrm{x}\mathbf{i} \mp \mathrm{u}'_\mathrm{z}\mathbf{k} \quad \text{for} \quad T < T_\mathrm{c},$$

permitting these two modes to mix. Considering that $\mathrm{u_x}$ and $\mathrm{u_z}$ constitute the basis of the irreducible representation of the symmetry element, the vector $\mathbf{u}'$ can be expressed as a linear combination;

$$\mathrm{u}' = c_\mathrm{x}\mathrm{u}'_\mathrm{x} + c_\mathrm{z}\mathrm{u}'_\mathrm{z}, \quad \text{where} \quad c_\mathrm{x}{}^2 + c_\mathrm{z}{}^2 = 1,$$

and, hence,

$$\mathbf{u}'^2 = c_\mathrm{x}^2\mathrm{u}'^2_\mathrm{x} + c_\mathrm{z}^2\mathrm{u}'^2_\mathrm{z}.$$

Using these notations, the potential V_c in the critical region can be assumed as

$$\begin{aligned} V_\mathrm{c} &= \tfrac{1}{2}A\mathbf{u}'^2 + \tfrac{1}{4}B\mathbf{u}'^2\mathrm{u}'^2_\mathrm{z}, \\ &= \tfrac{1}{2}A(c_\mathrm{x}^2\mathrm{u}'^2_\mathrm{x} + c_\mathrm{z}^2\mathrm{u}'^2_\mathrm{z}) + \tfrac{1}{4}B(c_\mathrm{x}^2\mathrm{u}'^2_\mathrm{x} + c_\mathrm{z}^2\mathrm{u}'^2_\mathrm{z})\mathrm{u}'^2_\mathrm{z} \\ &= c_\mathrm{x}^2V_\mathrm{c}(\mathrm{A_1}) + c_\mathrm{z}^2V_\mathrm{c}(\mathrm{B_{2u}}), \end{aligned}$$

where

$$V_\mathrm{c}(\mathrm{A_1}) = \tfrac{1}{2}A\mathrm{u}'^2_\mathrm{x} + \tfrac{1}{4}B\mathrm{u}'^2_\mathrm{x}\mathrm{u}'^2_\mathrm{z} \quad \text{and} \quad V_\mathrm{c}(\mathrm{B_{2u}}) = \tfrac{1}{2}A\mathrm{u}'^2_\mathrm{z} + \tfrac{1}{4}B\mathrm{u}'^2_\mathrm{z}\mathrm{u}'^2_\mathrm{z}$$

are the potentials for A_1 and B_{2u} modes mixed in quartic terms. However, we can consider that u'^2_z approaches to the mean-field average $\langle\mathrm{u}'^2_\mathrm{z}\rangle = -A/B$ with decreasing temperature. Accordingly, toward the noncritical region below T_c, these potentials approach

$$V(\mathrm{A_1}) = \tfrac{1}{4}A\mathrm{u}_\mathrm{x}'^2 \quad \text{and} \quad V(\mathrm{B_{2u}}) = -A^2/4B,$$

giving stability to the lattice modes $\mathrm{A_1}$ and $\mathrm{B_{2u}}$ below T_c. However, associated with pseudospins in B_{2u} symmetry, the frequency of u_z' mode becomes virtually zero at temperatures below T_c, while the frequency of the u_x' mode becomes higher with decreasing temperature. Accordingly, the parabolic potential $V(A_1)$ below T_c is characterized by the factor $\frac{1}{2}A$, in contrast to A of $V(B_{2u})$ above T_c. Therefore, the soft mode frequency below T_o is described as

$$\varpi(0, \Delta\omega)^2 = 2A'(T_o - T) \tag{4.20}$$

in the mean-field approximation. As compared with (4.19a) for $T > T_o$, the factor 2 in (4.20) for $T < T_o$ is well recognized in the experimental curves in Figs. 4.3 and 4.4, reflecting the particular symmetry change at the structural transitions.

4.5 Observation of Soft-Mode Spectra

At the threshold of a displacive phase transition, pseudospins are in collective motion, interacting with the lattice mode at a low excitation energy near a specific point in the Brillouin zone. Such an interaction in near phase is responsible for sinusoidal modulation of the pseudospin mode characterized by a small wavevector $\Delta\boldsymbol{q}$ and frequency $\Delta\omega$. Thermodynamically, specified by a minimum of the Gibbs potential, the equilibrium shifts with decreasing temperature by a quartic anharmonicity that has emerged at T_c. Hence, fluctuations δG at the minimum Gibbs potential can be expressed as related to $\Delta\boldsymbol{q}$ and $\Delta\omega$;

$$\delta G = \tfrac{1}{2}m\varpi(\Delta\boldsymbol{q}, \Delta\omega)^2\delta\eta^2, \tag{4.21}$$

where m is the effective mass, indicating harmonic variation with the characteristic frequency

$$\varpi(\Delta q, \Delta\omega)^2 = \varpi(0, \Delta\omega)^2 + \kappa\Delta\boldsymbol{q}^2 \tag{4.22}$$

for a small $\Delta\omega$ and $\Delta\boldsymbol{q}$. Here, the parameter κ corresponds to the kinetic energy of fluctuations. Considering symmetry change at the transition, the factors A' in (4.20) are not the same at temperatures above and below T_c. Therefore, rewriting these constants as $A'_>$, $\kappa_>$ and $A'_<$, $\kappa_<$, respectively, we have

$$\varpi_>(\Delta q, \Delta\omega)^2 = A'_>(T - T_o) + \kappa_>\Delta\boldsymbol{q}^2 \tag{4.22a}$$

and

$$\varpi_<(\Delta q, \Delta\omega)^2 = A'_<(T - T_o) + \kappa_<\Delta\boldsymbol{q}^2 \tag{4.22b}$$

It is noted that these expressions can be applied to neutron inelastic scattering experiments, where soft modes can be detected as anomalies in scattering intensities at a fixed value of $\Delta \boldsymbol{q}$. However, unless scattering angles are scanned, we expect no information for breaking spatial symmetry from intensity anomalies. Leaving the problem of spatial fluctuations to later discussions, here we outline how soft modes can be detected in usual scattering experiments at a constant $\boldsymbol{q}$.

The critical region is signified by long-wave fluctuations at $\Delta \boldsymbol{q}$. In a simple dielectric crystal as discussed by Cochran, we may consider $\boldsymbol{\sigma}_{\mathrm{m}} \approx \boldsymbol{p}_{\mathrm{m}} = e(\mathbf{u}_{+} - \mathbf{u}_{-})_{\mathrm{m}}$. In the phase transition at $\boldsymbol{G} = 0$, we may write simply $\Delta \boldsymbol{q} = \boldsymbol{q}$ for small fluctuations, and the Fourier transform $\boldsymbol{p}_{\pm q}$ can be considered as driven by the internal field $\boldsymbol{E}_{\pm q}$ due to correlations with distant $\boldsymbol{p}_{\mathrm{m}}$. In an applied field $\boldsymbol{E}$, responding to the effective field $\boldsymbol{E}'_{\pm q} = \boldsymbol{E} + \boldsymbol{E}_{\pm q}$, the singular behavior of $\boldsymbol{p}_{\pm q}$ can be studied from the susceptibility in the limit of $\boldsymbol{E} \rightarrow 0$. For such a dipolar oscillation in the crystal, the equation of motion of $\boldsymbol{p}_{\pm q}$ can be written as

$$\frac{\mathrm{d}^2 \boldsymbol{p}_{\pm q}}{\mathrm{d}t^2} + \gamma \frac{\mathrm{d}\boldsymbol{p}_{\pm q}}{\mathrm{d}t} + \varpi^2 \boldsymbol{p}_{\pm q} = \frac{e^2}{m} \boldsymbol{E}'_{\pm q} \exp(-i\omega t).$$

The susceptibility can then be defined as

$$\begin{aligned} \chi_{\pm q}(\omega) &= \chi'_{\pm q}(\omega) - i\chi''_{\pm q}(\omega) = \lim_{E \rightarrow 0} (\boldsymbol{p}_{\pm q}/\boldsymbol{E}'_{\pm q}) \\ &= \boldsymbol{p}_{\pm q}/\boldsymbol{E}_{\pm q} = (e^2/m)/(\varpi^2 - \omega^2 + i\gamma\omega), \end{aligned} \tag{4.23}$$

in which the wavevector $\boldsymbol{q}$ is a fixed parameter and implicit, exhibiting primarily the temporal behavior. The real and imaginary parts of $\chi_{\pm q}(\omega)$ are

$$\chi_{\pm q}'(\omega) = (e^2/m)(\varpi^2 - \omega^2)/\{(\varpi^2 - \omega^2)^2 + \gamma^2\omega^2\} \tag{4.23a}$$

and

$$\chi_{\pm q}''(\omega) = (e^2/m)\gamma\omega/\{(\varpi^2 - \omega^2)^2 - \gamma^2\omega^2\}. \tag{4.23b}$$

These are the basic formulas for dielectric analysis of the soft mode, whose characteristic frequency ϖ can be identified from the peak of $\chi''_{\pm q}(\omega)$ or from the inflection point of $\chi'_{\pm q}(\omega)$, that occurs at $\omega = \varpi$ if damping can be neglected, i.e. $\gamma < \varpi^{-1}$ (*underdamped*), otherwise these parts show a relaxational decay (*overdamped*).

For a phase transition at a nonlattice point $\boldsymbol{G}_{\mathrm{i}} \neq 0$, soft modes can normally be observed by neutron inelastic scattering, because the wavevector of thermal neutrons are comparable with lattice constants. In fact, neutrons are scattered by heavy nuclei (or magnetic spins) occupying lattice points, serving as ideal probes for phonon spectra. By virtue of a finite $\boldsymbol{G}_{\mathrm{i}}$ comparable in magnitude with the wavevector of neutrons, such a vector determines the scattering geometry as required by the conservation law of wavevectors. For instance, for scattering at a zone-boundary point $\boldsymbol{G}_{\mathrm{i}} = \frac{1}{2}\boldsymbol{G}$, we consider an exact scattering geometry of $\boldsymbol{K}_2 - \boldsymbol{K}_1 = \frac{1}{2}\boldsymbol{G}$, where the energy relation is given by $\varepsilon_2 - \varepsilon_1 = \varepsilon_o \mp \Delta\varepsilon$. Here $\boldsymbol{K}_1$, ε_1 and $\boldsymbol{K}_2$, ε_2 are wavevectors and energies of incident and scattered neutrons, respectively, and ε_o is the lattice excitation energy associated with $\frac{1}{2}\boldsymbol{G}$. At the fixed geometry, phase fluctuations are expected as related to loss and gain of the neutron energy $\mp\Delta\omega$ during inelastic scattering process. The scattering intensity is generally expressed by the time average of correlated amplitudes of scattered neutrons;

$$\begin{aligned} I\left(\tfrac{1}{2}\boldsymbol{G}, \Delta\varepsilon\right) &= \left\langle \boldsymbol{A}_{1/2G}{}^{*} \boldsymbol{A}_{1/2G} \right\rangle_{\mathrm{t}} = \left\langle \sum\nolimits_{\mathrm{mn}} A_{1/2G,\mathrm{m}}{}^{*} A_{1/2G,\mathrm{n}} \right\rangle_{\mathrm{t}} \\ &= I\left(\tfrac{1}{2}\boldsymbol{G}, 0\right) + \left\langle \sum\nolimits_{\mathrm{m}\neq\mathrm{n}} A_{1/2G,\mathrm{m}}{}^{*} A_{1/2G,\mathrm{n}} \right\rangle_{\mathrm{t}}, \end{aligned}$$

where

$$I(\tfrac{1}{2}\boldsymbol{G}, 0) = \left\langle \sum\nolimits_{\mathrm{m}} A_{1/2G,\mathrm{m}}{}^{*} A_{1/2G,\mathrm{m}} \right\rangle_{\mathrm{t}}.$$

The quantity $A_{1/2G,\mathrm{m}}$ is called the *scattering amplitude* from the nucleus at a site m and the total scattering amplitude is given by

$$\begin{aligned} \boldsymbol{A}_{1/2G} &\propto \sum\nolimits_{\mathrm{m}} \mathbf{u}_{\mathrm{m}} \exp i[-(\varepsilon_2 - \varepsilon_1 - \varepsilon_o \pm \Delta\varepsilon)t_{\mathrm{m}}/\hbar] \\ &= \sum\nolimits_{\mathrm{m}} \mathbf{u}_{\mathrm{m}} \exp i[(\omega_2 - \omega_1 - \omega_o)t_{\mathrm{m}}] \exp i(\pm\Delta\omega . t_{\mathrm{m}}), \end{aligned}$$

where these energies are expressed in frequencies $\omega_{1,2} = \varepsilon_{1,2}/\hbar$ and $\omega_o = \varepsilon_o/\hbar$. Writing

$$\mathbf{u}_{\mathrm{m}} \exp i[(\omega_2 - \omega_1 - \omega_o)t_{\mathrm{m}}] = \mathbf{u}_o,$$

we obtain

$$\boldsymbol{A}_{1/2G} \propto \mathbf{u}_o(t) \exp(-i\Delta\omega . t) + \boldsymbol{u}_o(t) \exp(i\Delta\omega . t).$$

Therefore, the scattering anomaly can be expressed as

$$\begin{aligned} \Delta I(\tfrac{1}{2}\boldsymbol{G}, \Delta\omega) &= I(\tfrac{1}{2}\boldsymbol{G}, \Delta\omega) - I(\tfrac{1}{2}\boldsymbol{G}, 0) \\ &= 2\Re\langle \mathbf{u}_o(t)^{*} . \mathbf{u}_o(t') \rangle_{\mathrm{t}}. \end{aligned}$$

Considering that $\mathbf{u}_o$ and $\mathbf{u}_o{}^{*}$ are driven by effective fields $\boldsymbol{F}_o \exp(\mp i\Delta\omega . t)$ originating from the coupling with pseudospins, the equations of motion can be written as

$$\frac{\mathrm{d}^2\mathbf{u}_o}{\mathrm{d}t^2} + \gamma\frac{\mathrm{d}\mathbf{u}_o}{\mathrm{d}t} + \varpi^2\mathbf{u}_o = \boldsymbol{F}_o \exp(-i\Delta\omega . t)$$

and

$$\frac{\mathrm{d}^2\mathbf{u}_o{}^{*}}{\mathrm{d}t^2} + \gamma\frac{\mathrm{d}\mathbf{u}_o{}^{*}}{\mathrm{d}t} + \varpi^2\mathbf{u}_o = \boldsymbol{F}_o{}^{*} \exp(i\Delta\omega t)$$

where the steady solutions are determined from

$$(-\Delta\omega^2 + \varpi^2 \mp i\gamma\Delta\omega)|\mathbf{u}_o| = |\boldsymbol{F}_o|.$$

Accordingly,

$$\begin{aligned} \Re\langle \mathbf{u}_o{}^{*} . \mathbf{u}_o \rangle_{\mathrm{t}} &= |\mathbf{u}_o|^2 \langle \cos\Delta\omega(t - t') \rangle_{\mathrm{t}} \\ &= |\boldsymbol{F}_o|^2 \langle \cos\Delta\omega(t - t') \rangle_{\mathrm{t}} \gamma\Delta\omega / [(\varpi^2 - \Delta\omega^2) + \gamma^2\Delta\omega^2], \end{aligned}$$

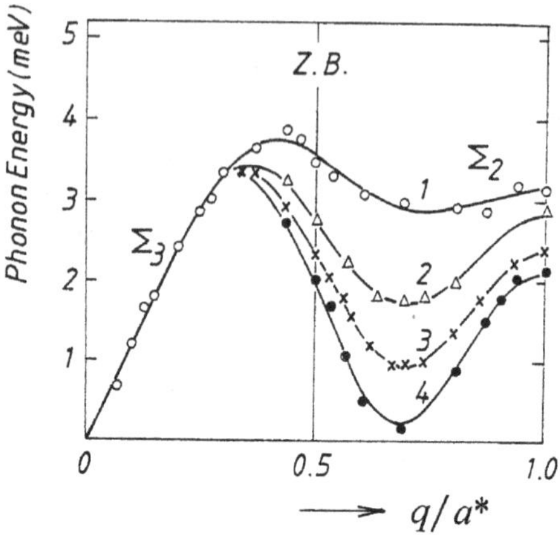

Fig. 4.5. Phonon energy in K_2SeO_4 measured by neutron inelastic scattering at $\boldsymbol{G}_\mathrm{i} = 0.7\boldsymbol{a}^*$. Curves 1, 2, 3 and 4 were obtained at 250, 175, 145 and 130K, respectively. (From M. Iizumi, J. D. Axe, G. Shirane and K. Shimaoka, Phys. Rev. B15, 4392 (1977).)

where $\langle \cos \Delta\omega(t - t')\rangle_\mathrm{t} = \sin(\Delta\omega.t_\mathrm{o})/(\Delta\omega.t_\mathrm{o})$ is the time correlation function Γ_t defined in (3.23), which is evaluated in the timescale t_o of observation. The value of Γ_t is close to 1 in the critical region, where $\Delta\omega.t_\mathrm{o} < 1$ for the impact time for neutron scattering. $\Re\langle \mathbf{u}_\mathrm{o}{}^*.\mathbf{u}_\mathrm{o}\rangle_\mathrm{t}$ is proportional to the imaginary part of the susceptibility $\chi''(\Delta\omega)$ defined in (4.23b), and hence the scattering anomaly at $\frac{1}{2}\boldsymbol{G}$ is expressed by

$$\Delta I(\tfrac{1}{2}\boldsymbol{G}, \Delta\omega) = 2|\boldsymbol{F}_\mathrm{o}|^2 \chi''_{1/2G}(\Delta\omega), \qquad (4.24)$$

allowing one to identify the soft frequency from the peak of scattering anomalies that occur when $\Delta\omega = \varpi$.

Although neutron inelastic scattering at the Brillouin-zone boundary was discussed in the above, the argument is also valid for scattering at an arbitrary point $\boldsymbol{G}_\mathrm{i}$. Figures 4.5 and 4.6a show typical examples of such soft modes, which were observed for orthorhombic K_2SeO_4 crystals at $\boldsymbol{G}_\mathrm{i} = 0.7\boldsymbol{a}^*$, as well as scattering results at the zone boundaries in $SrTiO_3$ and $KMnF_3$. Another example of a soft mode shown in Fig. 4.6b is ferroelectric anomalies observed from dielectric spectra of $\varepsilon'(\omega)$ in TSCC by Sawada and Horioka, which were interpreted as mixed with zero-frequency anomalies as discussed in the next section.

The susceptibility represents the linear response of the order variable mode to the mean field that grows with increasing correlations. Soft modes should always be observed when approaching the critical temperature of a "continuous" structural change, where a quartic potential emerges originating from the fluctuating lattice potential.

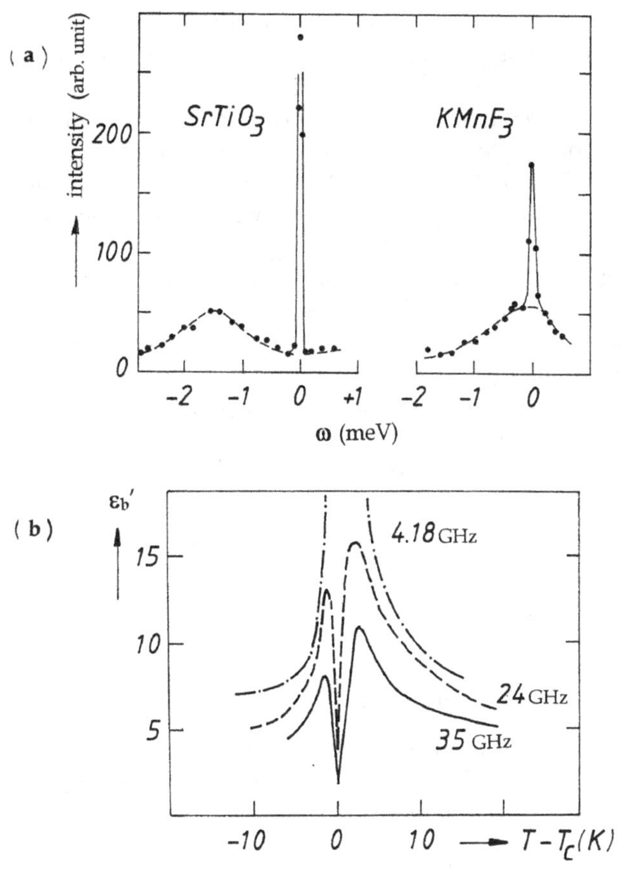

Fig. 4.6. (a) Soft-mode spectra from $SrTiO_3$ and $KMnO_3$. (From S. M. Shapiro, J. D. Axe, G. Shirane and T. Riste, Phys. Rev. B6, 4332 (1972).); (b) Oscillator-relaxator behavior in the dielectric response from TSCC near $T_c = 130K$. (From A. Sawada and M. Horioka, Jpn. J. Appl. Phys. 24-2, 390 (1985).)

4.6 The Central Peak

It was discovered that phonon susceptibility curves observed in practical crystals exhibited another anomalous absorption near zero frequency, in addition to a temperature-dependent soft mode, as shown in Fig. 4.6a. Being significantly sharp, such a zero-frequency peak, called the *central peak*, attracted many investigations in spite of its unidentified origin. Riste and his coworkers [36] discovered such a central peak in the phonon spectrum from the cell-doubling phase transition of $SrTiO_3$ crystals at temperatures close to T_c. Shapiro et al. [37] further investigated the central peak phenomenon in other systems as well, and published the spectra as shown in Fig. 4.6a. Such an absorption line at zero frequency signifies a relaxation to the lattice, which is likely due to some imperfections, although the origin cannot be positively identified from such a featureless decay. Besides, the measured relaxation time is typically of the order of 10^{-9} sec, which is often the limit of instrumental resolution. In dielectric studies of the ferroelectric phase transition in TSCC, Sawada and Horioka [38] analyzed the dielectric spectra in terms of a coupling between the soft lattice mode and the central peak, thereby interpreting the anomalies as a decay of the lattice mode when the frequency approaches to zero.

Damping occurs normally with anharmonic potentials of an odd power in the lattice potential, which are associated with strains in the crystal. In contrast, the decay at zero frequency signifies the presence of another mechanism, due presumably to lattice imperfections. We can therefore consider the damping mechanism for the lattice mode $\mathbf{u}_q$, due not only to damping to lattice strains but also to an additional relaxational mode $\mathbf{v}_q$, in the equation of motion driven by an effective driving field $\boldsymbol{F}_q$, namely

$$\frac{\mathrm{d}^2\mathbf{u}_q}{\mathrm{d}t^2} + \gamma\frac{\mathrm{d}\mathbf{u}_q}{\mathrm{d}t} + \gamma'\frac{\mathrm{d}\mathbf{v}_q}{\mathrm{d}t} + \varpi^2\mathbf{u}_q = F_q\exp(-i\omega t),$$

where the mode $\mathbf{v}_q$ obeys the relaxational equation

$$\frac{\mathrm{d}\mathbf{v}_q}{\mathrm{d}t} + \frac{\mathbf{v}_q}{\tau} = \boldsymbol{F}_q\exp(-i\omega t),$$

where τ is the relaxation time of $\mathbf{v}_q$. Assuming that the coupling between these modes is simply given by $\mathbf{v}_q = c\mathbf{u}_q$, the steady-state solutions of these equations can be expressed as

$$\mathbf{u}_{qo}(-\omega^2 - i\omega\gamma + \varpi^2) - i\omega\gamma'\mathbf{v}_{qo} = \boldsymbol{F}_q,$$

Therefore, the susceptibility for $\mathbf{u}_q$ is given by

$$\chi_q(\omega) = 1/\{\varpi^2 - \omega^2 - i\omega\gamma - ic\gamma' F_q\omega\tau/(1 - i\omega\tau)\},$$

or by letting $c\gamma' F_q = \delta^2$ for convenience we have the formula for a so-called *coupled oscillator-relaxator*:

$$\chi_q(\omega) = 1/\{\varpi^2 - \omega^2 - i\omega\gamma - \delta^2\omega\tau/(1 - i\omega\tau)\}. \tag{4.25}$$

Particularly, if the conditions $\gamma \ll \delta^2\tau$ and $\varpi \gg \tau^{-1}$ are fulfilled [39], the imaginary part of (4.25) can be shown to be

$$\chi_q''(\omega) = \frac{\omega}{\varpi^2 - \omega^2}\frac{\delta^2}{\varpi^2}\frac{\tau'}{1 + \omega^2\tau'^2} + \left(1 - \frac{\delta^2}{\varpi^2}\right)\frac{\varpi\gamma}{(\varpi^2 - \omega^2)^2 + \omega^2\gamma^2}, \tag{4.26}$$

where

$$\tau'^{-1} = \tau^{-1}/(1 - \delta^2/\varpi^2).$$

The second term in (4.26) represents an absorption due to a soft mode at $\omega = \varpi$, whereas the first one shows a relaxation of Debye's type that becomes prominent at $\omega = \delta$. A notable feature of the formula (4.26) is that the soft mode is terminated at a nonzero frequency $\varpi = \delta$, which is then taken over by the relaxation mode. The dielectric dispersion spectra of TSCC in Fig. 4.6b are dominated by a relaxational mode. In $SrTiO_3$, from the observed plot shown in Fig. 4.7 it is not immediately evident if the linear extrapolation of ϖ^2 toward T_c indicates a small non-zero δ^2, however the estimated value agrees in the order of magnitude with the value obtained from the corresponding anomalies in EPR spectra. For TSCC, Sawada and Horioka reported that $\delta = 0.6\,\mathrm{cm}^{-1}$ and $\tau = 0.9$ s were estimated from dielectric measurements at $T_c + 6$K, whereas Fujimoto and his collaborators [21] evaluated the soft-mode frequency as the order of 20GHz. Although not in sufficient agreement, the soft mode appears to have a non-zero terminal frequency in these experimental results, indicating a finite coupling δ between the soft and relaxational modes. Thus, lattice imperfections are considered to play a significant role in structural transformations, although their mechanism cannot be revealed in detail from observed central peaks.

4.7 Symmetry-Breaking Fluctuations in Binary Phase Transitions

Thermodynamically, critical fluctuations can be described by (4.21), where a variation of the Gibbs potential δG occurs around the equilibrium, arising from momentum-energy exchanges between pseudospins and soft phonons. The critical fluctuations are thus sinusoidal in space-time characterized by $\Delta \boldsymbol{q}$ and $\Delta\omega$ or described by the fluctuating phase. Therefore, as signified in part by $\pm\Delta \boldsymbol{q}$, the spatial variation in the critical region can be revealed

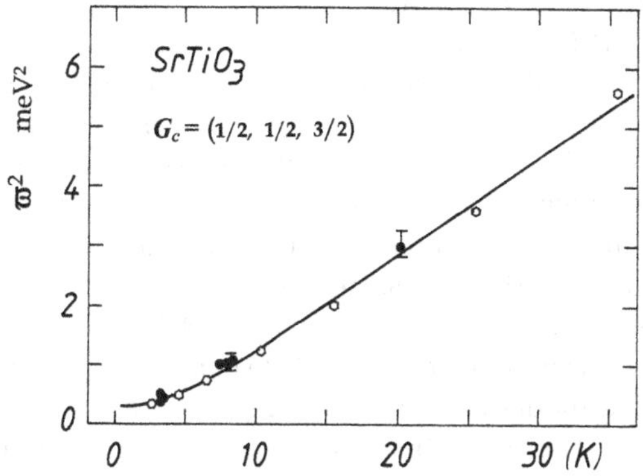

Fig. 4.7. A plot of the squared soft-mode frequency ϖ^2 vs. $T - T_c$, from neutron inelastic scattering from $SrTiO_3$ at $\boldsymbol{G}_i = (1/2, 1/2, 3/2)$.

by scanning $\boldsymbol{q}$ in scattering experiments or by sampling the condensate with magnetic resonance probes.

Caused by the coupling w with a lattice mode in near phase, binary pseudospins fluctuate around $(\pm\boldsymbol{q}, \mp\omega)$, for which the corresponding kinetic energies can be written as

$$\varepsilon(\boldsymbol{q} \pm \Delta\boldsymbol{q}) = (\hbar^2/2m)(\boldsymbol{q} \pm \Delta\boldsymbol{q})^2, \qquad \varepsilon(-\boldsymbol{q} \pm \Delta\boldsymbol{q}) = (\hbar^2/2m)(-\boldsymbol{q} \pm \Delta\boldsymbol{q})^2.$$

Here, $\pm\Delta\boldsymbol{q}$ and the corresponding energy variation $\varepsilon(\boldsymbol{q}\pm\Delta\boldsymbol{q})-\varepsilon(\boldsymbol{q}) = \mp\hbar\Delta\omega$ are due to the interaction with the lattice mode. Therefore, the pseudospin modes are described by

$$\begin{aligned}\boldsymbol{\sigma}_{q\pm\Delta q} &= \boldsymbol{\sigma}_q \exp i\{(\boldsymbol{q} \pm \Delta\boldsymbol{q}).\boldsymbol{r} - (\omega \pm \Delta\omega)t\},\\ \boldsymbol{\sigma}_{-q\pm\Delta q} &= \boldsymbol{\sigma}_{-q} \exp i\{(-\boldsymbol{q} \pm \Delta\boldsymbol{q}).\boldsymbol{r} + (\omega \pm \Delta\omega)t\},\end{aligned}$$

where the amplitudes are written as $\boldsymbol{\sigma}_q$ and $\boldsymbol{\sigma}_{-q}$, respectively.

A binary crystal system is signified by reflection symmetry on the mirror plane that is identifiable crystallographically. The pseudospin modes propagating in opposite directions are therefore reflected with respect to the mirror plane, and these amplitudes should be related by the condition $\boldsymbol{\sigma}(\boldsymbol{r}) \to -\boldsymbol{\sigma}(-\boldsymbol{r})$. Owing to the sinusoidal nature, the inversion $\boldsymbol{r} \to -\boldsymbol{r}$ in the crystal space is equivalent to the wavevector inversion $\boldsymbol{q} \to -\boldsymbol{q}$ in the reciprocal space. At temperatures below T_c, the inversion symmetry is violated by forming two opposite domains that are related to broken reflection symmetry for these pseudospin modes. Therefore, the inversion relation $\boldsymbol{\sigma}_q \to -\boldsymbol{\sigma}_{-q}$ should be applied to these amplitudes.

We note that these fluctuating kinetic energies are identical if $\Delta\boldsymbol{q} = 0$, i.e. $\varepsilon(\boldsymbol{q}) = \varepsilon(-\boldsymbol{q})$, and hence for breaking reflection symmetry it is necessary to identify asymmetrical fluctuations between $+\Delta\boldsymbol{q}$ and $-\Delta\boldsymbol{q}$. Also noted is that inversion of pseudospins is primarily independent of the hosting harmonic lattice, so that such fluctuations should be related to anharmonic interactions.

Writing that $\boldsymbol{K} = \pm\boldsymbol{q}\mp\Delta\boldsymbol{q}$ for convenience, the kinetic energies of fluctuations $\varepsilon(\pm\boldsymbol{K}) = \hbar^2\boldsymbol{K}^2/2m$ are plotted against $\pm\boldsymbol{K}$ in Fig. 4.8. Emphasized by the enlarged central portion, two parabolic curves for $\varepsilon(\boldsymbol{K})$ and $\varepsilon(-\boldsymbol{K})$ intersect at $\boldsymbol{K} = 0$, at which $\varepsilon(0) = \varepsilon(\boldsymbol{q}) = \varepsilon(-\boldsymbol{q}) = \hbar^2\boldsymbol{q}^2/2m$ for $\Delta\boldsymbol{q} = 0$. On the other hand, if a perturbing potential exists at $\boldsymbol{K} = 0$, these two states can no longer be independent, allowing fluctuations to describe by combined states. Being a familiar level-crossing problem, such a degeneracy at $\boldsymbol{K} = 0$ can be lifted by a perturbing anharmonic potential.

Using $\pm$ signs for $\Delta\boldsymbol{q}$ and $\Delta\omega$ defined as in (4.1), the perturbed pseudospin mode $\boldsymbol{\sigma}(x,t)$ in the vicinity of $\boldsymbol{K} = 0$ can be expressed by a linear combination of two propagating modes at $\boldsymbol{K} = \pm\boldsymbol{q}\mp\Delta\boldsymbol{q}$ and $\Delta\varepsilon = \hbar\Delta\omega$. Here, taking the direction of propagation x as perpendicular to the mirror plane, the fluctuating mode can be expressed as

$$\boldsymbol{\sigma}(x,t) = c_+\boldsymbol{\sigma}_o \exp i(Kx - \Delta\omega.t) + c_-\boldsymbol{\sigma}_o \exp i(-Kx + \Delta\omega.t)$$

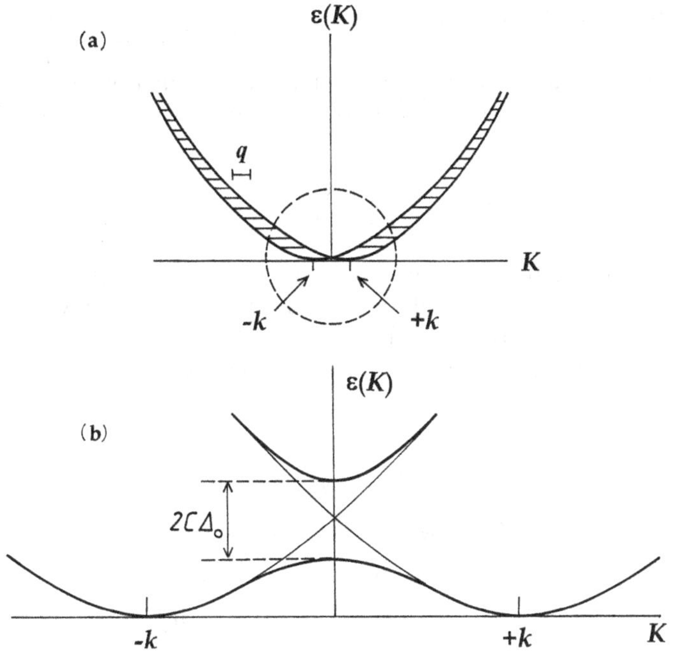

Fig. 4.8. (a) Critical spatial fluctuations near the minimum condensate energies at $K = \pm k$; (b) a magnified view of the circled part in (a) showing an energy gap at $K = 0$ due to a perturbing quartic potential.

or

$$\boldsymbol{\sigma}(\phi) = c_+ \boldsymbol{\sigma}_o \exp(i\phi) + c_- \boldsymbol{\sigma}_o \exp(-i\phi), \tag{4.27}$$

where $Kx - \Delta\omega.t = \phi$ is the phase of fluctuations in the vicinity of $K = 0$. The coefficients c_+ and c_- are mixing constants that are normalized as $c_+^2 + c_-^2 = 1$. For a small value of K, being a continuous function of x and t, the phase ϕ can be considered to take continuous angles in the range $0 \leq \phi \leq 2\pi$ in repetition.

Corresponding to the pseudospin mode, the fluctuating lattice energy can generally be expressed as

$$\delta U = \tfrac{1}{2} A \mathbf{u}_K \mathbf{u}_{-K} + \tfrac{1}{2}\kappa \left(\frac{\partial \mathbf{u}_K}{\partial x}\right)\left(\frac{\partial \mathbf{u}_{-K}}{\partial x}\right) + \tfrac{1}{4} B \sum\nolimits_{K,-K} \mathbf{u}_K \mathbf{u}_{-K} \sum\nolimits_{K',-K'} \mathbf{u}_{K'} \mathbf{u}_{-K'}, \tag{4.28}$$

where the first and second terms represent the potential and kinetic energies for harmonic distortion, and the third one is anharmonic in the fourth order. For such a quartic potential, we only need to consider the phonon scattering, $K + (-K) \to K' + (-K')$, to obtain a secular perturbation, as in the Cowley theory. If the phonon scattering $(K', -K')$ is regarded as independent from the scattering $(K, -K)$, the factor $\sum_{K',-K'} \mathbf{u}_{K'} \mathbf{u}_{-K'}$ in the quartic energy can be replaced by the mean-field average $\langle \mathbf{u}_{K'}^2 \rangle$, which is given by $-A/B$ with coefficients in Landau's expansion. Known as the Wick approximation, such an approximation allows one to make the calculation simpler, reducing the perturbation to quardratic one. As the result, the quartic potential energy can be expressed as

$$\delta U_p = -\tfrac{1}{4} A \sum\nolimits_{K,-K'} \mathbf{u}_K \mathbf{u}_{-K}, \tag{4.28a}$$

excluding all other terms in the vicinity of $K = 0$. It is noted that the same approximation was used in Subsection 4.4.2 for the symmetry change in the ferroelectric phase transition in TSCC.

Writing (4.28a) with indexes $K = q \pm \Delta q$ explicitly, we have

$$\begin{aligned}
\delta U_p = -\tfrac{1}{4} A[&\mathbf{u}_{q\pm\Delta q}{}^* \mathbf{u}_{q\pm\Delta q} + \mathbf{u}_{-q\pm\Delta q}{}^* \mathbf{u}_{-q\pm\Delta q} \\
&+ \mathbf{u}_{q+\Delta q}{}^* \mathbf{u}_{-q+\Delta q} + \mathbf{u}_{q-\Delta q}{}^* \mathbf{u}_{-q-\Delta q} \\
&+ \mathbf{u}_{q+\Delta q}{}^* \mathbf{u}_{-q-\Delta q} + \mathbf{u}_{q-\Delta q}{}^* \mathbf{u}_{-q+\Delta q} \\
&+ \mathbf{u}_{-q+\Delta q}{}^* \mathbf{u}_{q+\Delta q} + \mathbf{u}_{-q-\Delta q}{}^* \mathbf{u}_{q-\Delta q} \\
&+ \mathbf{u}_{-q-\Delta q}{}^* \mathbf{u}_{q+\Delta q} + \mathbf{u}_{-q+\Delta q}{}^* \mathbf{u}_{q-\Delta q}],
\end{aligned}$$

where $\mathbf{u}_{q+\Delta q} = \mathbf{u}_o \exp i\{(q+\Delta q)x - (\varpi + \Delta\omega)t\}$, $\mathbf{u}_{q-\Delta q} = \mathbf{u}_o \exp i\{(q-\Delta q)x - (\varpi - \Delta\omega)t\}$, and so forth. Considering only those terms for $\pm q \mp \Delta q$ modes, the fluctuation potential energy can be expressed as

$$\begin{aligned}
\delta U_p = -\tfrac{1}{4} A\mathbf{u}_o^2 [2 + 4\cos(2qx) + 2\cos\{2(q+\Delta q)x - 2\Delta\omega.t\} \\
+ 2\cos\{2(q-\Delta q)x + 2\Delta\omega.t\}].
\end{aligned}$$

Clearly, only the last two terms can be effective for the phase matching with the perturbed pseudospin $\boldsymbol{\sigma}(x,t)$ of (4.27) in the vicinity of $K = 0$. For the pseudospin modes at $K = 0$, we can select the partial potential $V_p(2\phi)$ in phase with the quartic lattice potential δU_p signified by the phase $2\phi = 2(Kx - \Delta\omega.t)$. Thus, we arrive at the perturbing potential energy for $\boldsymbol{\sigma}(\phi)$ at $K = 0$:

$$V(\phi) = C\Delta_o \cos 2\phi, \tag{4.29}$$

for which $\Delta_o \cos 2\phi$ can be regarded as an effective displacement due to the quartic strains and C is a constant proportionality factor.

For the unperturbed modes $\boldsymbol{\sigma}(\phi)$ and $\boldsymbol{\sigma}(-\phi)$ with degenerated energies at $K = 0$, we calculate the matrix element of $V(\phi)$:

$$\begin{aligned}
\int_0^{2\pi} \boldsymbol{\sigma}(\phi)^* V(\phi) \boldsymbol{\sigma}(-\phi) \mathrm{d}\phi \Big/ \int_0^{2\pi} \mathrm{d}\phi &= \frac{C\Delta_o}{2\pi} \int_0^{2\pi} \exp(-2i\phi)\cos(2\phi)\mathrm{d}\phi \\
&= \frac{C\Delta_o}{2\pi} \int_0^{\pi} \cos^2(2\phi)\mathrm{d}\phi = \frac{C}{8\pi}\Delta_o.
\end{aligned}$$

Writing $C' = C/8\pi$ for brevity, for the energy $\varepsilon(\phi) = \varepsilon(-\phi) = \hbar^2 q^2/2m$ at $K = 0$, the degeneracy will be lifted as calculated with the secular equation

$$\begin{vmatrix} \varepsilon(\phi) - \varepsilon & C'\Delta_o \\ C'\Delta_o & \varepsilon(-\phi) - \varepsilon \end{vmatrix} = 0.$$

Solving this equation, we obtain

$$\begin{aligned} \varepsilon = \varepsilon_\pm &= \tfrac{1}{2}\{\varepsilon(\phi) + \varepsilon(-\phi)\} \pm \left[\tfrac{1}{4}\{\varepsilon(\phi) - \varepsilon(-\phi)\}^2 - (C'\Delta_o)^2\right]^{1/2} \\ &= \hbar^2 q^2/2m \pm C'\Delta_o, \end{aligned} \tag{4.30}$$

which gives an energy gap $\varepsilon_+ - \varepsilon_- = 2C'\Delta_o$ at $K = 0$. Corresponding to these energies $\varepsilon_\pm$ separated by $2C'\Delta_o$, the pseudospin modes of (4.27) are given by symmetric and antisymmetric combinations of $\boldsymbol{\sigma}(\pm\phi)$, i.e. $\frac{1}{2}\{\boldsymbol{\sigma}(\phi) \pm \boldsymbol{\sigma}(-\phi)\}$. It is noted that these energies are unchanged by reflection of these combined modes, where the mixing constants c_+ and c_- in (4.27) are ± 1, as determined by the normalization condition $c_+^2 + c_-^2 = 1$. Thus the normalized functions can be expressed as

$$\boldsymbol{\sigma}_\pm(\phi) = \tfrac{1}{2}\{\boldsymbol{\sigma}(\phi) \pm \boldsymbol{\sigma}(\pi - \phi)\} = \tfrac{1}{2}\{\boldsymbol{\sigma}(\phi) \mp \boldsymbol{\sigma}(-\phi)\}, \tag{4.31}$$

which are then assigned to the perturbed levels $\varepsilon_\pm$, respectively. The antisymmetric $\boldsymbol{\sigma}_-(\phi)$ represents the lower level ε_-, and symmetric $\boldsymbol{\sigma}_+(\phi)$ is for the upper level ε_+, being expressed as proportional to $\cos\phi$ and $\sin\phi$, which are traditionally called the phase and amplitude modes.

It is noted that the reflection symmetry constitutes a subgroup of the symmetry group of a binary crystal, whereas the lattice structure remains unchanged under the space-time reversal $(x, t) \to (-x, -t)$ or the phase reversal $\phi \to -\phi$. Being represented by combined $\boldsymbol{\sigma}$ and $\mathbf{u}$, energies $\varepsilon_\pm$ of the perturbed condensate should also be associated with the symmetric and antisymmetric combinations of lattice modes $\mathbf{u}_\pm$, i.e.

$$u_\pm(\phi) = \tfrac{1}{2}\{u(\phi) \pm u(-\phi)\} \propto \cos\phi \quad \text{and} \quad \sin\phi, \tag{4.32}$$

and, hence, are assigned to the combinations of $(\boldsymbol{\sigma}_-, u_+)$ and $(\boldsymbol{\sigma}_+, \mathrm{u}_-)$, which are characterized by $\cos\phi$ and $\sin\phi$, respectively. A condensate is a combined object, where $\boldsymbol{\sigma}_+$ and $\boldsymbol{\sigma}_-$ exchange momentum and energy with the lattice modes u_- and u_+, respectively. In neutron-scattering experiments, anomalies are primarily related to a symmetry change in the lattice mode $\mathrm{u}_\mp$ in the energy states $\varepsilon_\pm$, whereas those in magnetic resonance are due to breaking reflection symmetry exhibited by the pseudospin mode $\boldsymbol{\sigma}_\pm$. The mirror symmetry is not violated in $\boldsymbol{\sigma}_+$ and $\mathbf{u}_-$ of the $\sin\phi$ mode in the upper energy ε_+ that is not quite stable at $\phi = 0$. We might as well call these fluctuation modes of a condensate as $\cos\phi$- and $\sin\phi$-modes to avoid confusion with the traditional nomenclatures, phase and amplitude modes.

We note in the above argument that the state ε_+ of the $\sin\phi$ mode is unstable at $\phi = 0$, whereas ε_- of the $\cos\phi$ mode fluctuates between two minima at $\phi = \pm\frac{1}{2}\pi$. The argument is correct in idealized crystals, however in practice we cannot disregard other kinds of potential that may play a significant role in stabilizing condensates. In fact, in a uniform applied electric field, the asymmetrical $\sin\phi$ mode can be stabilized at $\phi = \pm\frac{1}{2}\pi$, thereby converting $\sin\phi$ to $\cos\phi$. Experimentally it is confirmed that such an asymmetrical potential originates from the internal long-range field in a ferroelectric domain or from applying externally electric field (see Section 5.2).

The collective pseudospin mode $\boldsymbol{\sigma}(\phi)$ is an internal variable, and the macroscopic properties of the critical region are generally specified by distributed phases ϕ as $f(\boldsymbol{\sigma})\mathrm{d}\phi$. It is more practical to express such a distribution of pseudospin amplitudes in the range between σ and $\sigma + \mathrm{d}\sigma$, instead of the phase between ϕ and $\phi + \mathrm{d}\phi$. For a cos-mode, the variable ϕ can be converted to σ by the relation $\mathrm{d}\sigma = \sigma_o(-\sin\phi)\mathrm{d}\phi$, hence by letting $\sigma/\sigma_o = \xi = \cos\phi$, we can write $\mathrm{d}\phi = -\mathrm{d}\xi/(1-\xi^2)^{1/2}$. In this context, the density function is expressed as $f(\sigma)\mathrm{d}\xi/(1-\xi^2)^{1/2}$, which becomes infinite as $\xi \to \pm 1$, or $\sigma \to \sigma_o$, thereby visualizing binary phase fluctuations between $\pm\frac{1}{2}\pi$. On the other hand, for the $\sin\phi$ mode $\sigma/\sigma_o = \sin\phi$, we have $\mathrm{d}\xi = \cos\phi\mathrm{d}\phi$ and $\mathrm{d}\phi = \mathrm{d}\xi/\xi$, where the density is centered in the vicinity of $\xi = 0$. In Part Two, we will discuss these fluctuations in detail in relation to observed quantities in practical experiments.

Such a spatial profile of the sinusoidal fluctuations is thus observable, if the characteristic time $2\pi/\Delta\omega$ are sufficiently long as compared with the timescale t_o of experiments. In fact, $\Delta\omega$ is typically of the order of 10^{11}Hz, which is slightly higher than conventional magnetic resonance frequencies, and the reciprocal impact time in neutron scattering is shorter than $\Delta\omega$, so that such spatial fluctuations can be explicit in these observations of critical states.

Neutron inelastic scattering experiments by Bernard and his coworkers [41] on the phase transition in β-$ThBr_4$ crystals at 81K (Fig. 4.9a) and biphenyl crystals at 41.5K are the examples among others, where two modes of fluctuations u_+ and u_- were clearly resolved in the critical regions. Fig. 4.9b shows another example for critical fluctuation modes observed near the ferroelectric phase transition in K_2SeO_4 crystals at 95K, which were identified in Raman scattering experiments by Wada and his group [41]. In their studies, the $\sin\phi$ mode of lattice fluctuations was not observed, because presumably, order variables were not associated with the Raman active mode in the range between 95 and 129K. The two fluctuation modes in phonon- and pseudospin measurements are compared in the sketch shown in Fig. 4.10. In spite of these experimental results, the origin of central peaks was not positively identified, although the condensate model is regarded as a valid model.

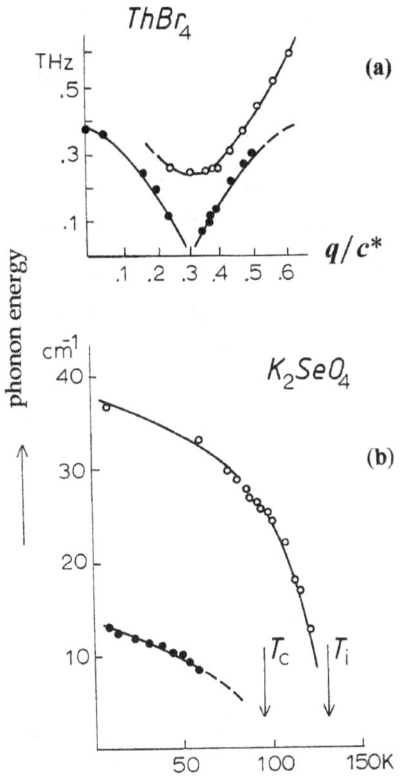

Fig. 4.9. (a) The phonon dispersion curve in $ThBr_4$ crystals at 81K, showing resolved amplitude and phase modes. (From L. Bernard, R. Currat, P. Delanoye, C. M. E. Zeyen, S. Hubert and Kouchkovsky, J. Phys. C16, 433 (1983).) (b) Phonon energy curve observed by Raman scattering from K_2SeO_3. The phase between T_i and T_c is incommensurate, and the phase below T_c is ferroelectric. (From M. Wada, H. Uwe, A. Sawada, Y. Ishibashi, Y. Takagi and T. Sakudo, J. Phys. Soc. Japan, 43, 544 (1977).)

4.8 Macroscopic Observation of a Binary Phase Transition; λ-anomaly of the Specific Heat

A significant feature of pseudospin condensates is their thermal stability, for which the soft mode of low damping is considered as responsible, whereas the collective pseudospin mode is observable in a short timescale $t_o \leq 2\pi/\varpi$. It is noted that near a phase transition at the order of 100K, such a low-energy excitation signifies mechanical strains in the lattice structure, which cannot be subjected to statistical arguments based on the *ergodic* hypothesis. Sharing primarily no thermal energies with the crystal, such a *nonergodic* excitation is not thermally accessible, although underdamped soft modes indicate a slow energy transfer between condensates and their surroundings. In this context,

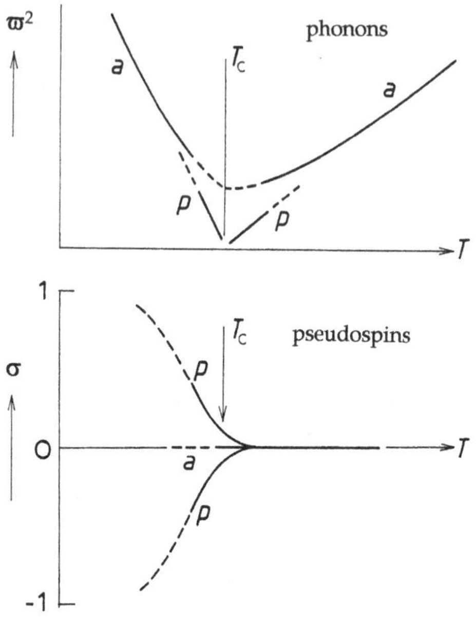

Fig. 4.10. Phonon dispersion curves and the corresponding pseudospin variation. The amplitude and phase modes are indicated by a and p.

the Gibbs free energy can be taken as a legitimate thermodynamic potential, because the volume of a crystal may not be constant under internal stresses. The threshold of a phase transition is dominated by short-range correlations, where the condensates are in mechanical equilibrium with the lattice strains. It is noted that the specific heat of such a system is measured with a considerably longer timescale than the characteristic time $2\pi/\varpi$ of condensates.

In Section 4.7, the fluctuating mode $\boldsymbol{\sigma}(\pm\phi)$ was considered as perturbed by spontaneous strains expressed by an effective displacement $\Delta(\phi) = \Delta_o \cos\phi$. As derived from the lattice strain energy in (4.28), the corresponding internal strain energy δU_{strain} must be considered as the "sink" for the ordering energy, although thermally almost isolated from the rest of the crystal. In terms of the strains $\Delta_K(\phi)$, the fluctuating macroscopic energy δU_{strain} can be written as

$$\delta U_{\text{strain}} = -\tfrac{1}{2}\alpha \sum\nolimits_K \Delta_K(\phi)\Delta_{K'}(\phi'), \tag{4.33}$$

where

$$\Delta_K = \Delta_o \cos 2\phi, \quad \Delta'_K = \Delta_o \cos 2\phi',$$
$$\phi = Kx - \Omega t \quad \text{and} \quad \phi' = Kx' - \Omega t'.$$

Here, $K = q \pm \Delta q$, $\Omega = \varpi \mp \Delta\omega$, and α is the proportionality constant, therefore,

$$\delta U_{\text{strain}} = -\tfrac{1}{2}\alpha\Delta_o^2 \cos 2\phi \cos 2\phi'.$$

As indicated by (4.31) and (4.32), critical space-time fluctuations are observed in two independent modes as $\cos\phi$ and $\sin\phi$ at constant Ω and constant K. In the trigonometric formula $2\cos 2\phi \cos 2\phi = \cos 2(\phi+\phi') + \cos 2(\phi-\phi')$, we notice that the first term depends on the average phase $\phi_o = \frac{1}{2}(\phi+\phi')$, that is however constant and hence insignificant. On the other hand, the second cosine term determined by $\cos 2\Omega(t-t')$ at $x = x'$ leads to the unvanishing time average $\frac{1}{2}$, if integrated over a long timescale, and for thermal observation the space-time average of δU_{strain} can be expressed as $\langle \delta U_{\text{strain}} \rangle_t = -(1/4)\alpha\Delta_o^2 + \text{const}$. Hence, we can write the corresponding strain energy of the lattice as

$$W_u = \tfrac{1}{4}A\Delta_o^2 + \text{const.}, \tag{4.34}$$

where A is a constant after adjusting the magnitude to the macroscopic W_u. We consider that the fluctuating pseudospin mode exchanges the energies $\pm\hbar\Delta\omega$ with W_u that can be interpreted as the internal "heat sink".

In (4.30), correlation energies of collective pseudospins are expressed in two separate modes of kinetic energies $\varepsilon_\pm(\phi)$. For the fluctuating wavevector $K = q_o \mp \Delta q$, (4.30) can be written as

$$\varepsilon_\pm(q_o \mp K, \Delta_o) = \varepsilon_\pm(\pm\Delta q, \Delta_o) = \tfrac{1}{2}(X_o + X) \pm (X_o X + C'^2\Delta_o^2)^{1/2},$$

where the abbreviations $X_o = \hbar^2 q_o^2/m$ and $X = \hbar^2 K^2/m$ are used for convenience. It is noted that this equation was derived at fixed space time coordinates (x,t), while the wavevector q or K fluctuates.

The Gibbs free energy of a system of pseudospins and the lattice strains can be written as

$$G_\pm = \int_{-q_o}^{+q_o} \langle \varepsilon_\pm(q_o \mp K, \Delta_o)\rangle_t \frac{dK}{2\pi} + W_u,$$

which can be minimized with respect to Δ_o independently for the two states of a condensate in thermal equilibrium with the surroundings. For the lower state, we can write

$$\frac{dG_-}{d\Delta_o} = 2\int_0^{q_o} (d\varepsilon_-/d\Delta_o)\frac{dK}{\pi} + \frac{dW_u}{d\Delta_o} = 0,$$

where

$$\frac{d\varepsilon_-}{d\Delta_o} = -\frac{C'^2\Delta_o}{(X_o X + C'^2\Delta_o^2)^{1/2}}.$$

Writing $X_o X = \xi^2$, where $\xi = (\hbar^2/m)K q_o$, we have $dK = (q_o/X_o)d\xi$, and so

$$\int_0^{q_o} \frac{dK}{(X_o X + C'^2\Delta_o^2)^{1/2}} = \frac{q_o}{X_o}\int_0^{X_o} \frac{d\xi}{(\xi^2 + C'^2\Delta_o^2)^{1/2}} = \frac{q_o}{X_o}\sinh^{-1}\frac{X_o}{C'\Delta_o},$$

and hence the equilibrium condition can be expressed as

$$\tfrac{1}{2}A\Delta_o - \left(\frac{2C'^2 m\Delta_o}{\pi\hbar^2 q_o}\right)\sinh^{-1}\frac{\hbar^2 q_o^2}{mC'\Delta_o} = 0.$$

From this relation, the strain amplitude Δ_o in equilibrium can be determined from

$$\frac{\hbar^2 q_o^2}{mC'\Delta_o} = \sinh\left(\frac{-\hbar^2 q_o \pi A}{4mC'\Delta_o}\right),$$

or

$$|C'|\Delta_o \approx \left(\frac{2\hbar^2 q_o^2}{m}\right)\exp\left(\frac{-\hbar^2 q_o \pi A}{4mC'^2}\right),$$

if the argument of the sinh-function is larger than 1. We can derive the same conclusion from minimizing the function G_+. Thus, the Gibbs free energy of a binary system exhibits a discontinuity $2C'\Delta_o$ at T_c.

The specific heat curve measured with varying temperature showed an anomaly characterized by a shape of the Greek letter "lambda", so called the λ-anomaly, as illustrated in Fig. 1.10. With the condensate model, a sharp rise of the curve at T_c, as approached from above, can be interpreted as "sudden" appearance of the discontinuity $\Delta G = 2C'\Delta_o$ due to the outset of correlated motion, and hence $(\Delta C_p)_{T_c} = \infty$, whereas a gradual tail below T_c signifies the slow ordering process in and below the critical region.

In the non-critical region below T_c, the ordering process should be consistent with the temperature-dependence of the order parameter for a long-range order, i.e., $\eta \propto (T_o - T)^{1/2}$. We may therefore consider the temperature-dependent Weiss field E_{int} as responsible for the gradual change in the specific heat curve. As will be analyzed in detail in Section 5.9, we can consider that the pseudospin energy is transferred stepwise to the lattice as $\Delta\sigma E_{\text{int}}(T) \propto \Delta T$ in the *soliton* potential, representing deformed levels of the structure. Therefore, the change in the Gibbs potential below T_c for a temperature step ΔT can be expressed as

$$\Delta G_< = -2C'\Delta - \alpha\left\{\frac{\Delta\sigma E_{\text{int}}(T)}{\Delta T}\right\}\Delta T,$$

and hence

$$G_<(T) = -2C'\Delta_o(T_o - T) - \alpha\left(\frac{\Delta\sigma}{\Delta T}\right)\int_{T_o}^{T} E_{\text{int}}\,\mathrm{d}T$$

where α is a constant related to previously defined α in (1.17). Assuming $E_{\text{int}} \propto (T_o - T)^{1/2}$, and writing k for the front factor of the integral to simplify the expression, we obtain

$$\Delta C_p = \left(\frac{\partial \Delta G_<}{\partial T}\right)_p = 2C'\Delta_o + \tfrac{1}{2}k/(T_o - T)^{1/2},$$

which agrees at least qualitatively with the gradual tail in observed curves, except that it becomes infinity as $T \to T_o$ instead of the transition temperature T_c. Presumably, in the critical region the distributed E_{int} is responsible for the anomaly between T_o and T_c. Nevertheless, using a empirical critical exponent

α', we may write $\Delta C_p \propto (T_c - T)^{-\alpha'}$ for observed anomalous curves, where $E_{int} \propto \eta(T_c - T)^{\beta}$ can be considered to cover the region between T_o and T_c with the exponents α' and β. Hence, we have a relationship $\beta - 1 = -\alpha'$, which gives specifically $\alpha' = \beta = \frac{1}{2}$ and $T_c = T_o$ in the mean-field approximation.

5

Dynamics of Pseudospins Condensates and the Long-Range Order

5.1 Imperfections in Practical Crystals

In the critical region, pseudospin condensates are in sinusoidal modes of fluctuations expressed as $\sigma_o \cos\phi$ and $\sigma_o \sin\phi$, where the amplitude σ_o is infinitesimal at the threshold, and the phase $\phi = \Delta q.x + \Delta\omega.t + \phi_o$ represents the propagation at a speed $v = \Delta\omega/\Delta q$ along a specific direction x in an anisotropic crystal. In idealized crystals, the phase constant ϕ_o is left undetermined, unless a boundary condition is imposed at space-time coordinates (x_o, t_o) of a lattice site. It is noted that observing small Δq and $\Delta\omega$ by light- and neutron-scattering experiments does not fully substantiate pseudospin condensates unless the scatterers are duly identified. By magnetic resonance sampling, on the other hand, collective pseudospins in slow motion can be visualized, yielding a credible image of condensates in the laboratory frame of reference.

Real crystals are by no means "perfect" because of the presence of surfaces and unavoidable imperfections, which violate the lattice periodicity. Of course, we may assume practical crystals to be perfect, if observing with probes at sufficiently higher kinetic energies than the depth of imperfection potentials. Nevertheless, for a low-energy condensate, such lattice imperfections can be significant obstacles for propagation, thereby immobilizing (*pinning*) condensates in their vicinity. Being stationary in crystals as pinned by imperfections, condensates can be observed in quality crystals with a sufficiently low defect density. The quality ferroelectric crystals of this kind, for example, can be evaluated by values of *coersive* force in the *hysteresis* curve for dielectric polarization.

Although existing in various types in practical crystals, the most significant role played by imperfections is that the lattice periodicity is only disrupted at their sites, as described by a model of a point imperfection, and in quality crystals, lattice defects can be considered primarily of this type. Although a perfect crystal is only a theoretical model, such point defects are essential for pinning condensates, which nevertheless allows us to study the nature of condensates in relation to a given lattice structure. Pinned condensates constitute primarily a subject for experimental investigation of structural phase transitions.

5.2 The Pinning Potential

In this section, we consider only the critical region where the propagating mode of collective pseudospins is sinusoidal in character. A stationary point defect at a lattice site $\boldsymbol{r}_\mathrm{i}$ can be represented by local field $F(\boldsymbol{r}-\boldsymbol{r}_\mathrm{i})$ at a point $\boldsymbol{r}$ close to $\boldsymbol{r}_\mathrm{i}$, reflecting symmetry at the defect site, constituting a subgroup of the point group of the crystal. We assume that such a field $F(\boldsymbol{r}-\boldsymbol{r}_\mathrm{i})$ interacting with a pseudospin σ_j at a site $\boldsymbol{r}_\mathrm{j}$ ($\mathrm{j} \neq \mathrm{i}$) represents the local distortion of the lattice. Considering correlations between pseudospins along a particular direction x in an anisotropic crystal, the attractive potential $V(x,t;x_\mathrm{i})$ for a collective pseudospin mode at the defect site x_i can be expressed by

$$\mathrm{d}V(x,t;x_\mathrm{i}) = -\sigma(\phi)F(x-x_\mathrm{i})\mathrm{d}x.$$

Further, we consider that the field F is symmetrical with respect to the defect center x_i; that is

$$F(x-x_\mathrm{i}) = F(x_\mathrm{i}-x), \tag{5.1}$$

although the defect symmetry can be lower in general if the coordinate x_i can shift from the original lattice point in the vicinity of a vacant site. In addition, we may assume that the field is highly localized in the vicinity of x_i, and hence we can write

$$F(x-x_\mathrm{i}) = F\delta(x-x_\mathrm{i}), \tag{5.2}$$

where the delta function signifies a localized field at $x = x_\mathrm{i}$ with a strength F. We can then define the *pinning potential* at x_i and t by the integral $-\int\{\partial V(x,t;x_\mathrm{i})/\partial x\}\mathrm{d}x$. For binary sin- and cos-modes, we define pinning potentials

$$V_\mathrm{A}(x_\mathrm{i},t) = -\int \sigma_\mathrm{o}F\sin(\Delta q.x - \Delta\omega.t + \phi_\mathrm{o})\delta(x-x_\mathrm{i})\mathrm{d}x$$

and

$$V_\mathrm{P}(x_\mathrm{i},t) = -\int \sigma_\mathrm{o}F\cos(\Delta q.x - \Delta\omega.t + \phi_\mathrm{o})\delta(x-x_\mathrm{i})\mathrm{d}x,$$

respectively. Using (5.1), these pinning potentials can be expressed as

$$V_\mathrm{A}(x_\mathrm{i},t) = -\sigma_\mathrm{o}F\sin(\Delta q.x_\mathrm{i} - \Delta\omega.t + \phi_\mathrm{o})$$

and (5.3)

$$V_\mathrm{P}(x_\mathrm{i},t) = -\sigma_\mathrm{o}F\cos(\Delta q.x_\mathrm{i} - \Delta\omega.t + \phi_\mathrm{o}).$$

These phases ϕ_i are distributed in the crystal, since the defect coordinates x_i are randomly distributed. However, it is noted that the spatial phases $\Delta q.x_\mathrm{i}$

are virtually continuous because of small Δq and random x_i, so that ϕ_i can be regarded as a continuous variable in the whole angular range in repetition. Accordingly, instead of distributed ϕ_i in a system of pinned condensates we may use the continuous phase variable

$$\phi = \Delta q.x - \Delta\omega.t + \phi_o \tag{5.4}$$

where $0 \le \phi \le 2\pi$. With such a continuous ϕ as defined in (5.4), the pinning potentials can be expressed by

$$V_A(\phi) = -V_o \sin\phi \quad \text{and} \quad V_P(\phi) = -V_o \cos\phi, \tag{5.5}$$

where we set $V_o = \sigma_o F$, for brevity. Normally, for symmetrical defects expressed by (5.1), $V_P(0) = -V_o < 0$, where a condensate in the cos mode can be stablilized at $\phi = 0$, and the sin mode is unstable at $\phi = 0$, and $V_A(0) = 0$.

On the other hand, when pinned by *asymmetrical defects*, the pinning potentials are $V_A(\frac{1}{2}\pi) = -V_o$ and $V_P(\frac{1}{2}\pi) = 0$, and, hence, the sin mode can be stabilized at $\phi = \frac{1}{2}\pi$. In fact, the normal defect is represented by a symmetrical potential, whereas an *external* field E can be considered to provide such an asymmetric potential $-\sigma_o E$, hence giving an equilibrium at $\phi = \frac{1}{2}\pi$. Thus, an external electric field causes a significant effect for ferroelectric phase transitions, as polar ordering can be induced in addition to the spontaneous mechanism. It is interesting to note that the internal field of long-range order may provide such an asymmetrical potential, showing a behavior similar to E.

Dynamically, a pinned phase mode $\sigma_P(\phi)$ at symmetrical defects should fluctuate in an oscillatory motion around the equilibrium $\phi = 0$, for which the restoring force

$$f_R = -\frac{\partial V_P(\phi)}{\partial x} = -\Delta q \frac{\partial V_P(\phi)}{\partial \phi} = -\Delta q V_o \sin\phi$$

is responsible. Therefore, for a small phase variation $\delta\phi = \phi - 0$ in the vicinity of $\phi = 0$, the pinning potential can be written as

$$V_P(\delta\phi) \approx V_o + \tfrac{1}{2}V_o(\delta\phi)^2,$$

which can be responsible for dynamic fluctuations of a pinned condensate. Obviously, such a fluctuation $\delta\phi$ should be originated from the phonon interaction in the condensates.

Rice [42] discussed such an oscillatory motion of a charge-density-wave condensate in the presence of an applied electric field $E = E_o \exp(-i\omega t)$, and derived the susceptibility formula. Assuming that E_o represents the amplitude of an applied field at the wavevector q and of negligible damping for simplicity, the equation of motion in the potential $V_o = \frac{1}{2}m\omega_o^2(\delta\phi)^2$ can be written as

$$\frac{m}{\Delta q}\frac{\mathrm{d}^2(\delta\phi)}{\mathrm{d}t^2} + \frac{m\omega_o^2}{\Delta q}\delta\phi = eE_o \exp(-i\omega t),$$

where m and e are effective mass and charge of the CDW-condensate. Writing $\delta\phi = (\delta\phi)_o \exp(-i\omega t)$, the steady solution gives the susceptibility

$$\chi(\omega) = (\delta\phi)_o/E_o = (e/m)/(\omega_o^2 - \omega^2), \tag{5.6}$$

representing the dielectric response of the pinned condensate, which shows a singularity at $\omega = \omega_o$. Pawlacyk et al. [43] discovered a very low-frequency mode of fluctuations at $\omega_o = 0.1$GHz in the dielectric spectra from TSCC crystals near $T_c \sim 130$K, which was clearly different from the soft-mode frequency of the order of 20GHz.

Supported by experimental evidence, such a pinning potential as a function of the internal variation $\delta\phi$ permits a thermodynamical description of fluctuations in terms of a dynamic Gibbs function $G(\delta\phi)$, i.e.

$$G(\delta\phi) = \int_0^L \left[\tfrac{1}{2}\kappa\{\partial\sigma(\delta\phi)/\partial t\}^2 + V(\delta\phi)\right] (\mathrm{d}x/L), \tag{5.7a}$$

where κ is related to the kinetic constant proportional to the inverse mass of a condensate and L is the length of integration.

In the critical region of a ferroelectric phase transition, an applied static electric field E will modify the pinning scheme of condensates, where the sin mode $\sigma_A(\delta\phi)$ can also be pinned. In the vicinity of $\phi = 0$, the pinning potentials can be expressed as

$$V_P(\delta\phi, E) = -V_o\cos(\delta\phi) - \sigma_o \int E\cos(\delta\phi)\mathrm{d}x$$

and

$$V_A(\delta\phi, E) = -\sigma_o \int E\sin(\delta\phi)\mathrm{d}x.$$

For a uniform applied field $E = -\mathrm{d}V/\mathrm{d}x = -\Delta q\mathrm{d}V/\mathrm{d}\phi$, where the potential function V is antisymmetric with respect to ϕ and x, in contrast to symmetric defect potentials. Hence the first integral vanishes, because

$$\begin{aligned}\int E\cos(\delta\phi)\mathrm{d}x &= -\int \left(\frac{\mathrm{d}V}{\mathrm{d}\phi}\right)\cos(\delta\phi)\mathrm{d}\phi = -\int \mathrm{d}V\cos(\delta\phi)\\ &= -\Delta V\cos(\delta\phi) - (-\Delta V)\cos(-\delta\phi) = 0,\end{aligned}$$

whereas the second integral is not zero;

$$\int E\sin(\delta\phi)\mathrm{d}x = -\Delta V\sin(\delta\phi) - (-\Delta V)\sin(-\delta\phi) = -2\Delta V\sin(\delta\phi).$$

Here, $2\Delta V = -2(E/\Delta q)\delta\phi$ represents the potential difference between the phase limits $\pm\delta\phi$ in these integrals. Therefore, although $V_P(\delta\phi, E)$ is virtually unchanged by a weak field E on, these pinning potentials can be written as

$$V_A(\delta\phi, E) = -2\sigma_o\Delta V\cos(\tfrac{1}{2}\pi \pm \delta\phi) \quad \text{and} \quad V_P(\delta\phi, E) = -2\sigma_o\Delta V\cos(\delta\phi), \tag{5.8}$$

signifying that the pinning equilibrium for the sin mode is also established at $\phi = \frac{1}{2}\pi$. Such a sine mode behaves like a cos mode, after shifting the phase by $\frac{1}{2}\pi$. With increasing E, such a $V_A(\delta\phi, E)$ becomes indistinguishable from $V_P(\delta\phi)$, because these equilibrium phases $\frac{1}{2}\pi$ and 0 are both in the same range between 0 and 2π. In any case, σ_A can be identified with a weak applied field E, although its lineshape is featureless. The experimental detail for such a *field pinning* will be discussed in Chapter 9. The Gibbs function for fluctuating $\sigma_{A,P}$ in an applied field E can be written as

$$G_{A,P}(\delta\phi, E) = \int_0^L \left[\tfrac{1}{2}\kappa\left\{\frac{\partial\sigma_{A,P}(\delta\phi)}{\partial t}\right\}^2 + V_{A,P}(\delta\phi, E)\right]\frac{dx}{L}, \tag{5.7b}$$

For amplitude and phase modes, such Gibbs functions can be minimized independently, and where the dynamic fluctuations are described as a harmonic phase variation $\delta\phi$.

5.3 The Lifshitz Condition for Incommensurate Fluctuations

In Chapters 3 and 4, we discussed modulated structures of collective pseudospins, originating from competing short-range correlations, although it was uncertain if such a structure constitutes a macroscopic phase. Nevertheless, if a state specified by the single continuous variable $\boldsymbol{\sigma}(\phi)$ is stable at a given temperarure and pressure, the crystal should be considered in a thermodynamic phase [45], for which the Gibbs potential $G(\boldsymbol{\sigma}(\phi))$ is expressed as a function of $\boldsymbol{\sigma}, p$ and T. There are examples of modulated phases among real crystal systems, where pinned pseudospins exhibit a stationary modulated structure incommensurate with the lattice period. In principle, such a modulated structure is time dependent due to interactions with the lattice, but is observed as if steadily modulated in the crystal space in a short timescale.

Incommensurate fluctuations were first observed at microwave frequencies in the critical region of the ferroelectric phase transition of TSCC [21], whereas stable incommensurate crystal phases had been known in other systems, for which Lifshitz formulated the thermodynamical criterion. In this section, we discuss the Lifshitz condition for incommensurability, which can be derived from correlated pseudospins, although the origin for fluctuations was unspecified in the Lifshitz argument.

Normally, the variable $\boldsymbol{\sigma}(\phi)$ is subjected to phase fluctuations in crystals, when immobilized by a pinning potential. Apart from the origin, such fluctuations can be described as due to correlations among pseudospins themselves. Considering the sinusoidal pseudospin mode $\boldsymbol{\sigma}(\phi) = \boldsymbol{\sigma}_o \exp i\phi$, the correlations are predominantly between different phases ϕ_1 and ϕ_2, whereas the amplitude $\boldsymbol{\sigma}_o$ is unchanged under constant p and T. Thus, the fluctuating phase difference

$\phi_1 - \phi_2$ is significant in the specific direction x, whereas $\boldsymbol{\sigma}_o$ is virtually infinitesimal in the critical region. We calculate the corresponding density correlations due to significant interference $\boldsymbol{\sigma}^*(\phi_1)\boldsymbol{\sigma}(\phi_2)$ between "nonlattice points" x_1 and x_2, when observed in a timescale $t_o \leq t_1 - t_2$. Therefore, the free energy of pseudospins should be characterized by an additional term G_L determined by such density correlations. He proposed to consider $G_L \propto \langle \boldsymbol{\sigma}^*(\phi_1)\boldsymbol{\sigma}(\phi_2)\rangle_t$, which is calculated as the time average over the timescale t_o.

For a one-dimensional correlations along x, such correlations can be written as

$$\langle \boldsymbol{\sigma}^*(\phi_1)\boldsymbol{\sigma}(\phi_2)\rangle_t = \boldsymbol{\sigma}_o^2 \left\langle \sum\nolimits_{\pm} \exp\{(\pm i\Delta q)(x_2 - x_1)\} \exp\{(\mp i\Delta\omega)(t_2 - t_1)\} \right\rangle_t,$$

in which the time correlation factor can be expressed as

$$\Gamma_t = \langle \exp\{\mp i\Delta\omega(t_2 - t_1)\}\rangle_t = t_o^{-1} \int_0^{t_o} \cos(\Delta\omega.\tau)d\tau = \frac{\sin(t_o\Delta\omega)}{t_o\Delta\omega},$$

where the variable $\tau = t_2 - t_1$ is in the range $0 \leq \tau \leq t_o$

Because of time-reversal symmetry, the function Γ_t is real, and the value is almost equal to 1 if the condition $t_o\Delta\omega \ll 1$ is fulfilled. Signified as "slow" in this case, such binary correlations yields a quasi-static condition, being dominated by the symmetrical spatial correlation factor $\cos\{\Delta q(x_2 - x_1)\}$. It is noted however that such a spatial correlation function for regular lattice points vanishes if $x_2 - x_1 =$ integer $\times$ lattice constant, whereas it is nonzero if otherwise, and observable for nonlattice points, provided that $t_o\Delta\omega \ll 1$. In this case, the fluctuations are revealed as incommensurate.

Denoting small deviations from a regular lattice point x as $x_1 = x - \delta x$ and $x_2 = x + \delta x$, where $|\delta x| < a$, the lattice constant, the correlation function can be written as

$$\begin{aligned}\Gamma(\delta x) &= \langle\{\boldsymbol{\sigma}(x_1, t_1) - \boldsymbol{\sigma}(x, t)\}^*\{\boldsymbol{\sigma}(x_2, t_2) - \boldsymbol{\sigma}(x, t)\}\rangle_t \\ &= \left\langle \boldsymbol{\sigma}^*(x, t)\frac{\partial\boldsymbol{\sigma}(x, t)}{\partial x} - \boldsymbol{\sigma}(x, t)\frac{\partial\boldsymbol{\sigma}^*(x, t)}{\partial x} \right\rangle_t \boldsymbol{\delta} x,\end{aligned}$$

where the quantity in the brackets $< \ldots >$ must be nonzero for $\Gamma(\boldsymbol{\delta}x)$ to express nonvanishing correlations for $\boldsymbol{\delta}x \neq 0$. Obviously, if $\boldsymbol{\delta}x = 0$, $\Gamma(0) = 0$ regardless of the quantity in the bracket. Lifshitz has proposed that the incommensurability is assured if the Gibbs potential has an extra term G_L proportional to $\Gamma(\delta x)$, namely

$$G_L = (iD/2) \int_o^L \left\langle \boldsymbol{\sigma}^* \frac{\partial\boldsymbol{\sigma}}{\partial x} - \boldsymbol{\sigma}\frac{\partial\boldsymbol{\sigma}^*}{\partial x} \right\rangle_t \frac{dx}{L}, \tag{5.9}$$

and $G_L \neq 0$ is called the Lifshitz condition for incommensurability. Here, the coefficient $iD/2$ in (5.9) is defined to include δx, and the factor $\frac{1}{2}i$ is set for convenience. In the above, a continuum crystal is assumed, which is valid in the long-wave approximation.

Although the equilibrium can be specified by the free energy in Landau's expansion, the dynamic Gibbs function for the fluctuating state should consist of the kinetic energy for propagation as well as the correlation term G_{L}. Thus the dynamic Gibbs function can be expressed as

$$G(\boldsymbol{\sigma}) = G(0) + \int_0^L \left\langle \tfrac{1}{2}a|\boldsymbol{\sigma}|^2 + \tfrac{1}{4}b|\boldsymbol{\sigma}|^4 + \tfrac{1}{2}\kappa\left|\frac{\partial\boldsymbol{\sigma}}{\partial x}\right|^2 \right\rangle_{\mathrm{t}} \frac{\mathrm{d}x}{L} + G_{\mathrm{L}}, \tag{5.10}$$

where $\kappa = mc_{\mathrm{o}}^2$ and c_{o} the speed of propagation. Assuming that $\Gamma_{\mathrm{t}} = 1$ for brevity, (5.10) can be written for $\boldsymbol{\sigma} = \boldsymbol{\sigma}_{\mathrm{o}} \exp i\phi$ as

$$G(\boldsymbol{\sigma}_{\mathrm{o}}, \phi) = G(0) + \int_0^L \left[\tfrac{1}{2}a\boldsymbol{\sigma}_{\mathrm{o}}^2 + \tfrac{1}{4}b\boldsymbol{\sigma}_{\mathrm{o}}^4 + \tfrac{1}{2}\kappa\left(\frac{\mathrm{d}\boldsymbol{\sigma}_{\mathrm{o}}}{\mathrm{d}x}\right)^2 + \tfrac{1}{2}\kappa\boldsymbol{\sigma}_{\mathrm{o}}^2\left(\frac{\mathrm{d}\phi}{\mathrm{d}x}\right)^2 + D\boldsymbol{\sigma}_{\mathrm{o}}^2\left(\frac{\mathrm{d}\phi}{\mathrm{d}x}\right)\right] \frac{\mathrm{d}x}{L}.$$

The pseudospin system can be in equilibrium with the lattice excitation. When the lattice counterpart is expressed by G_{S}, the thermal equilibrium can be determined by $\mathrm{d}\{G(\boldsymbol{\sigma}) + G_{\mathrm{S}}\} = 0$, where G_{S} is primarily independent of $\boldsymbol{\sigma}$. Therefore, the equilibrium values of $\boldsymbol{\sigma}_{\mathrm{o}}$ and ϕ can be obtained by solving the equations $\partial G/\partial\boldsymbol{\sigma}_{\mathrm{o}} = 0$ and $\partial G/\partial\phi = 0$ simultaneously:

$$a\boldsymbol{\sigma}_{\mathrm{o}} + b\boldsymbol{\sigma}_{\mathrm{o}}^3 + \kappa\left(\frac{\mathrm{d}^2\boldsymbol{\sigma}_{\mathrm{o}}}{\mathrm{d}x^2}\right) + \kappa\boldsymbol{\sigma}_{\mathrm{o}}\left(\frac{\mathrm{d}\phi}{\mathrm{d}x}\right)^2 + 2D\boldsymbol{\sigma}_{\mathrm{o}}\left(\frac{\mathrm{d}\phi}{\mathrm{d}x}\right) = 0 \tag{i}$$

and

$$\left\{\kappa\boldsymbol{\sigma}_{\mathrm{o}}^2\left(\frac{\mathrm{d}\phi}{\mathrm{d}x}\right) + D\boldsymbol{\sigma}_{\mathrm{o}}^2\right\}\left(\frac{\mathrm{d}}{\mathrm{d}\phi}\right)\left(\frac{\mathrm{d}\phi}{\mathrm{d}x}\right) = 0. \tag{ii}$$

From the equation (ii), we see immediately that

$$\frac{\mathrm{d}\phi}{\mathrm{d}x} = -\frac{D}{\kappa} = q,$$

indicating that the wavevector is generally irrational, as D and κ are parameters unrelated to the lattice periodicity and primarily temperature dependent. The term $\kappa(\mathrm{d}^2\boldsymbol{\sigma}_{\mathrm{o}}/\mathrm{d}x^2)$ in (i) is very small and negligible, so that

$$\boldsymbol{\sigma}_{\mathrm{o}}^2 = -(a - D^2/\kappa)/b,$$

indicating that the amplitude $\boldsymbol{\sigma}_{\mathrm{o}}$ is temperature dependent and, indeed, a function of temperature and pressure in practical systems. Although the origin is unspecified in the above argument, the coupling with the soft mode is responsible for incommensurate fluctuations.

5.4 A Pseudopotential for Condensate Locking and Commensurate Modulation

Pseudospins represent active groups occupying regular lattice points, hence their collective motion should be sensitive to any subtle change deviating from the regular periodic structure. We discussed point defects disrupting translational symmetry in Section 5.1; there is another significant case of *pseudo structure* for a structural transformation. A pseudo structure may not be clearly detected by crystallographic observation, as is often too small to be resolved by X-ray diffraction. Nevertheless, pseudospin fluctuations may be stabilized by such a pseudo structure when the wavelength becomes comparable with the repeat unit in the pseudo structure, being responsible for a commensurate lattice modulation. Such transitions from an incommensurate phase to a commensurately modulated phase as occurring at particular temperatures T_i have been found among practical systems, which are characterized by a change of continuous phase variables ϕ to discrete angles.

A simple example of pseudo potentials is screw symmetry due to successive rotations of active groups along a direction. Typically, such screw symmetry is observed in twofold or threefold screw axis, along which constituent molecules rotate by π or $2\pi/3$ in succession over two or three unit cells, respectively. Accordingly, in these cases, the repeat unit of pseudo symmetry can be two or three times longer than the regular lattice constant. Such an *incommensurate-to-commensurate* phase transition is considered for phase matching between phases of a pseudospin mode and a screw potential in the same direction.

Obviously an excitation energy is involved in such a modulated structure, for which a pseudolattice potential U_m with an m-fold screw axis along a specific crystallographic axis, say the b axis, is considered to be responsible. In the pseudopotential U_m for active groups at each site p between two ends of the unit, transversal vectors $u_{\perp p}$ can be considered for successive rotation-translation along the axis, namely

$$u_{\perp p} = u_{\perp o} \exp i\theta_p, \quad \text{where} \quad \theta_p = \pm\frac{2\pi}{m}p \quad \text{and} \quad p = 1, 2, \ldots, m. \tag{5.11}$$

The pseudopotential for screw symmetry can therefore be expressed as

$$U_m \propto \sum\nolimits_{\pm p} u_{\perp p} = \sum\nolimits_{\pm p} u_o\{\exp(i\theta_p) + \exp(-i\theta_p)\},$$

or

$$U_m \propto 2u_o \sum\nolimits_p \cos\theta_p = U_o \sum\nolimits_p \cos(G_b x_p),$$

where

$$G_b = b^*/m \quad \text{and} \quad x_p = pb.$$

The potential U_m is characterized by m maxima as specified by successive rotations θ_p at x_p, as given by (5.11). Accordingly, applying the phase matching rule to this case, the phase ϕ of a pseudospin mode can be locked into the phase of U_{m}, when $\mathrm{m}\phi$ becomes equal to the phase $G_{\mathrm{b}}x_{\mathrm{p}}$, i.e. $\phi = 2\pi\mathrm{p}/\mathrm{m}$, $\mathrm{p} = 0, 1, \ldots, \mathrm{m}-1$.

For a collective pseudospin mode we consider a potential V_{m}, as related to the lattice distorting potential U_{m}, thereby phase matching between $\boldsymbol{\sigma}$ and V_{m} can take place at a specific temperature T_{i}. In this context, for exact phase matching we can consider the potential V_{m} as proportional to $\cos(\mathrm{m}\phi)$, and write

$$V_{\mathrm{m}}(\phi) = \rho\{\boldsymbol{\sigma}^{\mathrm{m}} + (\boldsymbol{\sigma}^{\mathrm{m}})^*\}/\mathrm{m} = \left(\frac{2\rho}{\mathrm{m}}\right)\sigma_{\mathrm{o}}^{\mathrm{m}}\cos(\mathrm{m}\phi), \tag{5.12}$$

where ρ is constant. The incommensurate pseudospin mode $\boldsymbol{\sigma}(\phi)$ is therefore perturbed by the pseudopotential $V_{\mathrm{m}}(\phi)$ as the temperature is lowered through T_{i}, where the dynamic Gibbs function for $\boldsymbol{\sigma}(\phi)$ can be expressed as

$$G(\boldsymbol{\sigma}) = \int \frac{\mathrm{d}x}{L}\left\{\tfrac{1}{2}a\boldsymbol{\sigma}(\phi)^2 + \tfrac{1}{4}b\boldsymbol{\sigma}(\phi)^4 + \tfrac{1}{2}\kappa\left|\frac{\partial\boldsymbol{\sigma}(\phi)}{\partial x}\right|^2 + V_{\mathrm{m}}(\phi)\right\}. \tag{5.13}$$

It is realized that such a collective pseudospin mode is a nonergodic variable, hence, the temperature dependence should be attributed to long-range correlations, as will be discussed in Section 5.6. Therefore, representing thermodynamic properties, the variables σ_{o} and ϕ in the Gibbs function should be related to the temperature, but implicit in (5.13). Nevertheless, the Gibbs potential of (5.13) can be written as

$$G(\sigma_{\mathrm{o}}, \phi) = \int \frac{\mathrm{d}x}{\mathrm{L}}\left\{\tfrac{1}{2}a\sigma_{\mathrm{o}}^2 + \tfrac{1}{4}b\sigma_{\mathrm{o}}^4 + \tfrac{1}{2}\kappa\left(\frac{\partial\sigma_{\mathrm{o}}}{\partial x}\right)^2 + \tfrac{1}{2}\kappa\left(\frac{\partial\phi}{\partial x}\right)^2 + \left(\frac{2\rho}{\mathrm{m}}\right)\sigma_{\mathrm{o}}^{\mathrm{m}}\cos(\mathrm{m}\phi)\right\}.$$

Using the variation principle, the function $G(\sigma_{\mathrm{o}}, \phi)$ can be minimized for thermal equilibrium against arbitrary variations $\delta\sigma_{\mathrm{o}}$ and $\delta\phi$, for which the equations $\partial G/\partial\sigma_{\mathrm{o}} = 0$ and $\partial G/\partial\phi = 0$ are to be solved simultaneously; that is,

$$a\sigma_{\mathrm{o}} + b\sigma_{\mathrm{o}}^3 + 2\rho\sigma_{\mathrm{o}}^{m-1}\cos(\mathrm{m}\phi) + \kappa\sigma_{\mathrm{o}}\left(\frac{\mathrm{d}\phi}{\mathrm{d}x}\right)^2 + \kappa\left(\frac{\mathrm{d}^2\sigma_{\mathrm{o}}}{\mathrm{d}x^2}\right) = 0$$

and

$$\kappa\sigma_{\mathrm{o}}^2\frac{\mathrm{d}^2\phi}{\mathrm{d}x^2} + 2\rho\sigma_{\mathrm{o}}^{\mathrm{m}}\sin(\mathrm{m}\phi) = 0. \tag{5.14a}$$

Using the abbreviations $\psi = \mathrm{m}\phi$ and $\zeta = (2\mathrm{m}\rho/\kappa)\sigma_{\mathrm{o}}^{\mathrm{m}-2}$, the second equation can be expressed as

$$\frac{\mathrm{d}^2\psi}{\mathrm{d}x^2} - \zeta\sin\psi = 0, \tag{5.14b}$$

which is known as the *sine-Gordon* equation. Integrating (5.14a) once, we have

$$\tfrac{1}{2}\kappa\sigma_{\mathrm{o}}^2\left(\frac{\mathrm{d}\phi}{\mathrm{d}x}\right)^2 + V_{\mathrm{m}}(\phi) = \mathrm{const.}, \tag{5.14c}$$

representing the energy relation between $\boldsymbol{\sigma}(\phi)$ and the pseudo-potential U_{m}. It is noted that the kinetic energy of phase fluctuation given by the first term in (5.14c) originates from an energy exchange between V_{m} and U_{m} at T_{i}, although remaining implicit in the above, and only ϕ is considered for the phase transition.

The sine-Gordon equation describes a nonlinear motion with a finite amplitude σ_{o}, depending on the magnitude ρ of the pseudopotential. Frank and van der Merwe [46] discussed such a dynamical problem, and their results can be used for the present problem of nonlinear fluctuations. Following Böttiger's textbook [47], we write (5.14c) in the form

$$\tfrac{1}{2}(\mathrm{d}\psi/\mathrm{d}x)^2 - \zeta\cos\psi = E, \tag{5.15}$$

where E is the integration constant of (5.14b), which can be determined by values of ψ and $\mathrm{d}\psi/\mathrm{d}x$ specified at a point $x = x_{\text{o}}$, analogous to initial conditions for a "pendulum." Here, E and ζ represent the energy and the potential height, respectively, in a reduced scale. Such a classical motion is oscillatory with a finite amplitude if $E < \zeta$, whereas it is nonoscillatory if $E \geq \zeta$. In the former case, the mode ψ is stabilized by the potential $-\zeta\cos\psi$ in phase with the pseudoperiod, whereas in the latter case, the potential is no obstacle for free propagation.

The solution of (5.15) can be expressed in terms of an elliptic integral

$$x - x_{\text{o}} = \int_0^{\psi} [2(E + \zeta\cos\psi)]^{-1/2}\mathrm{d}\psi,$$

which can be rewritten in the standard form as

$$x - x_{\text{o}} = \left(\frac{\zeta^{1/2}}{\kappa}\right)\int_{\text{o}}^{\phi}(1 - \kappa^2\sin^2\Theta)\mathrm{d}\Theta, \tag{5.16}$$

where

$$\kappa^2 = 2\zeta/(E + \zeta)$$

is the squared modulus κ of the elliptic integral and $\Theta = \frac{1}{2}\psi$. The lower limit of the integral corresponds to $\Theta = 0$, and the upper limit designated as φ is the value of Θ at a given x. The integral of (5.16) can also be expressed in the reversed form

$$\sin\varphi = \mathrm{sn}\frac{\kappa(x - x_{\text{o}})}{\zeta^{1/2}}. \tag{5.17}$$

It is clear from (5.16) and (5.17) that $x - x_{\text{o}}$ varies periodically, as the angle φ varies in $\sin\varphi$, if the modulus is in the range $0 < \kappa < 1$, otherwise $x - x_{\text{o}}$ is not periodic for $\kappa \geq 1$. Correspondingly, the lock-in phase transition at temperature T_{i} can be specified as occurring at $E = \zeta$ or $\kappa = 1$ between the phases described by $E < \zeta$ and $E > \zeta$. Denoting the specific angle φ at T_{i} as

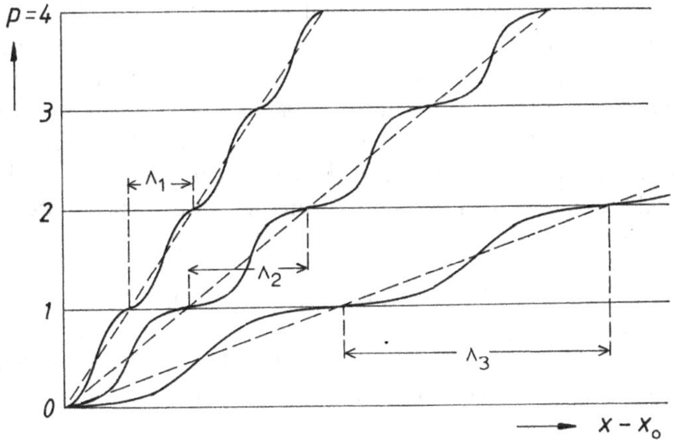

Fig. 5.1. Graphical illustration for kink solutions of the sine-Gordon equation plotted against $x - x_o$ for m = 5, p = 0, 1, 2, 3, 4. The "wavelength" Λ is defined as a distance between adjacent kinks with same derivatives.

θ, the phase transition can be described by the "kink" solution of (5.15); that is,

$$\sin\theta = \tanh\{(x - x_o)/\zeta^{1/2}\},$$

for which

$$\theta = \sin^{-1}[\tanh\{(x - x_o)/\zeta^{1/2}\} + \theta_p, \quad \text{where } \theta_p = p\pi, \quad p = 0, 1, 2, \ldots, m-1.$$

Therefore, as illustrated in Fig. 5.1, such transitions occur if $\boldsymbol{\sigma}(\phi)$ are trapped in the potential $V_m(\phi)$, as indicated by m kinks at $\theta_p = 0, \pi, 2\pi, \ldots, (m-1)\pi$, or by the relation $m\phi_p = \psi_p = 2\theta_p$ at $\phi_p = 2\pi p/m$. In fact, these discrete kinks $\sigma(\phi_p)$ were observed in K_2ZnCl_4 and Rb_2ZnCl_4 crystals as *discommensuration lines*, which can be explained by a simplified one-dimensional model at least qualitatively.

For such a modulated structure characterized by these kinks, it is useful to define the distance between kinks as $\Lambda(\kappa)$, which is expressed from (5.16) as

$$\kappa\Lambda(\kappa)/\zeta^{1/2} = 2\int_0^{\frac{\pi}{2}}(1 - \kappa^2\sin^2\Theta)d\Theta$$

The integral on the right, known as Jacobi's complete elliptic integral, is usually written as $K(\kappa)$ and, hence, $\kappa\Lambda(\kappa)/\zeta^{1/2} = 2K(\kappa)$. Here, the parameter $\Lambda(\kappa)$ is similar to the wavelength in a sinusoidal wave, representing the repeat length in the elliptical wave. Figure 5.1 shows numerical plots of the periodic elliptic function for p = 1, 2, 3 and 4, for the purpose of mathematical illustration.

In the above argument, we assumed exact phase matching between the pseudospin mode and the pseudolattice potential U_m for an incommensurate-to-commensurate phase transition. In practice, however, their coupling may not be exactly in phase, resulting in a small phase mismatch as observed in transition anomalies at T_i. In neutron inelastic scattering experiments, the pseudolattice potential, as perturbed by such a coupling with the pseudospin mode, exhibits a phase shift from the unperturbed potential $U_{mp} = U_o \cos G_b x_p$ as expressed by

$$U'_{mp} = U_o \cos(G_b x_p - \Delta\phi_p),$$

where the observed phase shift $\Delta\phi_p$ in the pth discommensuration line is usually written by the incommensurate parameter δ_p as

$$\Delta\phi_p = \phi_p \delta_p = (2\pi p/m)\delta_p. \tag{5.18}$$

Parameters δ_p determined from observed shifts represent incommensurate fluctuations effectively.

Although one-dimensional correlations is a valid assumption in anisotropic crystals, experimentally the model must be evaluated on sample crystals of good quality that are characterized by a small defect density. For phase-locking phase transitions in K_2ZnCl_4 and Rb_2ZnCl_4 crystals, Pan and Unruh [48] reported laminar patterns of discommensuration lines that were recorded by transmission electron microscopy (TEM). The photographs in Fig. 5.2a show patterns of *soliton stripes*, so called by these authors, of lines along the b axis in a plates of K_2ZnCl_4. Such patterns of discommensuration lines are interpretable primarily with the one-dimensional theory, although observed details, such as "splitting" and "vortex"-like behaviors of lines and so forth in dark fields of electron diffraction (Fig. 5.2c), require further explanation beyond the model of one-dimensional correlations. Among photographs published in ref. [48], it is also notable that isolated groups of three stripes (Fig. 5.2b) may be interpreted as arising from the pseudopotential $V_3(\phi)$. Further noted is that these lines are terminated at "vortexes," which may be attributed to unknown pinning mechanisms in the crystal. Also interesting in their observation is a fine doublet structure on each line, which appears as related to broken mirror symmetry at the threshold of the polar phase. Generally, in these experiments, pseudospin modes appear in variety of ways, depending on types and densities of the imperfections, while their intrinsic nature can only be revealed in high-quality crystals.

5.5 Propagation of a Collective Pseudospin Mode

The critical region of binary phase transitions is dominated by slow fluctuations in phase reversal $\phi \leftrightarrow -\phi$, whereas the collective pseudospins $\boldsymbol{\sigma}(\phi)$

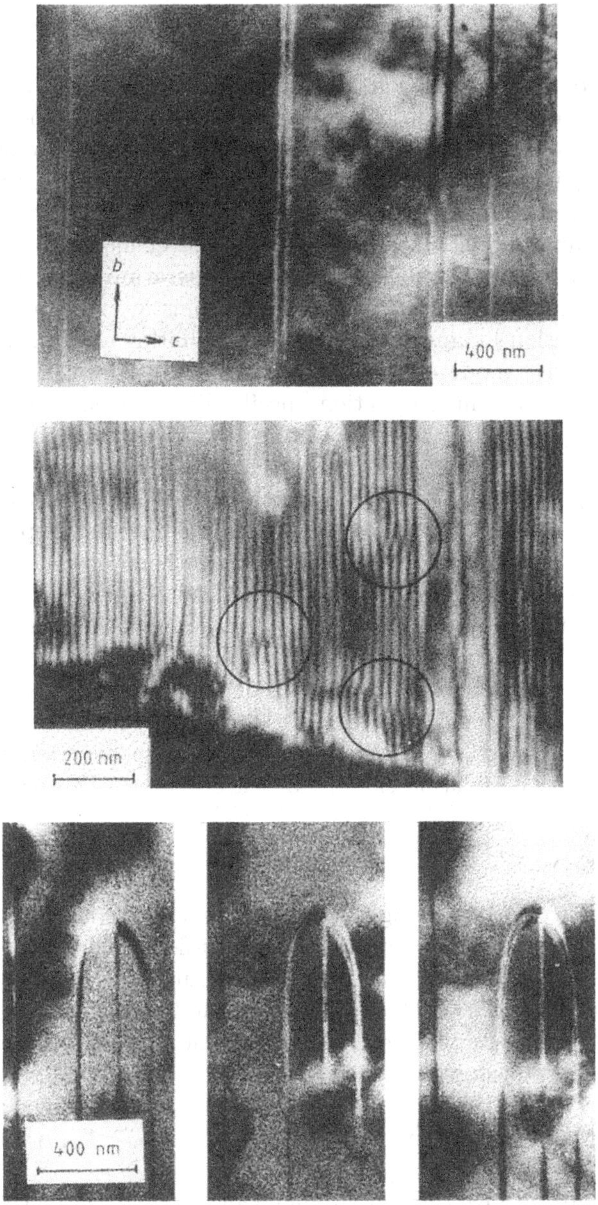

Fig. 5.2. (a) A dark-field image from satellite reflections from (100) plane of a K_2ZnCl_4 crystal at 205K showing discommensuration stripes parallel to the *b* direction. (b) A dark-field micrograph at 208K. In the circled area, discommensuration lines are evident in the pair structure. (c) The "vortex" of three pairs of discommensuration lines, where a splitting of outer pairs is visible, is seen in these photos displayed from right to left obtained with increasing time of electron irradiation. (From H.-G. Unruh, J. Phys. Cond. Matter 2, 323 (1990).)

represent a propagating mode in each domain, exhibiting a nonlinear character with decreasing temperature. While such a complex problem cannot be simply elucidated by a solution of one equation, we consider, as the first step, the problem of propagation of the collective pseudospin mode in a domain of a good quality crystal characterized with sufficiently low defect density, where no significant obstacles for propagation are present. Originating from minimum correlation energies at the transition threshold, a collective mode emerges at infinitesimal amplitude, which however increases to finite magnitudes with increasing correlations, as the temperature is lowered. At a given temperature below T_c, the collective pseudospins are in propagating motion in a low dimension in anisotropic crystals, exhibiting a nonlinear character.

The collective mode can generally be described by the expression $\boldsymbol{\sigma} = \boldsymbol{\sigma}_o f(\phi)$, propagating in a direction specified by the phase ϕ, although the amplitude $\boldsymbol{\sigma}_o$ and phase ϕ are constant at a given temperature. Observed thermodynamic quantities showed temperature dependences as expressed by empirical exponents on $(T_c - T)$, for which the responsible mechanism has not been verified as yet. Nevertheless, we discuss, as the first step, the dynamics of correlated pseudospins in one dimension with no obstructing potentials, using the long-wave approximation. At a given temperature, the motion is described primarily by the phase ϕ at a finite amplitude, while we leave the temperature-dependence to later discussions.

Here, for a displacive system, we consider that a pseudospin at each lattice site m is in a potential $V_m = \frac{1}{2}a\boldsymbol{\sigma}_m{}^2 + \frac{1}{4}b\boldsymbol{\sigma}_m{}^4$ perturbed by the correlation potential $-\sum_n J_{mn}\boldsymbol{\sigma}_m\boldsymbol{\sigma}_n$ in one-dimension along a direction x. It is noted that such binary correlation energies at the site m can be re-expressed as

$$\sum_{n\neq m} J_{mn}\boldsymbol{\sigma}_m\boldsymbol{\sigma}_n = \frac{1}{2}\sum_{n\neq m} J_{mn}(\boldsymbol{\sigma}_m - \boldsymbol{\sigma}_n)^2 - \sum_n J_{mn}\boldsymbol{\sigma}_n^2,$$

where the first term on the right is predominant, while the second one is just a constant at a given temperature in the mean-field accuracy. In this case, for correlations between the nearest neighbors m and n, we can consider that a displacement $\boldsymbol{\sigma}_m - \boldsymbol{\sigma}_n = (\partial\boldsymbol{\sigma}/\partial x)_m(x_m - x_n)$ in the long-wave approximation and, hence, the interaction behaves as if elastic. Therefore, disregarding the constant term, we can write

$$V_{m,m\pm1} = \frac{1}{2}C\sum_m\{(\boldsymbol{\sigma}_{m+1} - \boldsymbol{\sigma}_m)^2 + (\boldsymbol{\sigma}_m - \boldsymbol{\sigma}_{m-1})^2\}.$$

With such interactions, the Hamiltonian can be written for a chain crystal as

$$\mathbf{H} = \sum_n\left(\frac{p_m^2}{2m} + V_m + V_{m,m\pm1}\right),$$

where m is the effective mass of a pseudospin. Krumshansl and Schrieffer [49], Aubry [50] and many other investigators discussed the dynamics of an infinite number of particles m of a chain crystal with $\mathbf{H}$, which can be expressed in a long-wave approximation by an integral of the Hamiltonian density H, namely $\mathsf{H} = L^{-1}\int_o^L \mathrm{H}\mathrm{d}L$, where

$$\mathrm{H} = \frac{p(x,t)^2}{2m} + V\{\sigma(x,t)\} + \tfrac{1}{2}mc_o^2\{\partial\sigma(x,t)/\partial x\}^2,$$
$$V\{\sigma(x,t)\} = \tfrac{1}{2}a\sigma(x,t)^2 + \tfrac{1}{4}b\sigma(x,t)^4$$

and $c_o = (2LC/m)^{1/2}$ is the speed of propagation. Here, for the constants a and b, we consider $a > 0$ and $b = 0$ for $T > T_c$, whereas $a < 0$ and $b > 0$ for $T < T_c$, to be consistent with the Landau theory. Here, the momentum $p(x,t)$ is canonically conjugate to $\sigma(x,t)$ and related to the Hamiltonian density $\mathrm{H} = \mathrm{H}(\sigma, \partial\sigma/\partial x)$ by the canonical transformations

$$\frac{\mathrm{d}p}{\mathrm{d}t} = -\frac{\partial H}{\partial\sigma} - \frac{\partial}{\partial x}\left\{\frac{\partial \mathrm{H}}{\partial\left(\frac{\partial\sigma}{\partial x}\right)}\right\} \quad \text{and} \quad \frac{\mathrm{d}\sigma}{\mathrm{d}t} = \frac{\partial \mathrm{H}}{\partial p},$$

thereby obtaining the equation for propagation for $T < T_c$

$$m\left(\frac{\partial^2}{\partial t^2} - c_o^2\frac{\partial^2}{\partial x^2}\right)\sigma(x,t) = -\frac{\partial V}{\partial\sigma} = -a\sigma - b\sigma^3. \tag{5.19a}$$

This equation can be reduced to the ordinary differential equation

$$\frac{\mathrm{d}^2Y}{\mathrm{d}\phi^2} + Y - Y^3 = 0, \tag{5.19b}$$

by using rescaled variables,

$$Y = \frac{\sigma}{\sigma_o} \quad \text{and} \quad \phi = k(x - vt), \tag{5.19c}$$

where

$$\sigma_o = \left(\frac{|a|}{b}\right)^{1/2},$$
$$k^2 = \frac{|a|}{m(c_o^2 - v^2)} = \frac{k_o^2}{1 - v^2/c_o^2} \tag{5.19d}$$

and

$$k_o^2 = \frac{|a|}{mc_o^2}.$$

It is noted here that $v = c_o$ and $k_o = 0$ at the transition temperature T_c, whereas $v < c_o$ and $k \geq k_o$ characterize the phase below T_c. Writing $\omega = vk$ for the frequency, from the parameter k defined by the second expression in (5.19d) we obtain the *dispersion relation*

$$\omega^2 = c_o^2(k^2 - k_o^2). \tag{5.19e}$$

Such a dispersive property is a characteristic feature of a nonlinear propagation, playing an essential role for the collective pseudospins below T_c, as will be discussed for the soliton potential. Following Landau, we may consider that the parameter a changes signs when passing through the transition, where $a = 0$ specifies $T = T_c$. Hence, to be consistent with the soft-mode theory, $k = k_o$ corresponds to $\omega = 0$. It is interesting to note that such a nonzero k_o determines a initial modulation at T_c, otherwise $k = 0$ causes no modulated structure.

Although soluble analytically, (5.19b) can be simply solved for a small amplitude, ignoring Y^3. In this case, the linear equation

$$\frac{d^2Y}{d\phi^2} + Y = 0$$

has a sinusoidal solution

$$Y = Y_o \sin(\phi + \phi_o),$$

where ϕ_o is a phase constant, and Y_o is the infinitesimal amplitude.

On the other hand, (5.19b) can be analytically solved for a finite Y, using Jacobi's *elliptic function*. Integrating (5.19b) once, we obtain

$$2\left(\frac{dY}{d\phi}\right)^2 = (\lambda^2 - Y^2)(\mu^2 - Y^2), \tag{5.20}$$

where

$$\lambda^2 = 1 - (1 - \alpha^2)^{1/2} \quad \text{and} \quad \mu^2 = 1 + (1 - \alpha^2)^{1/2}.$$

Here the integration constant $\alpha = (dY/d\phi)_{\phi=0}$ represents the slope of $Y = Y(\phi)$ at $\phi = 0$, which can take a variety of values, depending on the amplitude σ_o, as seen from Fig. 5.3. Integrating (5.20) once more, the phase ϕ can be expressed by an integral called the *elliptic integral of the first kind*; that is,

$$\mu\phi/2^{1/2} = \int_0^{\xi_1} [(1 - \xi^2)(1 - \kappa^2\xi^2)]^{-1/2} d\xi, \tag{5.21a}$$

where $\xi = Y/\lambda$, and $\kappa = \lambda/\mu$ is the *modulus*, and the phase ϕ can be determined by the upper limit of the integral specified by $\xi = \xi_1$. The parameters λ, μ, κ and α all depend on $(dY/d\phi)_{\phi=0}$, which can be determined by σ_o, and so these are all temperature-dependent parameters. It is convenient to express λ and μ in terms of the modulus κ

$$\lambda = \frac{2^{1/2}\kappa}{(1 + \kappa^2)^{1/2}} \quad \text{and} \quad \mu = \frac{2^{1/2}}{(1 + \kappa^2)^{1/2}}.$$

The reverse form of (5.21a) is written as

$$\xi = \text{sn}(\mu\phi/2^{1/2}), \tag{5.21b}$$

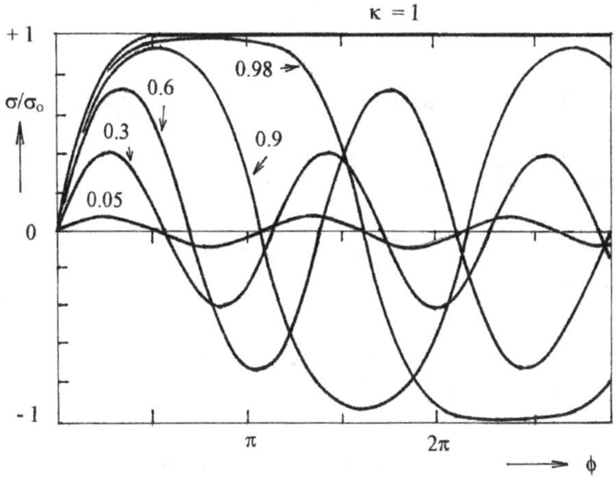

Fig. 5.3. Numerical plots of $Y = \lambda \mathrm{sn}(2^{-1/2}\mu\phi)$ for various values of the modulus κ.

which is the elliptic sn-function. Defining the angular variable Θ by $\xi = \sin\Theta$, (5.21a) can be written as

$$\mu\phi_1/2^{1/2} = \int_0^{\Theta_1} (1 - \kappa^2 \sin^2\Theta)\mathrm{d}\Theta,$$

where the upper limit Θ_1 is specified by the relation $\xi_1 = \sin\Theta_1 = \mathrm{sn}(\mu\phi_1/2^{1/2})$. Therefore, we can write the relation

$$\sigma_1 = \lambda\sigma_o \sin\Theta_1 = \lambda\sigma_o \mathrm{sn}\frac{\mu\phi_1}{2^{1/2}}, \tag{5.22}$$

which allows us to consider that σ_1 is the longitudinal component of the classical vector $\boldsymbol{\sigma}$ with an amplitude $\lambda\sigma_o$, making an angle $\frac{1}{2}\pi - \Theta_1 = \theta$ with the chain direction, as illustrated in Figs. 5.4a and 5.4b. In this interpretation, σ_1 grows as λ increases from 0 to 1 with temperature and the direction of $\boldsymbol{\sigma}$ rotates by θ while propagating along the x axis.

As shown in Fig. 5.3, Jacobi's sn-function is periodic for the modulus in the range $0 < \kappa < 1$, although it is not periodic in the specific case $\kappa = 1$. The period can be expressed as $4K(\kappa)$, where

$$K(\kappa) = \int_0^{\frac{\pi}{2}} (1 - \kappa^2 \sin^2\Theta)^{-1/2}\mathrm{d}\Theta \tag{5.23}$$

is the complete elliptic integral. In the process for κ to approach 0, $\kappa \to 0$ (or $\lambda \to 0$ and $\mu = 2^{1/2}$) corresponds to a periodic solution, whereas for the other extreme case of $\kappa = 1$ ($\lambda = \mu = 1$), (5.21b) takes the specific form

$$Y = \tanh\frac{\phi}{2^{1/2}}, \tag{5.24}$$

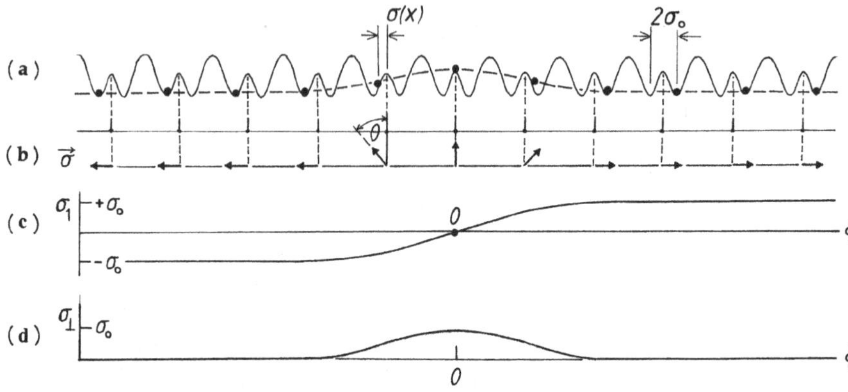

Fig. 5.4. (a) A pseudospin mode in a quasi-one-dimensional lattice; (b) pseudospin vectors in a collective mode; (c) longitudinal components σ_1 as given by $Y = \sigma_1/\sigma_o = \tanh(2^{-1/2}\phi)$; (d) transversal components $\sigma_\perp/\sigma_o = \mathrm{sech}^2(2^{-1/2}\phi)$.

which varies between -1 and $+1$ at $\phi = 0$ as shown in Fig. 5.4c, representing a *kink* of the pseudospin variable, that is consistent with mirror reflection on the plane of $\phi = 0$ perpenducular to the x axis. Such a plane at the kink may be considered a domain boundary.

The above theory falls short in other aspects of domain formation in real crystals. In order for a planar domain boundary to be represented by $\phi = 0$, interchain correlations should be considered as significant, which is, however, disregarded in the one dimensional theory. In addition, the pinning mechanism by random defects discussed in Section 5.1 should be revised for the present argument of domain boundaries.

On the other hand, for a classical displacement $\boldsymbol{\sigma}$ its transversal component should be considered as significant for interchain correlations. In addition to the longitudinal component $\sigma_1 = \lambda\sigma_o \cos\theta$ in (5.22), we consider the transversal component $\sigma_\perp = \lambda\sigma_o \sin\theta$ that is defined as

$$\sigma_\perp(\phi) = \lambda\sigma_o \mathrm{cn}\frac{\mu\phi}{2^{1/2}} \quad \text{for } 0 < \kappa < 1,$$

and

$$\sigma_\perp(\phi) = \sigma_o\,\mathrm{sech}\,\frac{\phi}{2^{1/2}} \quad \text{in the limit of } \kappa = 1.$$

Corresponding to σ_1 shown in Fig. 5.4c for $\kappa = 1$, the transversal $\sigma_\perp$ represents a *solitary* pulse as shown in Fig. 5.4d. Being prominent in the vicinity of $\phi = 0$ where ϕ changes from -1 to $+1$, the classical vector $\boldsymbol{\sigma}$ reverses its direction, for which a certain amount of energy is obviously required. Assuming that $\sigma_\perp$ can take all directions perpendicular to the x axis with an equal probability, the energy required for reversing $\boldsymbol{\sigma}$ can be expressed as proportional to $\pi \boldsymbol{k}^2{\sigma_\perp}^2$, where $\boldsymbol{k}$ is the wavevector of propagation. The field $F_s(\phi)$ or the corresponding potential $V_s(\phi) = -\mathrm{d}F_s/\mathrm{d}x$ proportional to $k^2{\sigma_\perp}^2$

should be involved in reversing the propagating pulse of $\sigma_\perp$. Hence, writing $V_s(\phi) \propto -k^2 \operatorname{sech}^2(\phi/2^{1/2})$, we obtain $F_s \propto \tanh(\phi/2^{1/2}) \propto \sigma_1$, which may be interpreted as an internal field defined in Chapter 3 by (3.31). Thus, known as the *soliton* in nonlinear dynamics, the potential $V_s(\phi)$ may be considered as representing the internal field due, essentially, to short-range correlations in the chain, but including those with distant pseudospins as well, if contributed by dipolar interactions as in ferroelectric crystals.

Generally, for $0 < \kappa < 1$, the direction of $\boldsymbol{\sigma}$ is reversed in a region of the narrow domain wall, for which we can consider a potential $V_s(\phi) = -\pi \boldsymbol{k}^2 \sigma_\perp{}^2 \propto -\mathrm{cn}^2(\mu\phi/2^{1/2})$ is responsible as related to $\sigma_\perp$, which is notably periodic and incommensurate in the lattice. In Sections 5.8 and 5.9, we will discuss the temperature dependence of $V_s(\phi)$ in conjunction with the long-range order.

5.6 A Hydrodynamic Model for Pseudospin Propagation

The linear chain model discussed in the previous section is capable of explaining dynamical aspects of correlated pseudospins, however their temperature dependent amplitudes cannot be explained, unless additional long-range correlations are taken into account. The quartic potential $\frac{1}{4}b\sigma^4$ emerging at the outset of a phase transition at T_c is considered in this model, although representing only short-range interactions. As a result, the finite amplitude is expressed in terms of a undetermined value of the modulus κ, wheras the continuous phase ϕ in the range $0 \le \phi \le 2\pi$ describes propagation through a crystal, being characterized by a constant speed v of propagation at a given temperature.

Equation (3. 31) derived by the variation principle is valid for the pseudospin $\boldsymbol{\sigma}(\phi)$ at any space-time (x, t) in the crystal, for which the relation $\lambda\boldsymbol{\sigma}(\phi) = \boldsymbol{F}(\phi)$ signifies the in-phase relation between $\boldsymbol{\sigma}(\phi)$ and the corresponding internal field $\boldsymbol{F}(\phi)$. On the other hand, if the pseudospins are responsive to an applied field $\boldsymbol{E}$, this relation should be revised as

$$\lambda\boldsymbol{\sigma}(\phi') = \boldsymbol{F}(\phi) + \boldsymbol{E}, \tag{5.25}$$

where the phase ϕ is shifted to ϕ'. Therefore, we can write

$$\boldsymbol{E} = \lambda\{(\boldsymbol{\sigma}(\phi') - \boldsymbol{\sigma}(\phi)\} = \lambda\left(\frac{\partial\boldsymbol{\sigma}}{\partial\phi}\right)\delta\phi, \tag{5.26}$$

implying that an external field $\boldsymbol{E}$ changes the amplitude by $\lambda\delta\boldsymbol{\sigma}$, accompanying with the phase shift $\delta\phi$.

In a polar phase, the ordered state at a temperature T is characterized by an internal electric field $\boldsymbol{E}_{\mathrm{int}}$ of long-range dipolar order. Assuming such a local field $\boldsymbol{E}_{\mathrm{int}}$ as if applied externally, we may expcct a phase shift in $\boldsymbol{\sigma}(\phi)$ as in (5.26), if $\boldsymbol{E}_{\mathrm{int}}$ is substituted for $\boldsymbol{E}$. Although $\boldsymbol{E}_{\mathrm{int}}$ is not distinguishable from an applied $\boldsymbol{E}$ experimentally, the temperature dependent amplitude of $\boldsymbol{\sigma}(\phi')$ in the polar state must be attributed to the net effective field $\boldsymbol{F}(\phi)+\boldsymbol{E}_{\text{int}}$ where $\boldsymbol{E}_{\text{int}}$ is temperature dependent. At this point, we note that the phase velocity in ϕ' cannot be the same as in the phase ϕ, and is temperature dependent.

We considered in Section 5.5 that the transversal component $\sigma_\perp$ is associated with the soliton potential, and that the nonlinear character of a pseudospin vector $\boldsymbol{\sigma}(\phi)$ arises from correlations in the chain as well as between neighboring chains. With this in mind, we proceed to a hydrodynamical description with the pseudospin density, $\boldsymbol{\sigma}^*.\boldsymbol{\sigma} = \sigma_1{}^2 + \sigma_\perp{}^2 = \rho_o$.

While ρ_o is constant at a given temperature, the components $\sigma_1 = \sigma_1(x,t)$ and $\sigma_\perp = \sigma_\perp(z,t;x)$ vary along the x direction, depending on the effective internal field. In this case, the effective field $\boldsymbol{F}(\phi') = \boldsymbol{F}(\phi) + \boldsymbol{E}_{\text{int}}$ can be expressed by longitudinal and transversal components, $F_\parallel$ and $F_\perp$ per volume in the hydrodynamical model, exerting these component forces on the corresponding pseudospin densities $\rho_\parallel = \sigma_1{}^2$ and $\rho_\perp = \sigma_\perp{}^2$, respectively. We consider a steady flow of "fluid" of a density ρ_o through a cylindrical pipe of a flexible cross section, where the density is modulated by the internal field $\boldsymbol{F}(\phi')$.

First, the speed of propagation $v = v(x,t)$ cannot be constant in the presence of $F_\parallel(x,t)$ and is modified by the force proportional to $-\partial F_\parallel/\partial x$;

$$\frac{\partial v}{\partial t} + v\frac{\partial v}{\partial x} = -\left(\frac{1}{\rho_o}\right)\frac{\partial F_\parallel}{\partial x}. \tag{i}$$

Second, the law of continuity should be applied to the longitudinal flow, because there is no transversal flow. Therefore, we have the equation of continuity

$$\frac{\partial \rho_\parallel}{\partial t} + \frac{\partial}{\partial x} v\rho_\parallel = 0. \tag{ii}$$

Third, for the transversal deviation, we assume that the density $\rho_\perp$ is in restoring motion with $F_\perp$ along a perpendicular direction z at any point x, for which the equation of motion for $\rho_\perp(z)$ at a given x can be written as

$$\frac{\partial^2 \rho_\perp}{\partial t^2} + \alpha(\rho_\perp - \rho_o) = F_\perp, \tag{iii}$$

where α is an elastic restoring constant in transversal directions.

Equation (i) is typically nonlinear because of the term $v(\partial v/\partial x)$, whereas (ii) and (iii) are linear equations. We assume that the nonlinearity is a weak perturbation, which can therefore be ignored in the first approximation where the equation (i) is linearized. In the following, analogous to a classical example of fluid through a flexible pipe discussed in Lamb's textbook [51], we can derive the Korteweg-deVries equation from (i), (ii) and (iii) combined. In the classical vector model, $|\boldsymbol{\sigma}|$ is constant at a given temperature while distributed densities between x and z directions vary along the x axis, as restricted by $\rho_\parallel + \rho_\perp = \rho_o$. In this case, we may consider that $\rho_\parallel = \rho(x)$ and $\rho_\perp = \rho(z)$ are independent functions of x and z, respectively, whereas $|\rho_\parallel| = |\rho_\perp| = \rho_o$

if calculated at x and $x + \frac{1}{2}\lambda$. Accordingly, we can write $F_{\parallel} = F(x)$ and $F_{\perp} = F(z)$, whose amplitudes are the same F_o in order to be consistent with (3.31). Using reduced variables, $x' = \alpha^{1/2}x, t' = \alpha^{1/2}t$, $\rho' = (\rho - \rho_o)/\rho_o$ and $F' = F/\rho_o$, the linearized equations can be written as

$$\frac{\partial \rho'}{\partial t'} + \frac{\partial v}{\partial x'} = 0,$$
$$\frac{\partial v}{\partial t'} - \frac{\partial F'}{\partial x'} = 0 \tag{iv}$$

and

$$\frac{\partial^2 \rho'}{\partial t'^2} + \rho' = F'.$$

Considering that the unperturbed propagation in the reduced space-time (x', t') is sinusoidal and proportional to $\exp i(kx' - \omega t')$, we can obtain from the set of equations (iv) the *dispersion* relation

$$\omega^2 = \frac{k^2}{(1 + k^2)},$$

where the frequency ω is approximately expressed for a small k as

$$\omega \approx k - \tfrac{1}{2}k^3 \tag{v}$$

In this case, the factor $\exp i\{k(x' - t') - \frac{1}{2}k^3 t'\}$ allows us to consider a monochromatic variation in the first approximation, which is then modified by the factor $\exp(-i\frac{1}{2}k^3 t')$. Defining new variables $\xi = k(x' - t')$ and $\tau = \frac{1}{2}k^3 t'$, the space-time coordinates (x', t') can be transformed to the variables (ξ, τ) by performing the differentiations

$$\frac{\partial}{\partial x'} = k\frac{\partial}{\partial \xi} \quad \text{and} \quad \frac{\partial}{\partial t'} = -k\frac{\partial}{\partial \xi} + \tfrac{1}{2}k^3\frac{\partial}{\partial \tau}, \tag{vi}$$

which will be used for calculating the nonlinear perturbation.

For the perturbed flow, the nonlinear terms are retained in the reduced form in equations (i) and (ii), which are written as

$$\frac{\partial v}{\partial t'} + v\frac{\partial v}{\partial x'} + \frac{\partial F'}{\partial x'} = 0 \tag{i'}$$

and

$$\frac{\partial \rho'}{\partial t'} + \frac{\partial v}{\partial x'} + \frac{\partial (\rho' v)}{\partial x'} = 0, \tag{ii'}$$

respectively. The linear equation (iii) is also simplified with reduced variables as

$$\frac{\partial^2 \rho'}{\partial t'^2} + \rho' = F'. \tag{iii'}$$

Equations (i′), (ii′) and (iii′) can now be transformed to (ξ,τ) by (vi), resulting in

$$-\frac{\partial v}{\partial \xi}+\tfrac{1}{2}k^2\frac{\partial v}{\partial \tau}+v\frac{\partial v}{\partial \xi}+\frac{\partial F'}{\partial \xi}=0, \qquad \text{(i}''\text{)}$$

$$-\frac{\partial \rho'}{\partial \xi}+\tfrac{1}{2}k^2\frac{\partial \rho'}{\partial \tau}+\frac{\partial v}{\partial \xi}+\frac{\partial(\rho' v)}{\partial \xi}=0 \qquad \text{(ii}''\text{)}$$

and

$$F'=\rho'+k^2\frac{\partial^2\rho'}{\partial \xi^2}-k^4\frac{\partial^2\rho'}{\partial \xi\partial \tau}+\tfrac{1}{4}k^6\frac{\partial^2\rho'}{\partial \tau^2}. \qquad \text{(iii}''\text{)}$$

Although ρ' and F' are the variables emerging at the transition threshold, the speed v starts from the threshold value v_o, all varying as functions of ξ and τ with decreasing temperature. Therefore, the nonlinearity developing with decreasing temperature can be described by ρ', F' and v' that are expressed in power series of k^2. Assuming that

$$\rho'=k^2\rho'_1+k^4\rho'_2+\ldots\ldots\ldots,$$
$$F'=k^2F'_1+k^4F'_2+\ldots\ldots\ldots$$

and

$$v=v_o+k^2v_1+k^4v_2+\ldots\ldots\ldots,$$

we expand the equations (i″), (ii″) and (iii″) into a series of k^2, where the relations among these *asymptotic* coefficients can be obtained from terms of the same power.

Comparing, first, the factors proportional to k^2, we obtain

$$-\frac{\partial \rho'_1}{\partial \xi}+\frac{\partial v_1}{\partial \xi}=0,$$
$$-\frac{\partial v_1}{\partial \xi}+\frac{\partial F'_1}{\partial \xi}=0$$

and

$$F'_1=\rho'_1.$$

Integrating these equations, we have the relation

$$F'_1=\rho'_1=v_1+\varphi(\tau), \qquad \text{(vii)}$$

where $\varphi(\tau)$ is an arbitrary function of τ. Comparing, next, terms proportional to k^4, we obtain

$$-\frac{\partial \rho'_2}{\partial \xi}+\frac{1}{2}\frac{\partial \rho'_1}{\partial \tau}+\frac{\partial v_2}{\partial \xi}+\frac{\partial(\rho_1 v_1)}{\partial \xi}=0,$$
$$-\frac{\partial v_2}{\partial \xi}+\frac{1}{2}\frac{\partial v_1}{\partial \tau}+v_1\frac{\partial v_1}{\partial \xi}+\frac{\partial F_2}{\partial \xi}=0$$

and

$$F_2' = \rho_2' + \frac{\partial^2 \rho_1'}{\partial \xi^2}.$$

We can eliminate ρ_2', v_2 and F_2' from these equations, arriving at the equation for v_1 and ρ_1':

$$\left(\frac{\partial v_1}{\partial \tau} + 3v_1\frac{\partial v_1}{\partial \xi} + \frac{\partial^3 v_1'}{\partial \xi^3}\right) + \varphi\frac{\partial v_1}{\partial \xi} + \frac{\partial \varphi}{\partial \tau} = 0$$

and

$$\left(\frac{\partial \rho_1'}{\partial \tau} + 3\rho_1'\frac{\partial \rho_1'}{\partial \xi} + \frac{\partial^3 \rho_1'}{\partial \xi^3}\right) - \varphi\frac{\partial \rho_1'}{\partial \xi} - \frac{\partial \varphi}{\partial \tau} = 0.$$

Because of (vii), the two terms in the brackets on the left of these equations become equal and opposite, so that both v_1 and ρ_1' satisfy the same differential equation

$$\frac{\partial V_1'}{\partial \tau} + 3V_1'\frac{\partial V_1'}{\partial \zeta} + \frac{\partial^3 V_1'}{\partial \zeta^3} = 0, \tag{5.27}$$

where, for convenience, we use the same symbol V_1' to represent v_1 or ρ_1', and arrive at (5.27) that is known as the *Korteweg-de Vries equation.* Here, in (5.27) the variable ξ is replaced by ζ, as necessitated by another requirement for the function $\varphi(\tau)$ that satisfies the equation

$$\varphi\frac{\partial V_1'}{\partial \xi} + \frac{\partial \varphi}{\partial \tau} = 0.$$

This can be satisfied by another transformation $\zeta = \xi + 3\int \varphi(\tau)d\tau$. It is noted from (vii) that the internal field F_1' can also be determined as a solution of (5.27).

We have shown in the above that these quantities developed as proportional to k^2 are all determined by the Korteweg-deVries equation that can be solved analytically, as discussed in the next section. The solution shows a particle-like behavior, and therefore is called a *soliton.* Although derived mathematically, the Korteweg-deVries equation gives physically significant solutions for nonlinear propagation up to the order k^2, as described by the potential k^2V_1', the density deviation $k^2\rho_1'$ and the corresponding speed of propagation $v_o + k^2v_1$ in the asymptotic approach. Applying to condensates in crystals, the lattice is stressed by the potential k^2V_1, which should be incorporated in the Born-Huang theory. We shall therefore refer to the corresponding lattice potential as the *soliton potential.*

It is noted that such a nonlinearity may also arise from a *dissipative* mechanism, instead of a *dispersive* character. However, the above approximation suffices for the present problem of correlations and, therefore, we discuss the Korteweg-deVries equation in the following sections, leaving the more general nonlinear problems to reference books on nonlinear physics (see, e.g., refs. [51]).

5.7 The Korteweg-deVries Equation

5.7.1 General Derivation

The soliton theory is a relatively recent topic in nonlinear physics. Although oscillatory phenomena are essentially nonlinear, the excitation in small amplitude is described in most cases by a linear equation of the Sturm-Liouville type. In general, the nonlinearity becomes significant with increasing excitation energy, as described by propagation in a soliton potential. Although derived for correlated pseudospins, the Korteweg-deVries equation can be obtained for many other applications, sharing common features among them. Mathematically, the problem can be characterized by finding a potential for keeping the eigenvalues unchanged. In this subsection, we discuss in general mathematical terms how such a potential as governed by the Korteweg-deVries equation evolves in a nonlinear process.

Consider a function $y(x)$ that satisfies a differential equation of the Strum-Liouville type:

$$\mathbf{D}^2 y(x) = \varepsilon_o y(x),$$

where $\mathbf{D} = \partial/\partial x$ is a differential operator and ε_o is a eigenvalue. In this case, the eigenfunction $y(x)$ represents a propagating wave corresponding to the eigenvalue ε_o. On the other hand, in the presence of a potential $V(x,\tau)$, where τ is the evolving parameter, the differential equation is

$$\mathbf{L}y(x,\tau) = [\mathbf{D}^2 - V(x,\tau)]y(x,\tau) = \varepsilon(\tau)y(x,\tau), \tag{5.28}$$

where $\mathbf{L} = \mathbf{D}^2 - V(x,\tau)$ is the operator in the potential, for which the eigenvalue $\varepsilon(\tau)$ varies as a function of the parameter τ and should normally be different from ε_o. Here, we consider the problem of finding such a potential $V(x,\tau)$ that keeps the eigenvalue unchanged, i.e. $\varepsilon = \varepsilon_o$ or $\partial\varepsilon/\partial\tau = 0$. Applying this to a physical problem, the variable τ may be interpreted as "real time" for the potential $V(x,\tau)$ to evolve; however, the origin for evolution is left unspecified in the meantime. In the following discussion, we express differential coefficients in the indexed form, e.g. $y_\tau = \partial y/\partial\tau$, $f_{xx} = \partial^2 y/\partial x^2$, etc., while retaining the symbol $\mathbf{D}y = \partial y/\partial x$ for a specific spatial derivative for convenience.

We first consider for (5.28) to be modified with varying τ in such a way that the condition $\varepsilon_\tau = 0$ is satisfied. In this case, the function $y(x)$ is assumed to change in space, corresponding to y_τ, in such a manner that

$$y_\tau = \mathbf{B}y, \tag{5.29}$$

where the operator $\mathbf{B}$ is primarily proportional to $\mathbf{D}$ in the absence of $V(x,\tau)$, representing a linear propagation, but will be modified as higher-order operators $\mathbf{D}^2, \mathbf{D}^3$ and so forth are included to obtain nonlinear propagation. Differentiating (5.28) with respect to τ, we obtain

$$\frac{\partial(\mathbf{L}y)}{\partial\tau} = \mathbf{L}_\tau y + \mathbf{L}y_\tau = -V_\tau y + \mathbf{LB}y,$$

while

$$\frac{\partial}{\partial \tau}(\lambda y) = \varepsilon_\tau y + \varepsilon y_\tau = \varepsilon_\tau y + \varepsilon \mathbf{B} y = \varepsilon_\tau y + \mathbf{BL} y.$$

Therefore,

$$(-V_\tau + [\mathbf{L}, \mathbf{B}])y = \varepsilon_\tau y, \quad \text{where} \quad [\mathbf{L}, \mathbf{B}] = \mathbf{LB} - \mathbf{BL},$$

from which we obtain the basic equation to be solved for the potential V; that is, for $\varepsilon_\tau = 0$,

$$(-V_\tau + [\mathbf{L}, \mathbf{B}])y = 0. \tag{5.30}$$

Consider $\mathbf{B}_1 = c\mathbf{D}$ in (5.29) and (5.30), as the simplest example for linear propagation,

$$\begin{aligned}[\mathbf{L}, \mathbf{B}_1]y &= \mathbf{L}(\mathbf{B}_1 y) - \mathbf{B}_1(\mathbf{L}y) = (\mathbf{D}^2 - V)(c\mathbf{D}y) - (c\mathbf{D})(V_{xx} - Vy) \\ &= 2c_x \mathbf{D}^2 y + c_{xx}\mathbf{D}y + cV_x y,\end{aligned}$$

which is just equal to $cV_x y$ if c is a constant. In this case, (5.30) can simply be written as $-V_\tau + cV_x = 0$, so that the solution of (5.30) is given as $V = V(x - c\tau)$ and $y = y(x - c\tau)$. Therefore, V and y remain unchanged by a translation $x \to x - c\tau$, so that this result is of no particular interest, as it represents only linear translation.

Next, we consider an operator $\mathbf{B}_2 = a\mathbf{D}^2 + c\mathbf{D} + b$, where a and b are functions of x and τ, while c remains constant as in $\mathbf{B}_1$. By a similar calculation however, we obtain from $\mathbf{B}_2$ the same potential $V = V(x - c\tau)$ as derived from $\mathbf{B}_1$. Therefore, to deal with a nonlinear effect, we should consider $\mathbf{B}_3 = a\mathbf{D}^3 + c\mathbf{D} + b$ for a possible evolving potential V with τ:

$$[\mathbf{L}, \mathbf{B}_3]y = (2c_x + 3aV_x)\mathbf{D}^2 y + (c_{xx} + 2b_x + 3aV_{xx})\mathbf{D}y + (b_{xx} + aV_{xxx} + cV_x)y.$$

For such a potential V, we set the coefficients of $\mathbf{D}^2 y$ and $\mathbf{D}y$ equal to zero and obtain differential equations that c and b can satisfy. Assuming a to be constant, we integrate these, arriving at

$$c = -\frac{3}{2}aV + C \quad \text{and} \quad b = -\frac{3}{4}aV_x + B,$$

where C and B are constants of integration. As the result, we can write that

$$[\mathbf{L}, \mathbf{B}_3]y = \left[\tfrac{1}{4}a(V_{xxx} - 6VV_x) + CV_x\right] y.$$

Using this result in (5.30),

$$\tfrac{1}{4}a(V_{xxx} - 6VV_x)y + (CV_x - V_\tau)y = 0,$$

where C can be selected to be zero by transforming (x, τ) to $(x', \tau) = (x - \alpha\tau, \tau)$, so that the second term on the left is expressed essentially as $-V_\tau y$.

Thus, we arrive at the following differential equation for a potential $V(x-\alpha\tau)$; that is, by letting $a=-4$, we have

$$V_\tau - 6VV_x + V_{xxx} = 0, \tag{5.31}$$

which is a standard form of the Korteweg-deVries equation.

Equation (5.29) describes the evolving function $y(x,\tau)$ in the Korteweg-deVries potential $V(x,\tau)$;

$$y_\tau = (-4\mathbf{D}^3 + 6V\mathbf{D} + 3V_x + B)y,$$

where the constant B can be eliminated by a suitable choice of α in the phase $x-\alpha\tau$ for further simplification. It is noted that the modified propagation is characterized by the eigenvalue ε_o and, hence, the wave equation for the function $y(x-\alpha\tau)$ is written as

$$y_{xx} - Vy = \varepsilon_o y. \tag{5.32}$$

Clearly, the wavevector k for propagation depends on the value of α, implying that the speed of propagation in the potential field V is different from the "potential-free space." Although a higher-order evolution of $V(x-\alpha\tau)$ is possible in the above derivation, the operator $\mathbf{B}_3$ is adequate for most nonlinear problems, for which the Korteweg-deVries potential plays an essential role.

5.7.2 Solutions of the Korteweg-deVries Equation

It is significant that the equation (5.31) is analytically soluble for a one-dimensional soliton potential. First, the Korteweg-deVries equation has been derived for a propagating potential $V=V(x-\alpha\tau)$, which is noted to be closely associated with the wave equation (5.32). Therefore, $V_\tau=-\alpha V_x$ and the equation (5.31) can be written as

$$-\alpha V_x - 6VV_x + V_{xxx} = 0.$$

Reexpressing it in the form

$$-\alpha V_x - \tfrac{6}{2}\frac{\mathrm{d}V^2}{\mathrm{d}x} + \frac{\mathrm{d}V_{xx}}{\mathrm{d}x} = 0,$$

which is integrated immediately as

$$V_{xx} = 3V^2 + \alpha V + \tfrac{1}{2}a,$$

where $\frac{1}{2}a$ is the constant of integration. Multiplying by V_x, we can integrate it once more and obtain

$$V_x^2 = 4V^3 + \alpha V^2 + aV + b,$$

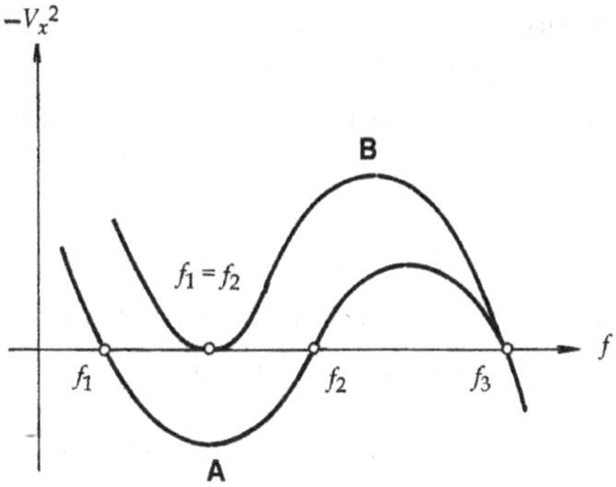

Fig. 5.5. Schematic curves for $-V_x^2$ vs. f, showing an oscillatory solution (A) and a solitary solution (B) of the Korteweg-deVries equation.

where b is another integration constant that should be chosen as positive. The algebraic expression on the right side can be rewritten by using a new variable defined by $V = -f$, for convenience, as

$$V_x^2 = -4(f - f_1)(f - f_2)(f - f_3),$$

where f_1, f_2 and f_3 are real roots of the cubic equation $V_x^2 = 0$ and, hence, $f_1 + f_2 + f_3 = \frac{1}{4}\alpha$. Assuming $f_1 < f_2 \leq f_3$, positive V_x^2 is restricted to the range $f_1 \leq f \leq f_2$, as illustrated in Fig. 5.5. To deal with this range of V_z^2, we use another variable g defined by $f_3 - f = g$ and

$$V_x^2 = 4g(f_3 - f_1 - g)(f_3 - f_2 - g).$$

Since $g > 0$ in this range, we can introduce a new variable ξ by writing $g = (f_3 - f_2)\xi^2$, and obtain

$$\xi_x^2 = (f_3 - f_1)(1 - \xi^2)(1 - \kappa^2\xi^2),$$

where

$$\kappa^2 = \frac{f_3 - f_2}{f_3 - f_1}. \tag{5.33}$$

The front factor $(f_3 - f_1)$ in ξ_x^2 can be dropped by further redefining the phase variable $\phi = (f_3 - f_1)^{1/2}(x - \alpha\tau)$, and obtain

$$\xi_\phi^2 = (1 - \xi^2)(1 - \kappa^2\xi^2),$$

where

$$\phi = \int_0^\xi [(1 - \xi^2)(1 - \kappa^2\xi^2)]^{1/2} d\xi \quad \text{and} \quad \xi = \mathrm{sn}(\phi, \kappa)$$

are the Jacobi elliptical integral and elliptic function, respectively, characterized by the modulus κ given by (5.33). By definition, the modulus is in the range $0 \leq \kappa \leq 1$, where at the limits $\kappa = 0$ and 1 that are determined by $f_3 = f_2$ and $f_2 = f_1$, respectively, the elliptic functions become specifically $\sin\phi$ and $\tanh\phi$. Except for the limit $\kappa = 1$, the sn-function is periodic with the period

$$4K(\kappa) = 4\int_0^1 [(1-\xi^2)(1-\kappa^2\xi^2)]^{-1/2}d\xi.$$

The potential as a solution of the Korteweg-deVries equation is therefore written in terms of the variable $z = x - \alpha\tau$ as

$$-V(z,\kappa) = f(z) = f_3 - g(z) = f_3 - (f_3 - f_2)\mathrm{sn}^2[(f_3 - f_1)^{1/2}z, \kappa], \quad (5.34a)$$

which is oscillatory between f_2 and f_3 with period $4K(\kappa)/(f_3 - f_2)^{1/2}$. Excluding the constant term, the potential $V(x,\kappa)$ can be expressed simply as

$$V(x,\kappa) = k^2\kappa^2\mathrm{sn}^2(kx - v\tau, \kappa), \quad (5.34b)$$

where the wavevector $k = (f_3 - f_1)^{1/2}$ is used, with the modulus κ and the speed of propagation $v = \alpha k$. The potential is periodic with the period $4K(\kappa)/\kappa k$, but becomes nonoscillatory in the limit $\kappa = 1$; that is,

$$V(z) = k^2\,\mathrm{sech}^2(kx - v\tau, 1).$$

As shown in Fig. 5.5, $\kappa = 1$ occurs when $f_1 = f_2$, where the curve of V_x^2 has a tangent along the horizontal axis. On the other hand, if $\kappa \to 0$, $f_3 - f_2 = k^2\kappa^2$ becomes infinitesimal, and the potential approaches zero, namely

$$V(x,\tau) = k^2\kappa^2\cos^2(kx - v\tau) \to 0.$$

Returning to the problem of hydrodynamic flow of the density $\rho = \boldsymbol{\sigma}^*.\boldsymbol{\sigma}$ in Section 5.6, the soliton potential is actually given by k^2V_1', where V_1' satisfies the Korteweg-deVries equation. The factor $f_3 - f_2$ is proportional to k^2, representing the amplitude of the potential $V_o(k)$, which can, therefore, be expressed as

$$V(z,\kappa) = -V_o(k)\mathrm{cn}^2(kx - v\tau, \kappa) \quad \text{for } 0 < \kappa < 1, \quad (5.35)$$

except for the constant term.

5.8 Soliton Potentials and the Long-Range Order

It is noted that the foregoing mathematical results are compatible with what was physically discussed for pseudospin propagation in Section 5.5. As remarked, the intrinsic Weiss field $\boldsymbol{F}$ due to short-range and long-range correlations signify the ordering process, for which the variable τ in the Korteweg-deVries equation can be considered as the *real time* for evolving order. However, such a time τ may remain as conjectural unless growing order is delineated in terms of decreasing temperature. In equilibrium, the pseudospin variable $\boldsymbol{\sigma}(x-\alpha\tau)$ should be in phase with the internal field $\boldsymbol{E}_{\text{int}}$, and hence the propagation should be dictated by the soliton potential V that represents $\boldsymbol{E}_{\text{int}}$. Therefore, the wave equation for $\boldsymbol{\sigma}$ can be written as

$$\left(\frac{\partial^2}{\partial\tau^2}-\alpha^2\frac{\partial^2}{\partial x^2}-\varepsilon_{\text{o}}\right)\boldsymbol{\sigma}(x-\alpha\tau)=-\text{grad}_{\sigma}V(x-\alpha\tau),$$

Here, the potential $V(x-\alpha\tau)$ generally expressed as proportional to $\text{cn}^2(z,\kappa)$, where $z=x-\alpha\tau$, is determined by the Korteweg-deVries equation.

The functions $\text{cn}^2(z,\kappa)$ and $\text{sn}^2(z,\kappa)$ are periodic, consisting of infinite number of identical potential valleys periodically separated by $2K(\kappa)$, as shown in Fig. 5.7. Although approximately sinusoidal for a small κ, such a potential curve for $0<\kappa<1$, called the *cnoidal* potential, consists of periodically repeated peaks that are generally incommensurate with the lattice period. The separation between these peaks increases as $\kappa\to 1$, and the cnoidal curve approaches a single-peak potential expressed as proportional to $\text{sech}^2(z,1)$ in the limit $\kappa\to 1$. In such an incommensurate potential $V(z,\kappa)$ for $0<\kappa<1$, the pseudospin wave can be "stabilized" in phase with the cnoidal period at a discrete negative eigenvalue, whereas it is free propagating for positive and continuous eigenvalues.

Mathematically, the function $\text{sn}^2(z,\kappa)$ can be expressed by a series

$$2\kappa^2\text{sn}^2(z,\kappa)=-2a\sum\nolimits_{m=-\infty}^{m=+\infty}\text{sech}^2(a^{1/2}x-cx_{\text{m}})+\text{const}, \tag{5.36}$$

where

$$a=\pi^2/4K'(\kappa)^2,\quad c=\pi K(\kappa)/K'(\kappa)\quad\text{and}\quad K'(\kappa)=K[(1-\kappa)^{1/2}],$$

indicating that each well of the periodic $\text{sn}^2(z,\kappa)$ at $x=x_{\text{m}}$ can be replaced approximately by such a sech^2 curve. For the derivation of (5.36), interested readers are referred to a standard textbook on the elliptic functions. The formula in (5.36) is quoted from the textbook by Toda "Introduction to elliptic functions" [52].

For a given $\kappa<1$, there are some overlaps between neighboring sech^2 peaks, whereas the overlaps diminish as κ approaches 1, i.e. $a\to 1$, $c\to\infty$. In this context, the cnoidal potential can be logically replaced by a periodic lattice of sech^2 potentials, where the finite overlaps are perturbations, resulting in a band structure of eigenvalues for stabilized pseudospin waves.

Physically, in a polar crystal, we realize that the dipolar field of long-range order $\boldsymbol{E}_{\text{dip}}$ may drive $\boldsymbol{\sigma}(z,\kappa)$ stabilized in the potential $V(z,\kappa)$ out to $\boldsymbol{\sigma}(z',\kappa')$ in another potential $V(z',\kappa')$ by shifting the phase and increasing the amplitude, as described by (5.25) and (5.26). We can consider that the energy $-\Delta\boldsymbol{\sigma}.\boldsymbol{E}_{\text{dip}}$ is transferred to the surroundings of the condensate if the temperature is lowered, thus reducing the pseudospin energy by the internal field $\boldsymbol{E}_{\text{dip}}$. It is noted that this postulate is consistent to the classical argument for work to magnetize a magnet (See Becker's textbook, *Theory of Heat*, pp. 9–11, ref. [14]). The wave $\boldsymbol{\sigma}(z,\kappa)$ can be more stabilized with decreasing temperature by periodic potentials with wider separations $2K(\kappa)$ at a larger κ and, thus, ordering can progress with increasing amplitudes of $\boldsymbol{\sigma}(z,\kappa)$ and $\boldsymbol{E}_{\text{dip}}$. The field $\boldsymbol{E}_{\text{dip}}$ can be assumed as arising from random dipolar orientations of neighboring pseudospin chains, and so the temperature dependence of an ordering process can be explained by the statistical principle. Using (5.25), such an energy transfer at a given T can be expressed as proportional to $-\boldsymbol{\sigma}.\Delta\boldsymbol{\sigma} \approx -\Delta(\frac{1}{2}\boldsymbol{\sigma}_{\text{dip}}{}^2)$, where $\boldsymbol{\sigma}_{\text{dip}} \propto \boldsymbol{E}_{\text{dip}}$ represents the dipolar contribution to $\boldsymbol{\sigma}$. With this interpretation, the basic mathematical problem can be reduced to obtaining eigenstates in a sech^2 potential well.

5.9 Mode Stabilization by the Eckart Potential

The soliton potential $V(z,\kappa)$ evolves as a consequence of nonlinear ordering, where the modulus κ varies as a function of temperature as related to dipolar order. Equation (5.36) shows that the soliton potential $V(z,\kappa) = -2\kappa^2\text{sn}^2(z,\kappa)$ is equivalent to a periodic array of potential $-2a\,\text{sech}^2(a^{1/2}x - cx_{\text{m}})$, so that the problem can be reduced to each elemental sech^2 potential known as the Eckart potential.

The Eckart potential is generally expressed in the form

$$V(z) = -V_{\text{o}}\,\text{sech}^2(z/d), \tag{5.37}$$

where the parameter d is introduced to express the width $2d$ of the peak, and V_{o} represents the depth of the potential.

The wave equation for a pseudospin mode $\sigma(z)$ can be written as

$$\frac{\mathrm{d}^2\sigma(z)}{\mathrm{d}z^2} + \{\varepsilon - V(z)\}\sigma(z) = 0, \tag{5.38}$$

where the eigenvalue ε should be negative for a stable mode and, hence, it is convenient to write $\varepsilon = -\mu^2$. Replacing z/d in (5.37) by z, μ and V_{o} can be further simplified with the relations $\beta^2 = \mu^2 d^2$ and $v_{\text{o}} = V_{\text{o}}d^2$, and (5.38) can be rewritten as

$$\frac{d^2\sigma(z)}{\mathrm{d}z^2} + (-\beta^2 + v_{\text{o}}\,\text{sech}^2 z)\sigma(z) = 0, \tag{5.39}$$

which is a differential equation familiar in Mathematical Physics, and the solution can be expressed in a *hypergeometric series.*

Following the book by Morse and Feshbach [53], we transform (5.39) to the standard form of the hypergeometric equation. Letting $\sigma(z) = Ay(z)\,\text{sech}^\beta z$, (5.39) can be expressed in terms of the function $y(z)$, i.e.

$$\frac{\mathrm{d}^2y}{\mathrm{d}z^2} - 2\beta(\tanh z)\frac{\mathrm{d}y}{\mathrm{d}z} + (v_{\text{o}} - \beta^2 - \beta)(\text{sech}^2 z)y = 0,$$

which is further rewritten with another argument $\zeta = \frac{1}{2}(1 - \tanh z)$ as

$$\zeta(1-\zeta)\frac{d^2y}{d\zeta^2} + (1+\beta)(1-2\zeta)\frac{dy}{d\zeta} + (v_o - \beta^2 - \beta)y = 0.$$

The hypergeometric equation is expressed with parameters **a**, **b** and **c** that are defined by the relations

$$\mathbf{a} + \mathbf{b} = 2\mathbf{c} - 1,$$

where

$$\mathbf{c} = 1 + \beta \quad \text{and} \quad \mathbf{ab} = -v_o + \beta^2 - \beta, \tag{i}$$

or solving these for **a**, **b** and **c**, we have

$$\mathbf{a}, \mathbf{b} = \tfrac{1}{2} + \beta \pm (v_o + \tfrac{1}{4})^{1/2}. \tag{ii}$$

With these parameters **a**, **b** and **c**, the hypergeometric equation in the standard form is

$$\zeta(1-\zeta)\frac{d^2y}{d\zeta^2} + [\mathbf{c} - (\mathbf{a}+\mathbf{b}+1)\zeta]\frac{dy}{d\zeta} - \mathbf{ab}y = 0. \tag{5.40}$$

We are, in fact, interested in a finite solution for $\zeta \to 0$ or $z \to \infty$, for which the function $y(\zeta)$ can be expanded into a series:

$$y(\zeta) = F(\mathbf{a}, \mathbf{b}, \mathbf{c}; \zeta) = 1 + (\mathbf{ab}/1!\mathbf{c})\zeta + [\mathbf{a}(\mathbf{a}+1)\mathbf{b}(\mathbf{b}+1)/2!\mathbf{c}(\mathbf{c}+1)]\zeta^2 + \ldots,$$

Among the solution of (5.39) given by

$$\sigma(z) = A \operatorname{sech}^{\beta} z F(\mathbf{a}, \mathbf{b}, \mathbf{c}; \zeta),$$

we are interested in the limiting cases for $\zeta \to 0$ and $\zeta \to 1$ for physical interpretation. In the former case, corresponding to $\zeta \to 0$, we have $z \to \infty$ and $y(\zeta) \to 1$ and, hence

$$[\sigma(z)]_{z\to+\infty} \to A \times 2^{\beta} \exp(-\beta z),$$

whereas in the other limit of $\zeta \to 1$, we have $z \to -\infty$ and the hypergeometric function converges to 1. For these solutions, the following identity formula can be used conveniently:

$$\begin{aligned} F(\mathbf{a}, \mathbf{b}, \mathbf{c}; \zeta) &= [\Gamma(\mathbf{c})\Gamma(\mathbf{c}-\mathbf{a}-\mathbf{b})/\Gamma(\mathbf{c}-\mathbf{a})\Gamma(\mathbf{c}-\mathbf{b})]F(\mathbf{a}, \mathbf{b}, \mathbf{a}+\mathbf{b}-\mathbf{c}+1; \zeta) \\ &\quad + (1-\zeta)^{c-a-b}[\Gamma(\mathbf{c})]\Gamma(\mathbf{a}+\mathbf{b}-\mathbf{c})/\Gamma(\mathbf{a})\Gamma(\mathbf{b})] \\ &\quad \times F(\mathbf{c}-\mathbf{a}, \mathbf{c}-\mathbf{b}, \mathbf{c}-\mathbf{a}-\mathbf{b}+1; 1-\zeta), \end{aligned}$$

where $\Gamma(\ldots)$ are so-called *gamma* functions. Here, we only quote necessary mathematical results, referring interested readers to refs. [51] and [53] for the detail.

Noting that $\mathrm{sech}^{\beta} z \approx 2^{\beta} \exp(+\beta z)$, $(1-\zeta)^{c-a-b} \approx \exp(-\beta z)$ and $F(\ldots;1-\zeta) \to 1$,

$$\begin{aligned}[\sigma(z)]_{z\to-\infty} &\to A \times 2^{\beta} \times [\{\Gamma(\mathbf{c})\Gamma(\mathbf{c}-\mathbf{a}-\mathbf{b})/\Gamma(\mathbf{c}-\mathbf{a})\Gamma(\mathbf{c}-\mathbf{b})\}\exp(+\beta z)] \\ &\quad +\{\Gamma(\mathbf{c})\Gamma(\mathbf{a}+\mathbf{b}-\mathbf{c})/\Gamma(\mathbf{a})\Gamma(\mathbf{b})\}\exp(-\beta z)] \\ &\propto \{\Gamma(\mathbf{c})\Gamma(\mathbf{a}+\mathbf{b}-\mathbf{c})/\Gamma(\mathbf{a})\Gamma(\mathbf{b})\} \times \exp(+ikz) \\ &\quad +\{\Gamma(\mathbf{c}-\mathbf{a}-\mathbf{b})\Gamma(\mathbf{a})\Gamma(\mathbf{b})/\Gamma(\mathbf{a}+\mathbf{b}-\mathbf{c})\Gamma(\mathbf{c}-\mathbf{a})\Gamma(\mathbf{c}-\mathbf{b})\} \\ &\quad \times \exp(-ikz)].\end{aligned}$$

From this result, with regard to the waves $\exp(\pm ikz)$, we can define the reflection and transmission coefficients as

$$R = \Gamma(\mathbf{c}-\mathbf{a}-\mathbf{b})\Gamma(\mathbf{a})\Gamma(\mathbf{b})/\Gamma(\mathbf{a}+\mathbf{b}-\mathbf{c})\Gamma(\mathbf{c}-\mathbf{a})\Gamma(\mathbf{c}-\mathbf{b})$$

and

$$T = \Gamma(\mathbf{a})\Gamma(\mathbf{b})/\Gamma(\mathbf{c})\Gamma(\mathbf{a}+\mathbf{b}-\mathbf{c}),$$

respectively. Here, the factor $\Gamma(\mathbf{c}-\mathbf{a})\Gamma(\mathbf{c}-\mathbf{b})$ in the denominator of R plays a significant role for reflection, because by the definitions (i) and (ii), that is

$$\Gamma(\mathbf{c}-\mathbf{a})\Gamma(\mathbf{c}-\mathbf{b}) = \Gamma(\tfrac{1}{2}+(v_o+\tfrac{1}{4})^{1/2})\Gamma(\tfrac{1}{2}-(v_o+\tfrac{1}{4})^{1/2}) = \pi/\cos[\pi(v_o+\tfrac{1}{4})^{1/2}],$$

and, hence,

$$R \propto \cos[\pi(v_o+\tfrac{1}{4})^{1/2}].$$

Thus, the reflection coefficient R can be equal to zero if

$$(v_o+\tfrac{1}{4})^{1/2} = \mathrm{n}+\tfrac{1}{2}, \quad \text{or} \quad v_o = \mathrm{n}(\mathrm{n}+1), \tag{iii}$$

where $\mathrm{n} = 1, 2, \ldots$. On the other hand, the gamma functions $\Gamma(\mathbf{a})$ and $\Gamma(\mathbf{b})$ have singularities at $\mathbf{a}, \mathbf{b} = 0, -1, -2, \ldots$, as illustrated in Fig. 5.6, and, hence, both R and T can be signified by "poles" in the complex plane when

$$\tfrac{1}{2}+\beta \pm (v_o+\tfrac{1}{4})^{1/2} = -\mathrm{m} \quad \text{where} \quad \mathrm{m} = 0, 1, 2, \ldots$$

Because there is no reflected wave under this condition, we must consider only poles above the real axis, namely

$$\beta = ikd = (\mathrm{n}+\tfrac{1}{2}) - (\mathrm{m}+\tfrac{1}{2}) = \mathrm{n}-\mathrm{m}, \tag{iv}$$

for which values of m should be limited to

$$\mathrm{m} = 0, 1, \ldots, \mathrm{n}-1. \tag{v}$$

Summarizing the above argument, the differential equation (5.38) for $R = 0$, representing the pseudospin wave stabilized by the potential $V(z)$, can be written as

$$\frac{d^2\sigma(z)}{dz^2} + [-(\mathrm{n}-\mathrm{m})^2 + \mathrm{n}(\mathrm{n}+1)\,\mathrm{sech}^2 z]\sigma(z) = 0, \tag{5.41}$$

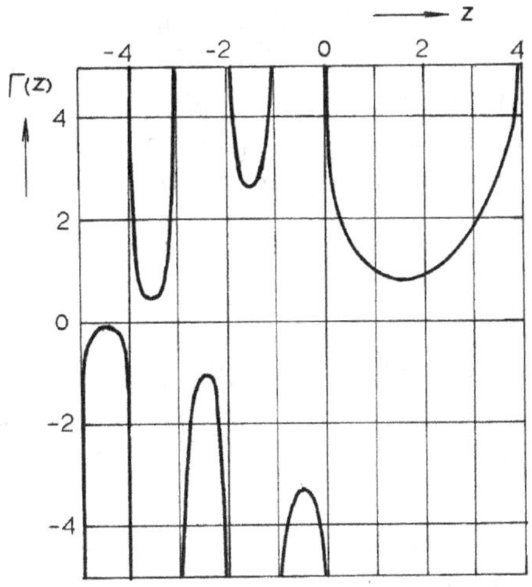

Fig. 5.6. Curves of the gamma function $\Gamma(z)$ showing discontinuities at $z = 0, -1, -2, \ldots$.

in which the eigenvalues and the potential depth are both discrete, as specified by (iv) and (v), and (iii). The wave $\sigma(z)$ is stable at discrete eigenvalues $\mathrm{n}-\mathrm{m}$, for which the potential depth is given by $-\mathrm{n}(\mathrm{n}+1)\,\mathrm{sech}^2 z$.

Obviously, $\mathrm{n} = 0$ represents no stabilized wave, for which no Eckart's potential is required, but signifying the transition threshold at T_c, at which the dipolar field of long-range order is yet insignificant. For $\mathrm{n} > 0$, (5.41) can be modified as

$$\frac{\mathrm{d}^2\sigma(z)}{\mathrm{d}z^2} + \{-\beta_\mathrm{p}{}^2 + \mathrm{n}(\mathrm{n}+1)\,\mathrm{sech}^2 z\}\sigma(z) = 0, \tag{5.42}$$

where $\beta_\mathrm{p} = \mathrm{n} - \mathrm{m} = \mathrm{n}, \mathrm{n}-1, \ldots$ from (v); hence the eigenvalues β_p can be specified by $\mathrm{p} = \mathrm{n}, \mathrm{n}-1, \ldots$. For instance, $\beta_1 = 1$ for $\mathrm{n} = 1$, $\beta_2 = 2, 1$ for $\mathrm{n} = 2$, and so on, signifying energy levels of $\sigma(z)$ stabilized in the potential $-\mathrm{n}(\mathrm{n}+1)\,\mathrm{sech}^2 z$. Expressing the pseudospin mode stabilized at a state p as $\sigma_\mathrm{p}(z)$, we have equations

$$\frac{\mathrm{d}^2\sigma_\mathrm{p}}{\mathrm{d}z^2} + \{-\beta_\mathrm{p}^2 + \mathrm{n}(\mathrm{n}+1)\,\mathrm{sech}^2 z\}\sigma_\mathrm{p} = 0$$

and

$$\frac{\mathrm{d}^2\sigma_{\mathrm{p}+1}}{\mathrm{d}z^2} + \{-\beta_{\mathrm{p}+1}^2 + (\mathrm{n}+1)(\mathrm{n}+2)\,\mathrm{sech}^2 z\}\sigma_{\mathrm{p}+1} = 0.$$

In these equations, the soliton potential increases by $(\Delta V_\mathrm{o})_{\Delta\mathrm{p}=1} = -2(\mathrm{n}+1)\,\mathrm{sech}^2 z$, as β_p is lowered by $\Delta\mathrm{p} = 1$ from a substate p to $\mathrm{p}+1$. The total number of such jumps is $\frac{1}{2}\mathrm{n}$, in order for the mode at n to be stabilized in the potential $-V(\mathrm{n})$. Figure 5.7 illustrates potentials $V(\mathrm{n})$ for $\mathrm{n} \approx 0$ and $\mathrm{n} = 1$,

2 and 3, where the eigenvalues of stabilized σ_p are shown. The soliton potential $V(n)$ can therefore be considered as composed of $\frac{1}{2}n$ fragmental potentials $(\Delta V_o)_{\Delta p=1}$, and we obtain $-V(n) = (\Delta V_o)_{\Delta p=1} \times n = -n(n+1)\,\mathrm{sech}^2 z$. Therefore, the eigenvalue decreases stepwise, when the energy $(\Delta V_o)_{\Delta p=1}$ is fragmentally transferred to the lattice, resulting in the stepwise increase in the local Weiss field.

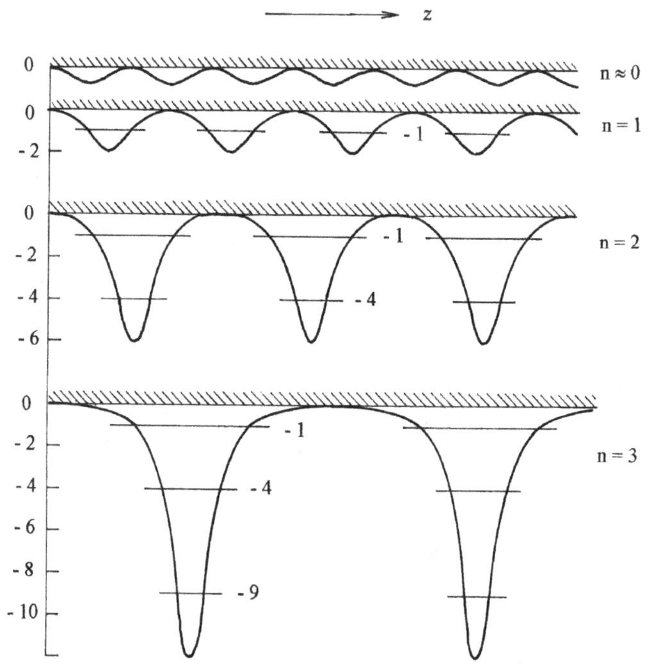

Fig. 5.7. Incommensurate *cnoidal* soliton potentials $V(z,n) = -n(n+1)\mathrm{sn}^2 z$, stabilizing pseudospin waves $\sigma(z,n)$ at negative eigenvalues. The potential is nearly sinusoidal at $n \sim 0$, while each well at $n > 0$ can be replaced by Eckart potentials $-n(n+1)\,\mathrm{sech}^2 z$ approximately. In this approximation, eigenvalues $p = n - m$, where $m = 0, 1, \ldots, n-1$, are shown here by horizontal levels in these wells, but should be broadened into bands due to overlaps between Eckart potentials.

It is logical to consider that stepwise changes between levels given by the soliton potential $(\Delta V_o)_{\Delta p=1}$ is due to "de-excitation" by the field $\boldsymbol{E}_{\mathrm{dip}}$, thereby transferring the energy $(\Delta V_o)_{\Delta p=1} = \Delta\boldsymbol{\sigma}.\boldsymbol{E}_{\mathrm{dip}}$ to the lattice, where $\Delta\boldsymbol{\sigma} \approx \Delta\boldsymbol{\sigma}_{\mathrm{dip}} = \boldsymbol{E}_{\mathrm{dip}}/\lambda$. Assuming random orientation of $\Delta\boldsymbol{\sigma}_{\mathrm{dip}}$, we can consider the statistical average of $\Delta\boldsymbol{\sigma}_{\mathrm{dip}}.\boldsymbol{E}_{\mathrm{dip}}$ that is proportional to $\langle\Delta\boldsymbol{\sigma}^2_{\mathrm{dip}}\rangle \propto k_B(T_p - T_{p+1})$, using the equipartion theorem for the ergodic quantity $\Delta\boldsymbol{\sigma}_{\mathrm{dip}}$, and write

$$(\Delta V_o)_{\Delta p=1} \propto c_p(T_p - T_{p+1}),$$

and

$$V(\mathrm{n}) = \sum\nolimits_{\mathrm{p}} (\Delta V_{\mathrm{o}})_{\Delta \mathrm{p}} \propto \sum\nolimits_{\mathrm{p}} c_{\mathrm{p}} (T_{\mathrm{p}} - T_{\mathrm{p}+1}). \tag{5.43}$$

Assuming c_{p} is constant of p, we obtain $V(\mathrm{n}) \propto T_{\mathrm{o}} - T$, which is consistent with $\langle \Delta \boldsymbol{\sigma}_{\mathrm{dip}} \rangle \propto (T_{\mathrm{o}} - T)^{1/2}$, using T_{o} in the Landau theory. It is noted that the mechanism postulated here is also consistent with Cowley's theory of phonon scattering by the quartic potential (subsection 4.4.1) at least in the region close to T_{o}. Although simplified by separating the dipolar part $\Delta \boldsymbol{\sigma}_{\mathrm{dip}}$ from $\boldsymbol{\sigma}$ in a polar crystal, the argument encompasses a nonpolar case where a stable modulated structure can exist. With this postulate, the actual transition temperature T_{c} defined for minimum correlations is clearly lower than T_{o}, indicating that $\boldsymbol{E}_{\mathrm{dip}}$ is meaningful only at temperatures lower than T_{o}.

In the soliton potential $V(z, \kappa)$ for $\kappa \leq 1$, levels n of stabilized $\sigma(z, \kappa)$ are broadened into a band structure due to overlap between adjacent Eckart wells. Hence, such a broadened structure in the cnoidal potentials allows a view for pseudospin modes to fluctuate thermally. Besides, considering planar domain boundaries signified by $\kappa \leq 1$ in practical crystals, the interchain correlations for $\kappa \leq 1$ should also be considered in order for $\boldsymbol{E}_{\mathrm{dip}}$ to fluctuate randomly. Assuming that $\Delta \boldsymbol{\sigma}_{\mathrm{dip}} . \boldsymbol{E}_{\mathrm{dip}}$ is the dominant temperature-dependent interaction energy in the non-critical region, the broad tail of the specific heat below T_{c} can be explained as previously discussed in Section 4.8. It appears to be correct that owing to the ergodic $\boldsymbol{E}_{\mathrm{dip}}$ ordering can progress thermally beyond the critical region.

One of the significant aspects of the soliton theory is that the unit represented by $(\Delta V_{\mathrm{o}})_{\Delta \mathrm{p}=1}$ behaves like an independent particle, so that the system may be viewed as a gas consisting of such soliton particles. Such a nature of soliton particles can be verified mathematically by the "two-soliton solution" of the Korteweg-deVries equation [48], however accepting this implication, we shall not go into the complex mathematical detail. Nevertheless, it is notable that such a soliton gas can certainly be regarded as a quasi-ergodic system that can be in thermal equilibrium with the lattice at a given T, for which the relation between $\boldsymbol{\sigma}(\phi)$ and T, if written properly, serves as the "equation of state" of the ordering system.

Part II

Experimental Studies

Structural phase transitions in crystals are complex phenomena, originating from an interplay between order variables and their hosting lattice. In the harmonic approximation, basic excitations in these subsystems are independent of each other, so that it may not be surprising to see that experimental results exclusively from these subsystems appear often to be incompatible with the other. For example, some ferroelectric phase transitions are displacive on the basis of soft-mode results, whereas dipolar ordering signifies the polar phase below the transition temperature T_c. However, neither view per se can deal with critical anomalies that are essentially due to interactions between order variables and soft phonons. In this context, such a traditional classification as displacive or order-disorder is not quite logical, unless critical anomalies are properly elucidated.

In Chapter 4, we discussed the order-variable condensate prevailing in the critical region. The condensate is a mobile object of a long life, owing to low damping of soft modes on both sides of T_c, whereas collective pseudospin variables at and just below T_c are signified by phase fluctuations between binary states in partial order. Such collective fluctuations are very slow in the timescale of microwave measurements for example, so that the crystal appears as quasi-statically modulated, whereas only the temporal profile of motion can be revealed from the phonon spectra.

Being represented by a pseudospin mode of a long wavelength, the condensate can be expressed in the form $\sigma_o f(\phi)$, where the amplitude σ_o and the phase ϕ are both functions of temperature. Therefore, experimental studies should be focused on the dynamics of the collective motion, whereas in thermal experiments only quantities averaged over the crystal can be detected in the long timescale. The nature of ϕ can be investigated by light and neutron inelastic scattering, whereas spatially distributed peudospin amplitudes can be visualized within short timescales of magnetic resonance sampling and of inelastic impacts of neutrons showing anomalies in the scattered intensity. Nuclear spin relaxation analysis can provide evidence for the coupling between pseudospins and soft phonons, and dielectric and Brillouin light-scattering experiments yield useful information about condensate dynamics in crystals. Needless to say, these data are complementary and should be combined to elucidate the nature of condensates in the critical region.

The internal Weiss field is a significant concept for interpreting ordering processes in noncritical regions, while remaining theoretical in most cases except for a few, yielding indirect information of the transition mechanism. On the other hand, we can consider it as the driving force for ordering, as formulated in the mean-field accuracy. Nevertheless, a more precise approach is required for dealing with the critical region.

In Part Two, principles of these basic measurements are outlined, and published experimental results from representative systems are discussed in light of the condensate model. Because there are many articles reviewing experimental results in the literature, we only need to discuss selected phase transitions where soft-mode and magnetic resonance studies were already carried out for critical anomalies. In Chapter 10, some structural phase transitions of different categories from the displacive mechanism are briefly discussed for comparison, which are nevertheless helpful for better understanding of the collective mechanism during structural changes.

6

Diffuse X-ray Diffraction and Neutron Inelastic Scattering from Modulated Crystals

6.1 Modulated Crystals

An idealized crystal of infinite arrays of ions and molecules can be considered as macroscopically uniform. In equilibrium at a given pressure p and temperature T, thermodynamical properties of a uniform crystal can be described by the Gibbs potential $G(p,T)$, for which the lattice structure can be specified only by an implicit parameter. Although surfaces and lattice defects are not entirely ignorable, bulk properties prevail in a large crystal, as signified by internal translational symmetry with periodic boundary conditions. In a crystal undergoing a structural change, on the other hand, crystal phases above and below the transition temperature T_c are primarily so idealized in the first approximation, whereas in the transition region, the crystal becomes spontaneously inhomogeneous due to locally violated lattice symmetry, for which no adequate thermodynamical description has so far been presented. Subjected to diffraction experiments however, such a spontaneously modified crystal shows a *modulated* structure.

The periodic structure of an idealized crystal is characterized by three basic translational vectors $\boldsymbol{a}_1$, $\boldsymbol{a}_2$ and $\boldsymbol{a}_3$ along the symmetrical axes, in which a continuous periodic function $f(\boldsymbol{r})$ at a position $\boldsymbol{r}$ is invariant under a basic translation

$$\boldsymbol{R} = n_1\boldsymbol{a}_1 + n_2\boldsymbol{a}_2 + n_3\boldsymbol{a}_3, \tag{6.1a}$$

where n_1, n_2 and n_3 are integers specifying a lattice point. For such a function, we have a relation

$$f(\boldsymbol{r}) = f(\boldsymbol{r} + \boldsymbol{R}). \tag{6.1b}$$

A perfect crystal can also be characterized by the Fourier transform $g(\boldsymbol{k})$ that is defined by

$$g(\boldsymbol{k}) = \int f(\boldsymbol{r})\exp(-i\boldsymbol{k}.\boldsymbol{r})\mathrm{d}^3\boldsymbol{r} \quad \text{or} \quad f(\boldsymbol{r}) = \int g(\boldsymbol{k})\exp(i\boldsymbol{k}.\boldsymbol{r})\mathrm{d}^3\boldsymbol{k}. \tag{6.2}$$

Combining (6.1b) and (6.2), we obtain the relation $\exp(i\boldsymbol{k}.\boldsymbol{R}) = 1$ at all lattice points $\boldsymbol{R}$, so that the vector $\boldsymbol{k}$ cannot be continuous, taking discrete values $\boldsymbol{k} = \boldsymbol{G}$ to satisfy $\boldsymbol{G}.\boldsymbol{R} = 2\pi \times \text{integer}$. Such a vector $\boldsymbol{G}$ represents a translation in the reciprocal lattice; that is,

$$\boldsymbol{G} = h\boldsymbol{a}_1{}^* + k\boldsymbol{a}_2{}^* + l\boldsymbol{a}_3{}^*, \tag{6.3a}$$

where

$$\boldsymbol{a}_1{}^* = (2\pi/\Omega)(\boldsymbol{a}_2 \times \boldsymbol{a}_3), \quad \boldsymbol{a}_2{}^* = (2\pi/\Omega)(\boldsymbol{a}_3 \times \boldsymbol{a}_1) \text{ and } \boldsymbol{a}_3{}^* = (2\pi/\Omega)(\boldsymbol{a}_1 \times \boldsymbol{a}_2)$$

are the basic translational vectors in the reciprocal lattice, and $\Omega = (\boldsymbol{a}_1, \boldsymbol{a}_2, \boldsymbol{a}_3)$ is the unit-cell volume in the crystal. Here h, k, and l are integral numbers, indexing lattice points in the reciprocal space. Thus, all unit cells are identical in a uniform crystal, where the macroscopic uniformity is represented by the invariance of functions $f(\boldsymbol{r})$ and $g(\boldsymbol{k})$ in the normal and reciprocal lattices, respectively. Corresponding to (6.1b) for the normal lattice, we can write

$$g(\boldsymbol{k}) = g(\boldsymbol{k} + \boldsymbol{G}) \tag{6.3b}$$

in the reciprocal lattice.

In a modulated crystal, the order variable cannot be a periodic function in the lattice translation, but its Fourier transform can be periodic with regard to a nonlattice point $\boldsymbol{G}_\mathrm{i}$ in the reciprocal lattice. Such a point $\boldsymbol{G}_\mathrm{i}$ cannot be specified by integral indexes, among which at least one, for example, h should be *irrational* in the unit of $\boldsymbol{a}_1{}^*$, indicating an incommensurate modulation along the symmetry axis $\boldsymbol{a}_1$. In this case, the lattice modulation can alternatively be specified by a vector $\boldsymbol{Q}$ from the nearest reciprocal lattice point $\boldsymbol{G}$:

$$\boldsymbol{Q} = \boldsymbol{G}_\mathrm{i} - \boldsymbol{G} = m\boldsymbol{a}_4{}^*, \tag{6.4}$$

where the index m represents a modulation at $\boldsymbol{G}_\mathrm{i}$ in a suitably defined unit $\boldsymbol{a}_4{}^*$. Thus, such a one-dimensional modulation can be expressed by four indexes (h, k, l, m) in the reciprocal lattice. De Wolff and his co-workers [54] have developed a group-theoretical method to deal with a multidimensional space with such a wavevector, which they called a *superspace*. Their supergroup theory is mathematically suitable for such *aperiodic* crystals that are reported in recent literature. Nevertheless, our primary objective in this monograph is to investigate the role played by disrupted translational symmetry in real crystals; hence, we stay in the traditional scheme composed of a reciprocal lattice $(\boldsymbol{a}_1{}^*, \boldsymbol{a}_2{}^*, \boldsymbol{a}_3{}^*)$ plus an additional modulation vector $\boldsymbol{a}_4{}^*$.

A lattice modulation at an irrational vector $\boldsymbol{Q}$ implies the presence of an excitation energy $\varepsilon(\boldsymbol{Q})$ in a given crystal. Considering that such an excitation occurs in an ordering process in the crystal, $\varepsilon(\boldsymbol{Q})$ must be offset by distorting the lattice structure under the equilibrium condition. Therefore, for a modulated crystal, it is important to obtain values of $\boldsymbol{Q}$ and the corresponding $\varepsilon(\boldsymbol{Q})$. In inelastic neutron scattering experiments, the scattering geometry can normally be arranged in such a way that $\boldsymbol{K}_o - \boldsymbol{K} = \boldsymbol{Q}$, where $\boldsymbol{K}_o$ and $\boldsymbol{K}$ are the wavevectors of incident and scattered neutron beams, respectively, and the value of $\boldsymbol{Q}$ can be determined for the maximum intensity of scattered neutrons. On the other hand, $\varepsilon(\boldsymbol{Q})$ can be obtained from the phonon spectra at $\boldsymbol{Q}$ in the scattering geometry. In contrast, from unmodulated crystals where $\boldsymbol{Q} = 0$, a collimated X-ray beam exhibits a diffraction pattern, reflecting from well-defined *crystal planes* specified by a rational $\boldsymbol{G}$, thus providing the method of structural analysis.

6.2 The Bragg Law of X-ray Diffraction

In this section, the principle of the Bragg diffraction from an ideal crystal is outlined, prior to discussing modulated crystals. A collimated X-ray beam, when falling onto a crystal, shows a diffraction pattern characteristic of the three-dimensional structure, which can then be analyzed with the concept of reflection by a large number of parallel crystal planes $\boldsymbol{G}$. The interaction between X-ray photons and orbiting electrons can be interpreted in classical terms as elastic collisions that signify no loss of X-ray energy upon impact; hence, the lattice structure remains virtually intact.

A regular crystal structure can be considered geometrically as composed of many sets of parallel planes of identical atoms. It is noted that such a group of parallel planes can be specified by the common normal vector $\boldsymbol{n}$ of a unit length, which can be shown as parallel to the reciprocal lattice vector $\boldsymbol{G}$. Denoting an arbitrary lattice point on a crystal plane by $\boldsymbol{R}$, and the distance between adjacent planes by $\boldsymbol{d} = d\boldsymbol{n}$, the equation of the plane can be expressed as

$$\boldsymbol{n}.(\boldsymbol{R} - \boldsymbol{d}) = 0;$$

hence,

$$\boldsymbol{n}.\boldsymbol{R} = \boldsymbol{n}.\boldsymbol{d} = d.$$

In this case, it is obvious that $\boldsymbol{n} \parallel \boldsymbol{G}$ and hence $G = (2\pi/d) \times \text{integer}$, because of the relation $\boldsymbol{G}.\boldsymbol{R} = 2\pi \times \text{integer}$, and the gap between adjacent planes can be calculated as

$$\frac{1}{d} = \left[\left(\frac{h}{a_1}\right)^2 + \left(\frac{k}{a_2}\right)^2 + \left(\frac{l}{a_3}\right)^2\right]^{1/2}. \tag{6.5}$$

The Bragg law of X-ray diffraction can be derived with the classical theory of radiation, using the conservation laws of energy and momentum for the photon impact with a crystal plane. These laws manifest that X-ray behaves like an optical beam obeying the reflection law, for which geometrical crystal planes are a useful concept. As is clear from the following theory, the vector $\boldsymbol{G}$ parallel to $\boldsymbol{n}$ represents a "crystal momentum", so that the Bragg law

$$\boldsymbol{K} - \boldsymbol{K}_o = \boldsymbol{G} \quad \text{where} \quad |\boldsymbol{K}| = |\boldsymbol{K}_o| \tag{6.6}$$

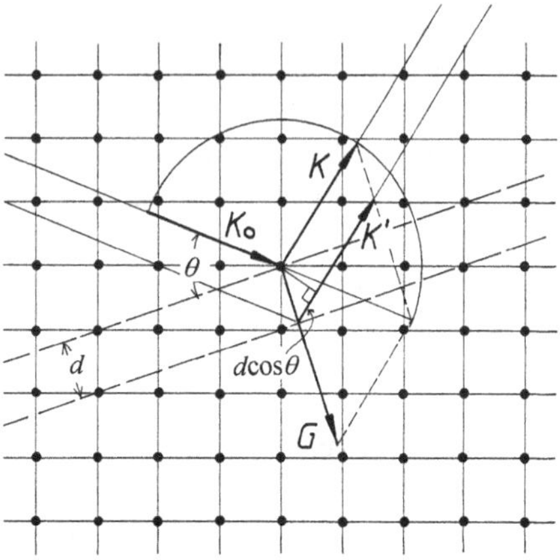

Fig. 6.1. A schematic diagram for the Bragg diffraction in two dimensions. Two diffracted rays of wavevectors $\boldsymbol{K}$ and $\boldsymbol{K}'$ with a path difference $d\cos\theta$. These rays interfere constructively, when $d\cos\theta = 1/2\lambda \times$ integer.

can be interpreted for momentum conservation for elastic impact. Typically the lattice constant is of the order of 5 Å; hence, the X-ray energy for diffraction estimated by (6.6) should be of the order of $10 \sim 50$ keV. While excited to higher atomic levels by impact, these electrons fall subsequently down to the ground level by emitting photons of the same energy as for the excitation. In contrast, heavy nuclei forming the lattice remain unchanged during impact. Figure 6.1 illustrates X-ray diffraction as interpreted with the concept of crystal planes.

The rigorous radiation theory is complex, but the result can be interpreted as simplified for a distant point of observation. The incident X-ray beam can be represented by a plane wave $E = E_\mathrm{o} \exp i(\boldsymbol{K}_\mathrm{o}.\boldsymbol{r} - \omega t)$, that interacts with a target ion located at a point $\boldsymbol{r}_\mathrm{o}$ and at a time t_o, inducing an oscillating electric dipole moment

$$\mathrm{d}\boldsymbol{p} \propto \rho(\boldsymbol{r}_\mathrm{o})\mathrm{d}^3\boldsymbol{r}_\mathrm{o} E_\mathrm{o} \exp i(\boldsymbol{K}_\mathrm{o}.\boldsymbol{r}_\mathrm{o} - \omega t_\mathrm{o})$$

in the volume element $\mathrm{d}^3\boldsymbol{r}_\mathrm{o}$ of the target charge cloud, where $\rho(\boldsymbol{r}_\mathrm{o})$ is the charge density. Such an induced oscillatory dipole moment radiates a spherical wave at the wavevector $\boldsymbol{K}$, whose amplitude observed at a distant point $\boldsymbol{r} \gg \boldsymbol{r}_\mathrm{o}$ is proportional to $\exp i\{\boldsymbol{K}.(\boldsymbol{r} - \boldsymbol{r}_\mathrm{o}) - \omega(t - t_\mathrm{o})\}/|\boldsymbol{r} - \boldsymbol{r}_\mathrm{o}|$. Therefore, the scattered amplitude at $\boldsymbol{r}$ can be expressed by

$$\begin{aligned} A_\mathrm{o} &\propto \int \mathrm{d}^3\boldsymbol{r}_\mathrm{o}\rho(\boldsymbol{r}_\mathrm{o})E_\mathrm{o} \exp i(\boldsymbol{K}_\mathrm{o}.\boldsymbol{r}_\mathrm{o} - \omega t_\mathrm{o}) \exp i[\boldsymbol{K}.(\boldsymbol{r} - \boldsymbol{r}_\mathrm{o}) - \omega(t - t_\mathrm{o})]/|\boldsymbol{r} - \boldsymbol{r}_\mathrm{o}| \\ &\approx E_\mathrm{o}\Big[\int \mathrm{d}^3\boldsymbol{r}_\mathrm{o}\rho(\boldsymbol{r}_\mathrm{o})\Big][\exp i(\boldsymbol{K}.\boldsymbol{r} - \omega t)/r] \exp i\{(\boldsymbol{K}_\mathrm{o} - \boldsymbol{K}).\boldsymbol{r}_\mathrm{o}\}, \end{aligned}$$

where we considered the approximation $|\boldsymbol{r}-\boldsymbol{r}_o| \sim r$ for $r \gg r_o$. The scattering amplitude relative to the incident E_o can then be defined as

$$A_o/E_o = \int \rho(\boldsymbol{r}_o) \exp i(\boldsymbol{K}_o - \boldsymbol{K}).\boldsymbol{r}_o \mathrm{d}^3\boldsymbol{r}_o,$$

to which the conservation law (6.6) gives a maximum amplitude that is nearly equal to 1 for a small $|\boldsymbol{r}_o|$, i.e. $\exp(-i\boldsymbol{G}.\boldsymbol{r}_o) \approx 1$ and, hence, the reflection law can be applied to a crystal plane signified by the vector $\boldsymbol{G}$, as if it is a rigid reflector.

In practice, a collimated beam strikes a finite area of the crystal plane, where there are a number of scatterers at $\boldsymbol{r}_{om}$ near lattice points $\boldsymbol{R}_{om}(\mathrm{m} = 1, 2, \ldots, \mathrm{p})$, reflecting X-ray photons in phase. Therefore, such reflections should be in constructive interference, as expressed by the total amplitude $\boldsymbol{A}_o = \sum_m A_{om}$, and the practical scattering amplitude is given by

$$\boldsymbol{A}_o/E_o = \sum\nolimits_m f_m(\boldsymbol{G}) \exp(-i\boldsymbol{G}.\boldsymbol{R}_m), \tag{6.7a}$$

where

$$f_m(\boldsymbol{G}) = \int \rho(\boldsymbol{r}_{om}) \exp(-i\boldsymbol{G}.\boldsymbol{r}_{om})\mathrm{d}^3\boldsymbol{r}_{om}. \tag{6.7b}$$

While $f_m(\boldsymbol{G})$ defined by (6.7b) is called the *atomic scattering factor*, we realize that these charge densities $\rho(\boldsymbol{r}_{om})$ are overlapped, giving a substantial overestimate of $\boldsymbol{A}_o$ when calculated with (6.7a). It is therefore logical to calculate reflections from all atoms spread in the target area. Using a delocalized coordinate vector $\boldsymbol{s}$, A_{om} is expressed as

$$\begin{aligned} A_{om}/E_o &= \int \rho(\boldsymbol{s}) \exp(-i\boldsymbol{G}.\boldsymbol{s})\mathrm{d}^3\mathrm{s} \\ &= \int \rho(\boldsymbol{s} - \boldsymbol{r}_{om}) \exp(-i\boldsymbol{G}.(\boldsymbol{s} - \boldsymbol{r}_{om})\mathrm{d}^3(\boldsymbol{s} - \boldsymbol{r}_{om}) \times \exp(-i\boldsymbol{G}.\boldsymbol{r}_{om}) \\ &= f_m(\boldsymbol{G}) \exp(-i\boldsymbol{G}.\boldsymbol{r}_{om}). \end{aligned}$$

In this case, the quantity

$$S(\boldsymbol{G}) = \sum\nolimits_m f_m(\boldsymbol{G}) \exp(-i\boldsymbol{G}.\boldsymbol{r}_{om}) \tag{6.8}$$

is called the *structural form factor*, and is expressed in terms of indexes h, k and l of the vector $\boldsymbol{G}$ representing the specific group of parallel crystal planes. Writing for instance $\boldsymbol{r}_{om} = x_m\boldsymbol{a}_1 + y_m\boldsymbol{a}_2 + z_m\boldsymbol{a}_3$,

$$S(\boldsymbol{G}) = \sum\nolimits_m f_m(\boldsymbol{G}) \exp\left\{\frac{2\pi i}{\Omega}(x_m h + y_m k + z_m l)\right\}. \tag{6.8a}$$

The electric field of scattered X-ray at a distance $r \gg r_o$ is expressed as

$$\boldsymbol{E}_G(\boldsymbol{r}) = r^{-1}E_o S(\boldsymbol{G}) \exp i(\boldsymbol{K}.\boldsymbol{r} - \omega t),$$

where the relative intensity $I(\boldsymbol{G})$ is given by

$$I(\boldsymbol{G})/I_o(\boldsymbol{G}) = \boldsymbol{E}_G{}^*(\boldsymbol{r})\boldsymbol{E}_G(\boldsymbol{G})/E_o^2 = r^{-2}S^*(\boldsymbol{G})S(\boldsymbol{G}), \tag{6.9}$$

indicating that the diffraction intensity from crystal planes $\boldsymbol{G}$ is determined by the structure factor $S(\boldsymbol{G})$.

Up to this point, crystal planes are considered as rigid reflectors for X-ray, but in reality the crystal is in thermal motion and its effects should be considered for observed diffraction. Assuming a simple harmonic vibration for each constituent, the scattered intensity (6.9) is modified by the Debye-Waller factor, as explained below.

Writing the position of a constituent scatterer as $\boldsymbol{r}_o(t) = \boldsymbol{r}_o + \boldsymbol{u}(t)$, where $\boldsymbol{u}(t)$ is the instantaneous displacement from the equilibrium position $\boldsymbol{r}_o$, $\boldsymbol{u}(t)$ obeys a harmonic oscillator equation in the Einstein model of solids. In this assumption, $\langle \boldsymbol{u}(t)\rangle = 0$ and $\langle \boldsymbol{r}(t)\rangle = \boldsymbol{r}_o$, and for small displacements the exponential factor in $\boldsymbol{E}_G(\boldsymbol{r})$ can be expanded as

$$\langle \exp i\boldsymbol{G}.\boldsymbol{r}_o(t)\rangle = \exp i\boldsymbol{G}.\boldsymbol{r}_o[1 - \tfrac{1}{2}\langle|\boldsymbol{G}.\boldsymbol{u}(t)|^2\rangle + \ldots.],$$

where the second term on the right side can be replaced by

$$\langle|\boldsymbol{G}.\boldsymbol{u}(t)|^2\rangle = \tfrac{1}{3}G^2\langle u(t)^2\rangle,$$

if the fluctuations can be assumed as isotropic in three-dimensional crystals. Hence the scattering intensity ratio between thermal and rigid crystals given by (6.9) is expressed by

$$I(\boldsymbol{G})/I_o(\boldsymbol{G}) = \exp\left\{-\tfrac{1}{3}G^2\langle u(t)^2\rangle\right\},$$

which is called the Debye-Waller factor. In the Einstein model, the average $\langle u(t)^2\rangle$ can be evaluated by the equipartition theorem that is applied to a harmonic oscillator of a mass M, i.e. $\frac{1}{2}M\omega^2\langle u(t)^2\rangle = (3/2)\mathrm{k_B}T$, and the Debye-Waller factor W is given by

$$W = \exp(-\mathrm{k_B}TG^2/M\omega^2). \tag{6.10}$$

This implies that the scattering intensity decreases with increasing temperature and that the diffraction from crystal planes at low G shows less thermal broadening than from those of higher G.

6.3 Diffuse Diffraction from Weakly Modulated Crystals

In a modulated crystal at an irrational vector $\boldsymbol{Q}$, the original lattice periodicity is modified; hence, unit cells are not all identical. Accordingly the vector $\boldsymbol{G}_\mathrm{i} = \boldsymbol{G} + \boldsymbol{Q}$ does not represent crystal planes. If, however, $|\boldsymbol{Q}| \ll |\boldsymbol{G}|$, X-ray diffraction exhibit anisotropically broadened patterns or *satellite* spots in some cases, where the concept of a crystal plane is acceptable approximately. On the other hand, for a phase transition at the Brillouin-zone boundary or at a nonlattice point, where $|\boldsymbol{Q}|$ is not small, the X-ray diffraction is not a logical method, and neutron inelastic scattering provides a direct method for studying such modulated crystals.

Binary structural changes at the center $\boldsymbol{Q} = 0$ or at the zone boundary $\boldsymbol{Q} = \frac{1}{2}\boldsymbol{G}$ in the Brillouin zone exhibit anomalies near transition temperatures that are interpreted as small fluctuations between $\boldsymbol{q}$ and $-\boldsymbol{q}$. As evidenced by soft modes observed in the critical region, such fluctuations are considered as arising from momentum and energy exchanges between order variables and soft phonons in condensates. Figures 6.2a and 6.2b illustrate scatterings of X-ray and neutrons at the zone center and boundary, respectively, where the direction of $\boldsymbol{q}$ is unspecified. In fact, such fluctuations as represented by a small $\boldsymbol{q}$ are significant in the critical region, and we first discuss diffuse X-ray diffraction experiments from weakly modulated crystals at $\boldsymbol{Q} = 0$, leaving strongly modulated cases to neutron-scattering studies in Section 6.5.

The conservation laws for X-ray scattering at $\boldsymbol{Q} = 0$ from a binary crystal can be written as

$$\boldsymbol{K}_\text{o} - \boldsymbol{K} = \boldsymbol{G} \pm \boldsymbol{q} \tag{6.11a}$$

and

$$\varepsilon(\boldsymbol{K}_\text{o}) - \varepsilon(\boldsymbol{K}) = \mp\Delta\varepsilon(\boldsymbol{q}), \tag{6.11b}$$

where $\varepsilon(\boldsymbol{G}) = 0$ represents the ground state of the crystal and $\varepsilon(\boldsymbol{K}_\text{o})$ and $\varepsilon(\boldsymbol{K})$ are X-ray energies before and after the impact. In the following, the Bragg theory in Section 6.2 is modified by writing $\omega = \varepsilon(\boldsymbol{K}_\text{o})/\hbar$, $\omega' = \varepsilon(\boldsymbol{K})/\hbar$ and $\omega - \omega' = \mp\Delta\omega$, replacing energies in (6.11b) by frequencies. In these notations, the scattered amplitude can be expressed as

$$\begin{aligned}
&A_\text{o}/E_\text{o}\\
&\propto \int \mathrm{d}^3\boldsymbol{r}_\text{o}\rho(\boldsymbol{r}_\text{o})\{\exp i(\boldsymbol{K}_\text{o}.\boldsymbol{r}_\text{o} - \omega t_\text{o})\}[\exp i\{\boldsymbol{K}.(\boldsymbol{r} - \boldsymbol{r}_\text{o}) - \omega'(t - t_\text{o})\}]/|\boldsymbol{r} - \boldsymbol{r}_\text{o}|\\
&\approx \int \mathrm{d}^3\boldsymbol{r}_\text{o}\rho(\boldsymbol{r}_\text{o})[\exp i(\boldsymbol{K}.\boldsymbol{r} - \omega' t)/r]\exp i\{(\boldsymbol{K}_\text{o} - \boldsymbol{K}).\boldsymbol{r}_\text{o} - (\omega - \omega').t_\text{o}\}.
\end{aligned}$$

Using the conservation laws (6.11ab), the field $E_G(\boldsymbol{r})$ of a scattered beam at a distant point $\boldsymbol{r}$ is expressed as

$$\begin{aligned}
E_G(\boldsymbol{r})/E_\text{o} \propto \left[\frac{\exp i(\boldsymbol{K}.\boldsymbol{r} - \omega' t)}{r}\right]&\\
\times \sum\nolimits_\text{m} [f(\boldsymbol{G} + \boldsymbol{q})\exp i(\boldsymbol{q}.\boldsymbol{R}_\text{m} - \Delta\omega.t_\text{o})&\\
+ f(\boldsymbol{G} - \boldsymbol{q})\exp i(-\boldsymbol{q}.\boldsymbol{R}_\text{m} + \Delta\omega.t_\text{o})],&
\end{aligned} \tag{6.12}$$

where

$$f(\boldsymbol{G} \pm \boldsymbol{q}) = \int \mathrm{d}^3\boldsymbol{r}_\text{o}\rho(\boldsymbol{r}_\text{o})\exp i(\boldsymbol{G} \pm \boldsymbol{q}).\boldsymbol{r}_\text{o}. \tag{6.13}$$

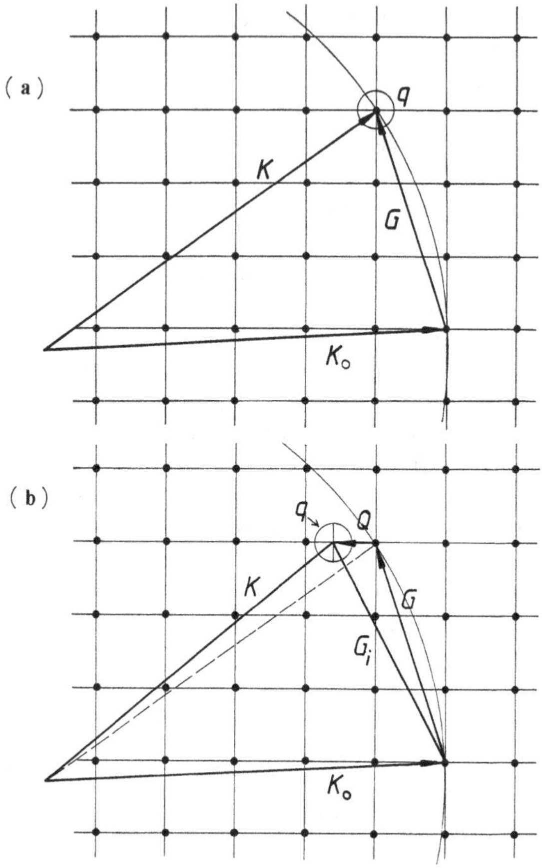

Fig. 6.2. (a) The Bragg diffraction $\boldsymbol{K}_o = \boldsymbol{K} + \boldsymbol{G}$, where $\boldsymbol{G}$ is a lattice translation vector. (b) Scattering by a nonlattice vector $\boldsymbol{G}_i = \boldsymbol{G} + \boldsymbol{Q}$, where $\boldsymbol{Q}$ in a modulation vector. The small vector $\boldsymbol{q}$ represents fluctuations in a modulated lattice.

Here, $f(+\boldsymbol{q}) = f(-\boldsymbol{q})$ at $\boldsymbol{G} = 0$, and $f(\boldsymbol{G}+\boldsymbol{q}) \approx f(\boldsymbol{G}-\boldsymbol{q}) \approx f(\boldsymbol{G})$ for $\boldsymbol{G} \neq 0$, and the scattering intensity can be expressed by

$$\begin{aligned} I(\boldsymbol{G} \pm \boldsymbol{q})/I_o &= r^{-2}|f(\boldsymbol{G})|^2 \sum\nolimits_{mn} [\exp i\{\boldsymbol{q}.(\boldsymbol{R}_m - \boldsymbol{R}_n) - \Delta\omega(t_{om} - t_{on})\} \\ &\qquad + \exp i\{-\boldsymbol{q}.(\boldsymbol{R}_m - \boldsymbol{R}_n) + \Delta\omega(t_{om} - t_{on})\}] \\ &= I(\boldsymbol{G})/I_o + 2r^{-2}|f(\boldsymbol{G})|^2 \\ &\quad \times \sum\nolimits_{m \neq n} \cos\{\boldsymbol{q}.(\boldsymbol{R}_m - \boldsymbol{R}_n) - \Delta\omega(t_{om} - t_{on})\}, \end{aligned} \tag{6.14}$$

where $\boldsymbol{q}.(\boldsymbol{R}_m - \boldsymbol{R}_n) - \Delta\omega(t_{om} - t_{on}) = \phi$ represents the phase of fluctuations on the crystal plane $\boldsymbol{G}$, and the second term is responsible for intensity anomalies. The observed intensity is determined by the average of (6.14) over a random time $t_{om} - t_{on} = \tau$ for photon impacts, resulting in zero anomalies if $\Delta\omega.\tau \gg 1$, otherwise unvanished for $\Delta\omega < 1/\tau$, where τ is distributed, whose limit at the long end may be regarded as the timescale t_o of measurement. In this case, $\Delta\omega < 1/t_o$ clearly determines observable fluctuations, where $\boldsymbol{R}_m - \boldsymbol{R}_n$ can specify a continuous spatial range along a specific direction x. Combining these space-time fluctuations, the phase ϕ is a continuous variable in the range between 0 and 2π in repetition. Taking such a time average of (6.14), we have

$$\frac{\langle \Delta I(\boldsymbol{G})\rangle_t}{I_o} = \frac{\langle I(\boldsymbol{G}\pm\boldsymbol{q}) - I(\boldsymbol{G})\rangle_t}{I_o} = 2r^{-2}|f(\boldsymbol{G})|^2(1/S)\int \langle\cos\phi(x,\tau)\rangle_t \mathrm{d}S, \tag{6.15}$$

where S is the effective area for X-ray impact on the plane of $\boldsymbol{G}$, hence the integral represents the spatial average of $\langle\cos\phi(x,\tau)\rangle_t$.

Assuming that S is a rectangular area L_xL_y, the average in (6.15) can be evaluated as

$$(1/\mathrm{S})\int\langle\cos\phi(x,\tau)\rangle_t\mathrm{dS} = (t_o^{-1}\int_0^{t_o}\mathrm{d}\tau)({L_x}^{-1}\int_0^{L_x}\cos\phi\mathrm{d}x)({L_y}^{-1}\int_0^{L_y}\mathrm{d}y),$$

where

$$L_x^{-1}\int\cos\phi\mathrm{d}x = (qL_x)^{-1}\int\cos\phi\mathrm{d}\phi = (qL_x)^{-1}(\sin\phi_2 - \sin\phi_1)$$
$$= (2/\Delta\phi)\sin(\Delta\phi/2)\cos\underline{\phi}$$

and

$${L_y}^{-1}\int_0^{L_y}\mathrm{d}y = 1.$$

Here, $\Delta\phi = \phi_2 - \phi_1 = qL_x$ and the average phase $\underline{\phi} = \frac{1}{2}(\phi_1+\phi_2)$ depend on the X-ray beam, whereas ϕ is random and continuous in the range $0 \le \phi \le 2\pi$. Letting $\underline{\phi} = qx - \Delta\omega.\tau$, where $0 \le \tau \le t_o$, the above time average can be calculated as

$$t_o^{-1}\int\cos\underline{\phi}\mathrm{d}\tau = \left[\frac{\sin\frac{1}{2}\Delta\omega.t_o}{(\frac{1}{2}\Delta\omega.t_o)}\right]\cos\left(qx - \tfrac{1}{2}\Delta\omega.t_o\right).$$

Therefore, the observable intensity anomaly in a diffraction pattern can be expressed as

$$\langle\Delta I(\boldsymbol{G})\rangle_t/I_o = \left[\frac{\sin\left(\frac{1}{2}\Delta\phi\right)}{\left(\frac{1}{2}\Delta\phi\right)}\frac{\sin\left(\frac{1}{2}\Delta\omega.t_o\right)}{\left(\frac{1}{2}\Delta\omega.t_o\right)}\right]\cos\varphi, \tag{6.16}$$

where $\varphi = qx - \frac{1}{2}\Delta\omega.t_o$ is a redefined phase angle in the range $0 \le \varphi \le 2\pi$. By virtue of the formula $\lim_{\theta\to 0}(\sin\theta/\theta) \to 1$, these front factors in the square brackets are practically equal to 1 for small values of $\Delta\phi$ and $\Delta\omega.t_o$. The condition $\Delta\omega.t_o < 1$ is significant in particular for such a spatial phase distribution of $\Delta\phi$ to be visualized in timescale t_o. Diffraction anomalies similar to the above can also be expected in phase transitions at zone-boundaries, but at arbitrary points in the Brillouin zone such anomalies are detectable only in neutron inelastic scattering experiments, as will be discussed in Section 6.5.

6.4 The Laue Formula and Diffuse Diffraction from Perovskites

Phase transitions at irrational points $\boldsymbol{G}_{\mathrm{i}}$ in the Brillouin zone cannot be subjected to X-ray studies in principle, whereas the diffraction method can be applied to some specific cases of crystals undergoing transitions to low-dimensional order. In perovskite crystals, structural phase transitions from cubic to tetragonal phases can be regarded as two-dimensional order in planes perpendicular to the tetragonal axis, if related to one-dimensional correlations along each of the two symmetry directions. If this view is correct, such a plane acts as a modulated crystal plane below the transition temperature, resulting in intensity anomalies in Bragg diffraction.

We first discussed pseudospin correlations in perovskites in Chapter 3, leading to two incommensurate directions perpendicular to the tetragonal axis in the low-temperature phase. Indeed, diffuse diffraction was observed in X-ray studies by Comes et al. [19], as shown by the diffraction photograph from $NaNbO_3$ at 700°C in Fig. 6.3. Müller and his co-workers [20] reported the corresponding anomalies in magnetic resonance lines from $SrTiO_3$ at 105K that can be interpreted as modulated in two independent directions. In this section, we show that such a diffuse diffraction pattern can be explained with

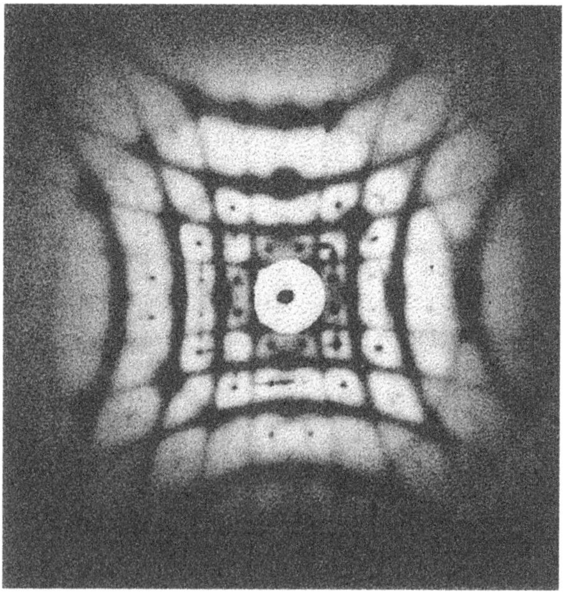

Fig. 6.3. A photograph of a two-dimensional diffuse-diffraction pattern from a perovskite $NiNbO_3$ at 700°C. (From R. Comes, R. Currat, F. Desnoyer, M. Lambert and A.M.Quittet, Ferroelectrics 12, 3 (1976).)

a model of two fluctuating one-dimensional condensates that are mutually perpendicular.

We first consider that correlated pseudospins are periodically placed along the axis of the repeat unit $\boldsymbol{a}_1$, being active for a structural change. In this model, for pseudospins located at $\boldsymbol{r}_\text{i} = n\boldsymbol{a}_1$ where n is an integer, we can define the reciprocal vector $\boldsymbol{G} = h\boldsymbol{a}_1{}^*$ where $\boldsymbol{a}_1{}^*.\boldsymbol{a}_1 = 2\pi$, which, however, does not represent crystal planes when these pseudospins are only linearly correlated. The scattered radiation originates from the dipole moments along $\boldsymbol{a}_1$, as induced by incident radiation. Considering the incident radiation perpendicular to $\boldsymbol{a}_1$, i.e. $\boldsymbol{K}_\text{o} \perp \boldsymbol{a}_1$, scattered beams can be observed in constructive or destructive interference in any direction $\boldsymbol{K}$ on the conical surface of an apex angle θ_1 from the axis, namely $\boldsymbol{K}.\boldsymbol{a}_1 = (2\pi a_1/\lambda)\cos\theta_1$, if θ_1 satisfies either $a_1\cos\theta_1 = \lambda$ integer or $a_1\cos\theta_1 = \frac{1}{2}\lambda \times$ integer, respectively. Such a scattered flux of X-ray beams can be detected on a photographic plate if placed perpendicular to $\boldsymbol{K}_\text{o}$, on which an image of symmetric hyperbola is obtained, as illustrated in Fig. 6.4. In three dimensions, hyperbola of scattered beams from two perpendicular lines of dipoles intersect on the plate, exhibiting a characterstic pattern of diffuse diffraction. In X-ray crystallography, this method of analysis is known as the Laue construction of diffraction patterns.

Using the Laue method for a perovskite crystal, the incident X-ray beam is set parallel to one of the cubic axes, e.g. $\boldsymbol{K}_\text{o} \parallel \boldsymbol{a}_3$, and diffraction cones are considered around the $\boldsymbol{a}_1$ and $\boldsymbol{a}_2$ axes individually. In perovskites, diffraction spots are generally identified by a set of indexes $(h + \frac{1}{2}, k + \frac{1}{2}, l)$, indicating that the phase transition occurs at the M-point of the Brillouin zone, and the diffuse diffraction spots and hyperbolic lines from $NaNbO_3$ crystals observed at 700°C are clear evidence for the two one-dimensional correlations that are primarily independent along the $\boldsymbol{a}_1$ and $\boldsymbol{a}_2$ axes. However, these are related by inversion symmetry, as implied by the magnetic resonance anomalies in [110] direction (ref. [20]).

The Laue condition for constructive interference can be written as

$$a_1\cos\theta_1 = h\lambda$$

for correlated pseudospins on the $\boldsymbol{a}_1$ axis, namely $(2\pi/\lambda)\cos\theta_1 = 2\pi h/a_1$, which can then be reexpressed as $|\boldsymbol{K}|\cos\theta_1 = ha_1{}^* = G_1$. Hence, the wavevector $\boldsymbol{K}$ has components $K_\parallel = G_1$ and $K_\perp \perp a_1$, and the scattering from such a linear chain in binary fluctuations can be signified by the structural form factor

$$f(G_1 \pm q) = \int \mathrm{d}^3 r_\text{o} \rho(r_\text{o}) \exp i(G_1 \pm q) r_\text{o}.$$

Here, assuming $q \ll G_1$ for weak modulation, we have $f(G_1 \pm q) \approx f(G_1)$. In Fig. 6.4b, such fluctuations of scattered X-rays are illustrated by shaded areas along the vectors $\boldsymbol{K}$ and $\Delta\boldsymbol{K}$. Considering temporal fluctuations as well, we can obtain the intensity anomalies of a beam scattered from a group of

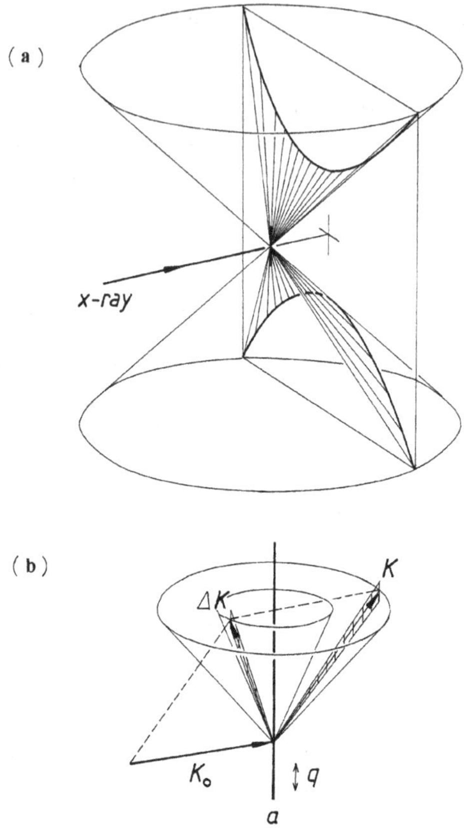

Fig. 6.4. (a) The Laue diffraction from a one-dimensional lattice along the $\boldsymbol{a}_1$ direction. X-ray beam $\perp \boldsymbol{a}_1$. (b) Diffreaction from a fluctuating one-dimensional lattice.

identical pseudospins as

$$I(G_1 \pm q) - I(G_1) \propto I_o \sum\nolimits_{m \neq n} \cos\{q(x_m - x_n) - \Delta\omega(t_{om} - t_{on})\},$$

where the phase $q(x_m - x_n) - \Delta\omega(t_{om} - t_{on})$ can be considered as continuous. Averaging this with respect to the distributed position $x = x_{om} - x_{on}$ at random time $t_{om} - t_{on} = \tau$ of impact in the range $0 < x < L$ and $-t_o < \tau < t_o$, respectively, the observed anomaly can be expressed as

$$\langle \Delta I(G_1) \rangle / I_o = \left[\frac{\sin(qL)}{qL}\right] \left[\frac{\sin(\Delta\omega.t_o)}{\Delta\omega.t_o}\right] \cos\varphi,$$

where $0 \leq \varphi \leq 2\pi$ and L represents the size of the X-ray beam. As in the previous argument, the condition $\Delta\omega.t_o \leq 1$ is essential for the anomaly to be observed in the timescale t_o.

6.5 Neutron Inelastic Scattering

Arising from a singularity in order variable correlations, a structural phase transition occurs at a specific temperature T_c, corresponding to minimum of the correlation energies. Such a singular behavior can occur not only at the center in the Brillouin zone $\boldsymbol{Q} = 0$ but also at a specific irrational wavevector $\boldsymbol{Q}_c \neq 0$, where the corresponding energy $\varepsilon(\boldsymbol{Q}_c)$ is nonzero. In the latter case, being likely associated with pseudosymmetry emerging at a particular temperature, fluctuations are observed from interactions in the condensate, and responsible for anomalies at wavevectors near $\boldsymbol{Q}_c$. The phase transition in K_2SeO_4 crystals at $T_c = 130K$ is a typical example in this category for $\boldsymbol{Q}_c \neq 0$, where so-called phonon-dispersion curves $\varepsilon = \varepsilon(\boldsymbol{Q})$ were observed at temperatures near 130K as shown in Fig. 4.5, where there is a notable dip at $\boldsymbol{Q}_c \sim 2\boldsymbol{a}^*/3$ when the temperature T_c is approached.

Such a transition vector $\boldsymbol{Q}_c$ is comparable with the wavevector of thermal neutrons, so that the energy fluctuation around $\varepsilon(\boldsymbol{Q}_c)$ can be studied by neutron inelastic scattering experiments. In neutron scattering, the wavevector and the related energy are observed as an excitation in the "phonon spectra", where the minimum of $\varepsilon = \varepsilon(\boldsymbol{Q}_c)$ signifies the phase transition. Hence, in the vicinity of $\varepsilon(\boldsymbol{Q}_c)$ critical anomalies can be interpreted as fluctuations due to interactions with pseudospins in the lattice, namely $\varepsilon(\boldsymbol{Q}) = \varepsilon(\boldsymbol{Q}_c) \mp \Delta\varepsilon$ and $\boldsymbol{Q} = \boldsymbol{Q}_c \pm \boldsymbol{q}$. On the other hand, incident neutrons can either lose or gain their kinetic energies during impact, so that scattered neutrons in the direction of $\boldsymbol{Q}$ at $\varepsilon(\boldsymbol{Q})$ should be modulated at $\mp\boldsymbol{q}'$ and $\pm\Delta\varepsilon'$. Denoted by "primes", these modulated quantities in scattered neutrons represent phonon fluctuations, and we can drop these primes from the following calculation to minimize the number of notations, but retaining these signs for the momentum-energy exchange in the critical region.

The conservation laws for neutron inelastic scattering can be expressed as

$$\boldsymbol{K}_o - \boldsymbol{K} = \boldsymbol{G}_i \pm \boldsymbol{q} \tag{6.17a}$$

and

$$E(\boldsymbol{K}_o) - E(\boldsymbol{K}) = \varepsilon(\boldsymbol{G}_i \pm \boldsymbol{q}) = \mp\Delta\varepsilon(\boldsymbol{q}) \tag{6.17b}$$

in the reciprocal space, where $\boldsymbol{G}_i = \boldsymbol{G} + \boldsymbol{Q}_c$ and $\Delta\varepsilon(\boldsymbol{q})$ can be expressed as $\frac{1}{2}\kappa\boldsymbol{q}^2$, representing the kinetic energy of fluctuations if $|\boldsymbol{q}| < |\boldsymbol{G}_i|$.

The basic experimental arrangement for neutron inelastic scattering is sketched in Fig. 6.5, which is called a *triple-axis spectrometer*. Thermally moderated neutrons from a nuclear reactor are admitted to a monochromator with a crystal (A) of a known lattice constant d, whereby the wavelength λ_o for the experiment can be selected by adjusting the angle θ of reflection in the Bragg constructive relation $2d\sin\theta = n\lambda_o$. A sample crystal is mounted on the rotatable goniometer that is placed in the cryostat (B). The wavelength λ of scattered neutrons can then be analyzed by the analyzer crystal (C) with a known lattice constant d'. Here, also by using the formula $n'\lambda = 2d'\sin\theta'$

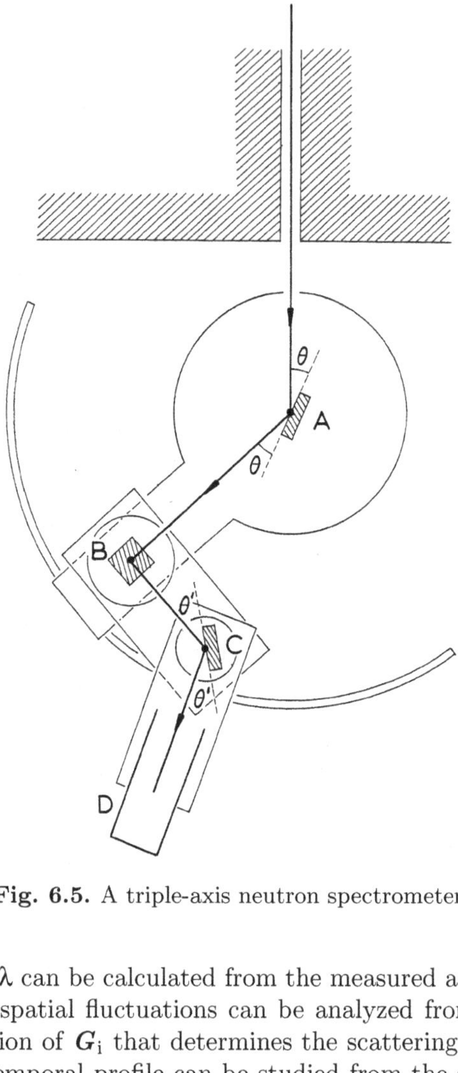

Fig. 6.5. A triple-axis neutron spectrometer.

for the crystal C, λ can be calculated from the measured angle θ'. At a given temperature, the spatial fluctuations can be analyzed from scattered intensities as the function of $\boldsymbol{G}_\mathrm{i}$ that determines the scattering geometry. On the other hand, the temporal profile can be studied from the scattered intensity at $\boldsymbol{G}_\mathrm{i}$, exhibiting the response to neutron impact.

Similar to the atomic scattering factor for X-ray diffraction, neutron inelastic scattering experiments determine the density fluctuation of in distributed scatterers, i.e., magnetic spins or heavy nuclei. We therefore write the scattering amplitude by a scatterer at a site m in the vicinity of $\boldsymbol{G}_\mathrm{i}$ as

$$\begin{aligned} n_\mathrm{m}(\boldsymbol{G}_\mathrm{i} \pm \boldsymbol{q}) &= n_\mathrm{o}(\boldsymbol{G}_\mathrm{i} \pm \boldsymbol{q}) \exp i\{(\boldsymbol{G}_\mathrm{i} \pm \boldsymbol{q}).\boldsymbol{r}_\mathrm{m} \mp (\Delta\varepsilon/\hbar)t_\mathrm{om}\} \\ &= (\exp i\,\boldsymbol{G}_\mathrm{i}.\boldsymbol{r}_\mathrm{m}) n_\mathrm{o}(\boldsymbol{G}_\mathrm{i} \pm \boldsymbol{q}) \exp i(\pm \boldsymbol{q}.\boldsymbol{r}_\mathrm{m} \mp \Delta\omega.t_\mathrm{om}), \end{aligned}$$

where $\Delta\omega = \Delta\varepsilon/\hbar$. Here, it is noted that for a small $|\boldsymbol{q}|$, $n_\mathrm{o}(\boldsymbol{G}_\mathrm{i} + \boldsymbol{q}) \approx n_\mathrm{o}(\boldsymbol{G}_\mathrm{i} - \boldsymbol{q})$, whereas in the critical region, as discussed in Chapter 4, $|\boldsymbol{q}|$ may not be small, and the binary fluctuations consist of symmetric and antisymmetric modes between $+\boldsymbol{q}$ and $-\boldsymbol{q}$. The scattering intensity can therefore be calculated as

$$I(\boldsymbol{G}_\mathrm{i} \pm \boldsymbol{q}) \propto \left\langle \sum_\mathrm{m} \sum_\mathrm{n} [n^*_\mathrm{m}(\boldsymbol{G}_\mathrm{i}+\boldsymbol{q})n_\mathrm{n}(\boldsymbol{G}_\mathrm{i}+\boldsymbol{q}) + n^*_\mathrm{m}(\boldsymbol{G}_\mathrm{i}-\boldsymbol{q})n_\mathrm{n}(\boldsymbol{G}_\mathrm{i}-\boldsymbol{q})] \right\rangle_\mathrm{t}$$
$$= \left\langle \sum_\mathrm{m} |n_\mathrm{m}(\boldsymbol{G}_\mathrm{i} \pm \boldsymbol{q})|^2 \right\rangle_\mathrm{t}$$
$$+ \left\langle \sum_{\mathrm{m}\neq\mathrm{n}} [n^*_\mathrm{m}(\boldsymbol{G}_\mathrm{i}+\boldsymbol{q})n_\mathrm{n}(\boldsymbol{G}_\mathrm{i}+\boldsymbol{q}) + n^*_\mathrm{m}(\boldsymbol{G}_\mathrm{i}-\boldsymbol{q})n_\mathrm{n}(\boldsymbol{G}_\mathrm{i}-\boldsymbol{q})] \right\rangle_\mathrm{t}.$$

Therefore, the second term in this expression should be responsible for anomalies from the normal intensities determined by the first one. The intensity anomalies are thus expressed as

$$\Delta I(\boldsymbol{G}_\mathrm{i} \pm \boldsymbol{q}) \propto \sum_{\mathrm{m}\neq\mathrm{n}} \exp i\boldsymbol{G}_\mathrm{i}.(\boldsymbol{r}_\mathrm{m} - \boldsymbol{r}_\mathrm{n})$$
$$\times [|n_\mathrm{o}(\boldsymbol{G}_\mathrm{i}+\boldsymbol{q})|^2 \exp i\{\boldsymbol{q}.(\boldsymbol{r}_\mathrm{m} - \boldsymbol{r}_\mathrm{n}) - \Delta\omega(t_\mathrm{om} - t_\mathrm{on})\}$$
$$+ |n_\mathrm{o}(\boldsymbol{G}_\mathrm{i}-\boldsymbol{q})|^2 \exp i\{-\boldsymbol{q}.(\boldsymbol{r}_\mathrm{m} - \boldsymbol{r}_n) + \Delta\omega(t_\mathrm{om} - t_\mathrm{on})\}]\rangle_\mathrm{t},$$

which should be symmetrical with regard to indexes m and n. Hence, for this expression we define variables $\boldsymbol{r} = \boldsymbol{r}_\mathrm{m} - \boldsymbol{r}_\mathrm{n}$ and $\tau = t_\mathrm{om} - t_\mathrm{on}$, which can be considered as continuous variables. Expanding the density factors as

$$|n_\mathrm{o}(\boldsymbol{G}_\mathrm{i} \pm \boldsymbol{q})|^2 = |n_\mathrm{o}(\boldsymbol{G}_\mathrm{i})|^2 \pm 2i\boldsymbol{q}.\,\mathrm{grad}_r\, |n_\mathrm{o}(\boldsymbol{G}_\mathrm{i})|^2,$$

we obtain the expression

$$\Delta I(\boldsymbol{G}_\mathrm{i}) \propto \left[\int |n_\mathrm{o}(\boldsymbol{G}_\mathrm{i})|^2 \cos(\boldsymbol{G}_\mathrm{i}.\boldsymbol{r}) \mathrm{d}^3\boldsymbol{r} \right] \langle \cos\phi \rangle_\mathrm{t}$$
$$+2 \left[\int \boldsymbol{q}.\,\mathrm{grad}_r\, |n_\mathrm{o}(\boldsymbol{G}_\mathrm{i})|^2 \cos(\boldsymbol{G}_\mathrm{i}.\boldsymbol{r}) \mathrm{d}^3\boldsymbol{r} \right] \langle \sin\phi \rangle_\mathrm{t}.$$

Using abbreviations A and P for these bracketed quantities, the anomalies can be written in a short form

$$\Delta I = \Delta I_A + \Delta I_P, \quad \text{where} \quad \Delta I_\mathrm{A} = A\langle \cos\phi \rangle_\mathrm{t} \quad \text{and} \quad \Delta I_\mathrm{P} = P\langle \sin\phi \rangle_\mathrm{t}. \tag{6.18a}$$

Here, the time-averages can be evaluated with respect to the timescale t_o of observation, as in (6.16):

$$\langle \sin\phi, \cos\phi \rangle_\mathrm{t} = \frac{\sin(\frac{1}{2}\Delta\omega.t_\mathrm{o})}{(\frac{1}{2}\Delta\omega.t_\mathrm{o})} (\sin\varphi, \cos\varphi), \quad 0 \le \varphi \le 2\pi. \tag{6.18b}$$

Writing time-averaged front factors in (6.18b) as A_t and P_t, (6.18a) can be reexpressed in terms of φ, instead of ϕ:

$$\Delta I_\mathrm{A} = A_\mathrm{t} \cos\varphi \quad \text{and} \quad \Delta I_\mathrm{P} = P_\mathrm{t} \sin\varphi. \tag{6.18c}$$

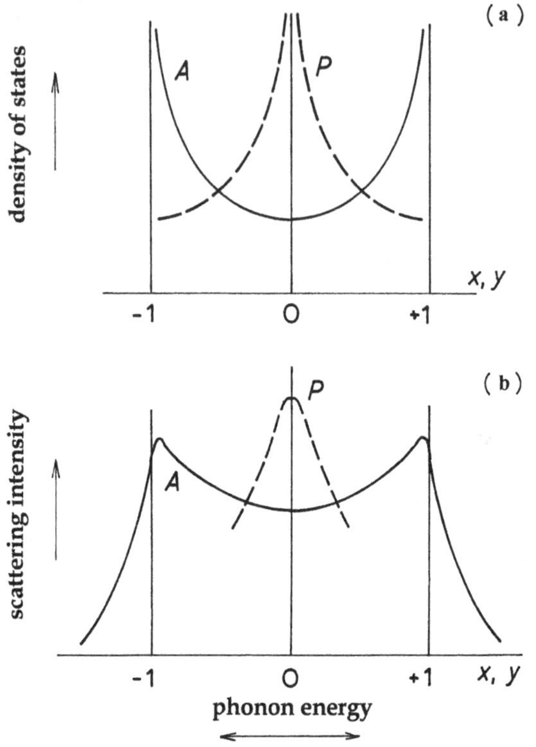

Fig. 6.6. Anomalous intensities of neutron inelastic scattering. A: amplitude mode, P: phase mode. (a) density of states; (b) intensity anomalies vs. phonon energy.

The anomalies in (6.18ac) exhibit characteric symmetry properties of critical fluctuations, that can be observed explicitly, if the condition $\Delta\omega.t_o \sim 1$ is fulfilled.

Expressed as functions of the phase φ, these ΔI_A and ΔI_P are distributed, and so the actual scattering intensities are described as $f(\Delta I_A)\mathrm{d}\varphi$ and $f(\Delta I_P)\mathrm{d}\varphi$ between φ and $\varphi + \mathrm{d}\varphi$, where the functional form f is unspecified. Therefore, we convert the variable φ to ΔI_A and obtain

$$f(\Delta I_A)\mathrm{d}\varphi = f(\Delta I_A)\left(\frac{\mathrm{d}\varphi}{\mathrm{d}\Delta I_A}\right)\mathrm{d}\Delta I_A = f(\Delta I_A)\mathrm{d}\Delta I_A/|\sin\varphi|,$$

and

$$f(\Delta I_P)\mathrm{d}\varphi = f(\Delta I_P)\mathrm{d}\Delta I_P/|\cos\varphi|.$$

Writing $\Delta I_A/A_t = x = \cos\varphi$ and $\Delta I_P/P_t = y = \sin\varphi$ for brevity, we express these fluctuation lineshapes in convenient forms for the amplitude and phase modes as

$$f(\Delta I_A)\mathrm{d}\varphi = f(A_t x)\mathrm{d}x/(1-x^2)^{1/2} \tag{6.19a}$$

and

$$f(\Delta I_P)d\varphi = f(P_t y)dy/(1-y^2)^{1/2} = f(P_t y)dy/x. \qquad (6.19b)$$

Figure 6.6 illustrates these lineshapes for amplitude and phase modes, which are simulated for experiments where the scattering was measured against the transfer energy $\pm\Delta\varepsilon$. Characterized by cos and sin functions, respectively, the distributed intensities of the former are spread between two edges, while peaked at the center $x = 0$ in the latter, and clearly distinguishable in some observed spectra. Figure 6.7 shows such intensity anomalies reported by Schulhof and his collaborators [55] near the Néel temperature T_N of antoferromagnetic MnF_2 crystals, whereas Fig. 6.8 was obtained from one-dimensional magnet $CsNiF_3$ by Pinn and Fender [56]. Two fluctuating modes are clearly evidenced in these results, although these authors called these sin and cos modes the phase and amplitude modes. However their assignments refer to the lattice, whereas neutrons are scattered from magnetic spins; hence, the semantic difference from our definition given in Chapter 4 is not conflict-

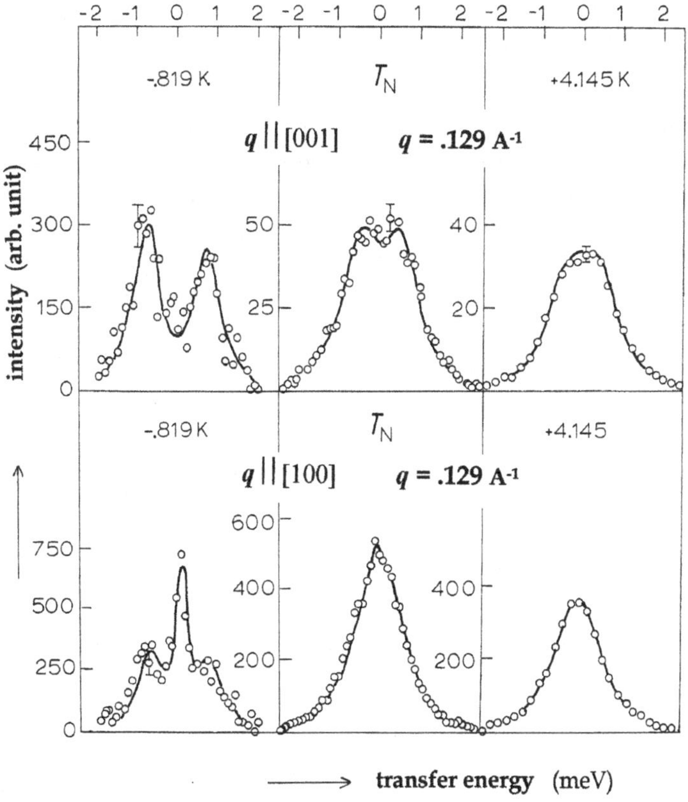

Fig. 6.7. Anomalous neutron scattering intensities from MnF_2 at the Neél temperature T_N: (a) $\boldsymbol{q} \parallel [001]$, (b) $\boldsymbol{q} \parallel [100]$.

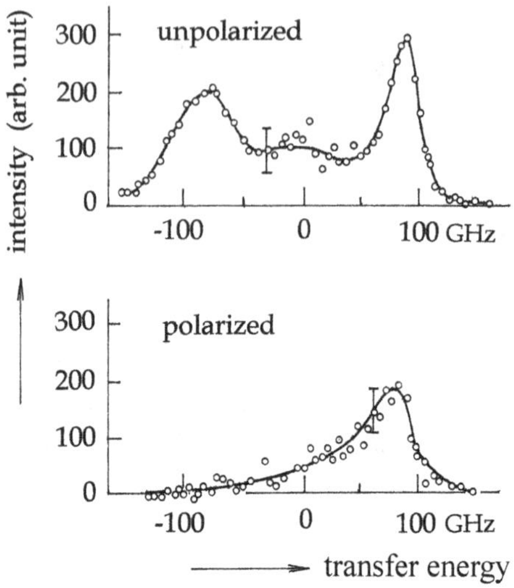

Fig. 6.8. Anomalous intensity distribution of unpolarized and polarized neutron scattering from magnetic spin waves in $CsNiF_3$ crystals. (From R. Pinn and B.E.F.Fender, Phys. Today, **38**, 47, (1985).)

ing in identifying the modes. It is noted that the intensity ratio between two modes depended on the scattering geometry with respect to crystal orientations, indicating that anisotropic fluctuations are signified by the directions of wavevectors $\pm\boldsymbol{q}$ and the gradient of the density square $|n_o(\boldsymbol{G}_i)|^2$. Further notable is that in $CsNiF_3$ the peak heights in the cos mode are unequal, for which asymmetric domain structure may be considered as responsible.

7

Light Scattering and Dielectric Studies on Structural Phase Transitions

7.1 Raman Scattering Studies on Structural Transitions

For structural phase transitions, neutron inelastic scattering is a straightforward method for investigating the energy-momentum exchange in pseudospin condensates in the critical region. In unmodulated crystals however, the scattering geometry $\boldsymbol{K}_o = \boldsymbol{K} \pm \boldsymbol{q}$ for a small $\boldsymbol{q}$ requires a very small scattering angle, making experiments extremely impractical. Instead, using intense coherent light beam from a laser oscillator, we can observe scattered light in enhanced intensity in the right-angle direction of scattering, i.e. $\boldsymbol{K}_o \perp \boldsymbol{K}$, thereby keeping the detector out of direct radiation. Such scattering occurs not only from optical transitions in specific ions interacting with optical phonons, but also with an acoustic excitation in a crystal as a whole at such a wavevector $\boldsymbol{Q}$ as in the relation $\boldsymbol{K}_o - \boldsymbol{K} = \pm\boldsymbol{Q}$, as shown in Fig. 7.1. In the former case, perturbed by optical lattice modes, the soft mode may be identified among transitions induced by the electric field of light if the ion is involved in a condensate. In the latter, critical fluctuations can be studied from the phonon wavevector $\boldsymbol{Q}$, if modulated by a specific $\boldsymbol{Q}_c$ at the transition threshold through a complex electro-elastic coupling in the crystal, although optical transitions are primarily induced by the electric field of light, independent of elastic properties.

For light-scattering experiments, it is a common practice to observe scattering in right-angle directions by scanning the frequency of light to reveal the temporal profile of critical fluctuations. Scattered light from a polar crystal is modulated by either optic or acoustic phonons, which are called *Raman or Brillouin scatterings*, respectively. In such light-scattering studies, the experimental objective is to determine the critical frequency ϖ of condensates, while in the Brillouin scattering, the characteristic wavevector $\boldsymbol{Q}_c$ can be determined from the specific scattering geometry. In these experiments, the sample crystal is rotated around a known axis for scattering in the spectrometer. We first review the principle of Raman scattering in this section, and then discuss the Brillouin scattering in Section 7.2.

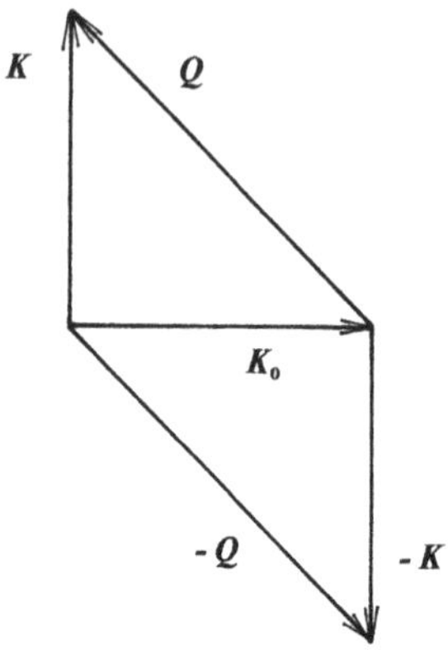

Fig. 7.1. The wavevector relation for right-angle scattering, $\pm\boldsymbol{K} = \boldsymbol{K}_\text{o} \mp \boldsymbol{Q}$.

Consider that an optical transition is induced by light between the ground state "0" and an excited state "1" of an ion associated with a pseudospin condensate. In the critical region, these ionic states can be perturbed by the soft mode, although always perturbed by an acoustic excitation at $\boldsymbol{Q}$ in the light scattering process $\Delta\boldsymbol{K} = \pm\boldsymbol{Q}$. Nevertheless, Placzek [57] discussed the general problem of ionic states perturbed by optic phonons in the lattice with adiabatic approximation, and his results can be directly transferred to the present Raman scattering problem. Following his theory, we write wavefunctions of these ionic states as

$$\psi_\text{o} = \varphi_\text{o}\chi_n \quad \text{and} \quad \psi_1 = \varphi_1\chi_n,$$

where φ_o and φ_1 are wavefunctions of unperturbed ionic states 0 and 1, respectively, and χ_n represents vibrational state n of the lattice. For the ionic Hamiltonian H, we have $\mathsf{H}\varphi_\text{o} = \varepsilon_\text{o}\varphi_\text{o}$ and $\mathsf{H}\varphi_1 = \varepsilon_1\varphi_1$, for which the lattice is considered as a perturbation represented by a harmonic oscillator at a single frequency ω in the Einstein model. Here, the frequency can be the characteristic frequency ϖ of the soft mode, if $\boldsymbol{Q} \parallel \boldsymbol{Q}_\text{c}$ near the transition threshold.

Assuming the lattice quantum $\hbar\omega$ is sufficiently smaller than the ionic excitation energy $\varepsilon_1 - \varepsilon_\text{o}$ at a given temperature, the external electric field $\boldsymbol{E}\cos\Omega t$ of coherent light polarizes the ionic states, inducing an electric dipole moment $\boldsymbol{p}$. We consider that the vector $\boldsymbol{E}$ of incident light has a direction fixed in the crystal and that such a linearly polarized field can be decomposed to two circularly polarized components in opposite directions:

$$\boldsymbol{E}\cos\Omega t = \tfrac{1}{2}\boldsymbol{E}_+\exp(i\Omega t) + \tfrac{1}{2}\boldsymbol{E}_-\exp(-i\Omega t).$$

For this problem, the Schrödinger equation can be written as

$$[\mathsf{H} - \tfrac{1}{2}\boldsymbol{p}.\boldsymbol{E}_+\exp(i\Omega t) - \tfrac{1}{2}\boldsymbol{p}.\boldsymbol{E}_-\exp(-i\Omega t)]\Psi = i\hbar\partial\Psi/\partial t, \tag{7.1}$$

where the perturbed wavefunction in the ground state is given by

$$\Psi_o = \psi_o \exp\{-i(\varepsilon_o + n\hbar\omega)t/\hbar\} \\ +\{\psi_{o+}\exp(i\Omega t) + \psi_{o-}\exp(-i\Omega t)\}\exp\{-i(\varepsilon_o + n\hbar\omega)t/\hbar\}.$$

Here, the first term is for the unperturbed ground state and the second one is for corrections due to the perturbation. In the first-order approximation, the functions $\psi_{o\pm}$ are related to ψ_o by

$$\{\mathsf{H} - (\varepsilon_o + n\hbar\omega) \pm \hbar\Omega\}\psi_{o\pm} = \boldsymbol{p}.\boldsymbol{E}_\pm\psi_o, \tag{7.2}$$

representing the ground state energies $\varepsilon_\pm = \varepsilon_o+(n\pm1)\hbar\omega$ that are polarized by circular fields $\boldsymbol{E}_\pm$ coupled with non-vanishing matrix elements $\langle\psi_{o\pm}|\boldsymbol{p}|\psi_o\rangle$. In the following calculation, we abbreviate these elements as $\langle\pm|\boldsymbol{p}|0\rangle$ for brevity. In this approximation $\varepsilon_\pm - \varepsilon_o = \pm\hbar\omega \ll \varepsilon_1 - \varepsilon_o$, so that the ionic state "0" can be considered as degenerate in the zero order, where the vibrational functions χ_n, χ_{n+1} and χ_{n-1} form a complete orthonormal set. Omitting n from χ_n,

$$\psi_o = \varphi_o\chi_o \quad \text{and} \quad \psi_{o\pm} = \varphi_o(c_+\chi_+ + c_-\chi_-) + c_o\psi_o,$$

where

$$c_\pm = \langle\pm|\boldsymbol{p}|0\rangle.\boldsymbol{E}_\pm/\hbar(\pm\omega \pm \Omega). \tag{7.3}$$

In the perturbed ground state, the time-dependent dipole moment induced by the field $\boldsymbol{E}\cos\Omega t$ can be calculated as

$$\begin{aligned}\int \Psi_o{}^*\boldsymbol{p}(t)\Psi_o \mathrm{d}v &= \int \psi_o{}^*\boldsymbol{p}(t)\psi_o \mathrm{d}v \\ &\quad + \exp(i\Omega t)\int[\psi_o\langle+|\boldsymbol{p}|0\rangle\psi_{o-}{}^* + \psi_o{}^*\langle0|\boldsymbol{p}|+\rangle\psi_{o+}]\mathrm{d}v \\ &\quad + \exp(-i\Omega t)\int[\psi_o{}^*\langle0|\boldsymbol{p}|-\rangle\psi_{o-} + \psi_o\langle0|\boldsymbol{p}|-\rangle\psi_{o+}{}^*]\mathrm{d}v \\ &= \langle0|\boldsymbol{p}|0\rangle c_o + \exp(i\Omega t)[\langle+|\boldsymbol{p}|0\rangle c_1{}^* + \langle0|\boldsymbol{p}|+\rangle c_+] \\ &\quad + \exp(-i\Omega t)[\langle0|\boldsymbol{p}|-\rangle c_- + \langle0|\boldsymbol{p}|-\rangle c_+{}^*],\end{aligned}$$

where $\mathrm{d}v$ is the volume element for integration. The first term in the last expression represents a permanent dipole moment if exists, whereas the second and third ones are induced dipole moments by $\boldsymbol{E}_\pm$, respectively. The components of the dipole moment can be written conveniently as

$$p_i(t) = \langle0|p_i|0\rangle + \sum\nolimits_j\{\alpha_{ij}(\Omega)E_{+j}\exp(i\Omega t) + \alpha_{ij}(-\Omega)E_{-j}\exp(-i\Omega t)\}, \tag{7.4a}$$

where

$$\begin{aligned}\alpha_{ij}(\Omega) &= \langle0|p_i|+\rangle\langle+|p_j|0\rangle/\hbar(\omega+\Omega) + \langle0|p_j|+\rangle\langle+|p_i|0\rangle/\hbar(\omega-\Omega) \\ &\quad + \langle0|p_i|-\rangle\langle-|p_j|0\rangle/\hbar(-\omega+\Omega) + \langle0|p_j|-\rangle\langle-|p_i|0\rangle/\hbar(-\omega-\Omega).\end{aligned} \tag{7.4b}$$

Here, these elements $\alpha_{ij}(\Omega)$ constitute the *polarizability tensor* with respect to coordinate axes i, j = x, y in the plane of $\boldsymbol{E}_{\pm}$. It is noted that these elements are related as

$$\alpha_{ij}(\Omega) = \alpha_{ij}{}^*(\Omega) \quad \text{and} \quad \alpha_{ij}(\Omega) = \alpha_{ji}{}^*(\Omega). \tag{7.5}$$

The first relation indicates that these components are all real quantities, whereas the second one states that the tensor $\alpha_{ij}(\Omega)$ is *Hermitian.*

Similar expressions can be written for the upper ionic state, i.e.

$$\Psi_1 = \exp\{-i(\varepsilon_1 + n\hbar\omega)/\hbar\}[\psi_{1+}\exp(i\Omega t) + \psi_{1-}\exp(-i\Omega t)],$$

where

$$\psi_{1\pm} = \langle\pm|\boldsymbol{p}|1\rangle.\boldsymbol{E}_{\pm}/\hbar(\pm\omega \pm \Omega).$$

For Raman transitions between states 0 and 1, we need to calculate matrix elements for Δn = ±1 and 0. We consider an induced emission from the state ψ_1 to the ground state Ψ_o, where the transition probability can be calculated from the matrix element

$$\begin{aligned}\int \Psi_o{}^*\boldsymbol{p}\psi_1 dv &= \exp(i\varepsilon_{10}t/\hbar)\int dv[c_o{}^*(\psi_o{}^*\boldsymbol{p}\psi_1) \\ &\quad + c_+{}^*(\psi_+{}^*\boldsymbol{p}\psi_1) + c_-{}^*(\psi_-{}^*\boldsymbol{p}\psi_1)) \\ &= c_o{}^*\exp(i\varepsilon_{10}t/\hbar)\int \psi_o{}^*\boldsymbol{p}\psi_1 dv \\ &\quad + [\langle 0|\boldsymbol{p}^*|\pm\rangle\langle\pm|\boldsymbol{p}|0\rangle/\hbar(\pm\omega + \Omega)]E_+\exp i(\Omega \pm \varepsilon_{10}t/\hbar) \\ &\quad + [\langle 0|\boldsymbol{p}^*|\pm\rangle\langle\pm|\boldsymbol{p}|0\rangle/\hbar(\pm\omega - \Omega)]E_-\exp i(-\Omega \pm \varepsilon_{10}t/\hbar),\end{aligned} \tag{7.6}$$

where $\varepsilon_{10} = \varepsilon_1 - \varepsilon_o$. The first term in (7.6) gives the transition matrix for $\Omega = \omega_{10} = \varepsilon_{10}/\hbar$, for which $\Delta n = 0$, whereas the second and third ones are for the Raman transitions $\Omega = \omega_{10} \pm \omega$, corresponding to $\Delta n = \pm 1$. It is noted that the quantities in the square brackets in the last expression are elements of the polarizability tensor $\alpha_{ij}(\Omega)$, which should be nonzero for observable Raman transitions. The satellites at $\omega_{10} - \omega$ and $\omega_{10} + \omega$ are traditionally referred to as Stokes and anti-Stokes lines, respectively, as illustrated in Fig. 7.2a. A typical Raman spectrometer is shown schematically in Fig. 7.2b, where the double (or triple) grating is the essential part to perform automatic frequency scanning in high resolution.

In the Placzek theory, it is too naïve to consider that the lattice frequency of the Einstein model represents the soft mode, whereas the phonon vector $\boldsymbol{Q}$ does not necessarily represent the soft mode. However, if Raman satellites show temperature-dependent frequency shifts; that is, if $\omega = |\Omega - \omega_{10}|$ is proportional to $(T - T_c)^{\beta}$ or $(T_c - T)^{\beta'}$, the corresponding scattering should be originated from the geometry for $\boldsymbol{Q} \parallel \boldsymbol{Q}_c$. To find such Raman lines near T_c, their temperature-dependence should be detected, implicating the soft mode, even if the active group is not positively identified. We have shown such

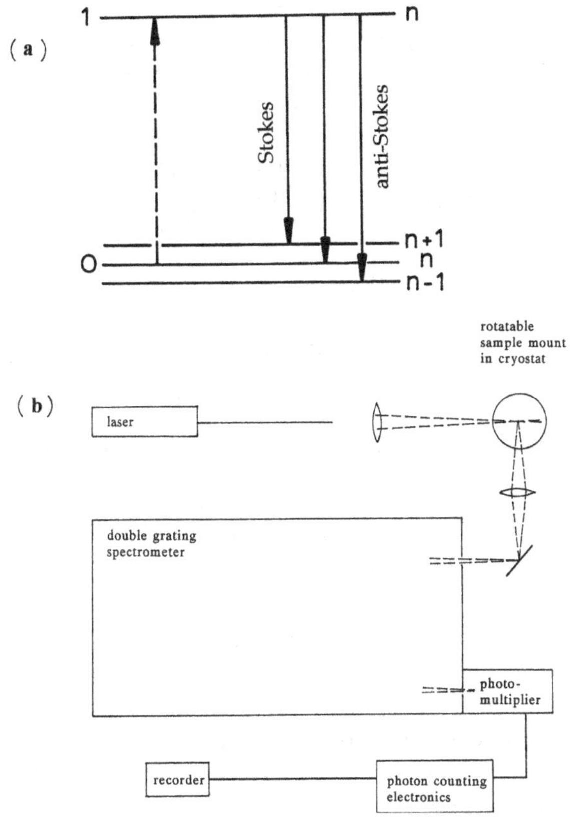

Fig. 7.2. (a) Raman scattering with an optical excitation $0 \to 1$. (b) A typical layout for a laser Raman spectrometer.

Raman observations of soft modes in Figs. 4.3 and 4.4 for phase transitions in TSCC and K_2SeO_4, respectively. Figure 7.3 shows another example of Raman measurements published by Toledano and his coworkers [58] on the ferroelastic phase transition in crystals of the lanthanum phosphate family.

The above argument can basically be for unmodulated crystals, unless critical fluctuations are considered for scattering at $\boldsymbol{Q}_c$. In this case, although much slower than the decay rate of optically excited ionic states, critical fluctuations could be explicitly observed in Raman spectra at temperatures very close to T_c, provided that these sidelines are separated from the main transition for $\Delta n = 0$.

Although often difficult to resolve in the critical region, the intensities of Raman lines at noncritical temperatures depend on the polarizability tensor α_{ij} at $\Omega = \omega_{10} \pm \varpi$, whose nonzero values signify the *Raman activities.* In a modulated crystal, scattered intensities are modified anisotropically with regard to the direction of modulation, but not easily analyzed from Raman

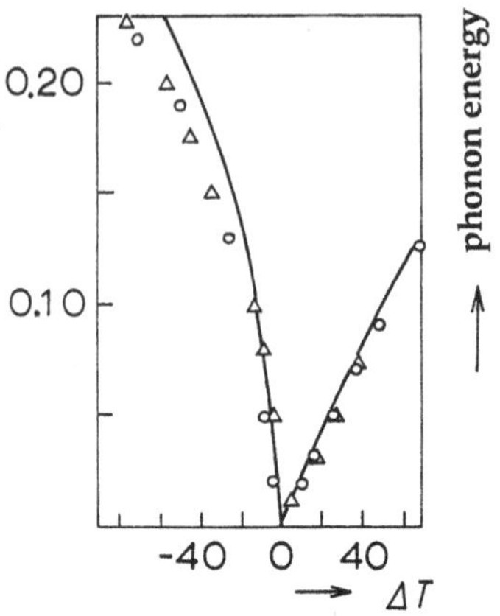

Fig. 7.3. An example of a structural phase transition as observed by a Raman scattering. The temperature dependence of soft-phonon energies in the ferroelastic phase transition of LaP_5O_{14} and $La_{0.5}Nd_{0.5}P_5O_{14}$ determined by Raman scattering. (From P.S.Peercy, *5th Int. Conf. Raman Spectroscopy*, p. 571 (1976); J.C. Toledano, E. Errandonea, and I.P. Jaguin, Solid state Comm. **20**, 905 (1976).)

spectra. The intensity anomalies should be expressed by those terms in (7.6), that are proportional to Raman-active products $p_i p_j$ in the critical region. Assuming a linear chain model for ordering, the induced dipole moment can be modulated as proportional to $\sigma(\phi)$; that is, $\delta p_1 \propto \sigma(\phi)$, whereas δp_2, $\delta p_3 = 0$, if assuming δp_1 parallel to Q_c, and $E_\pm$ are in the plane perpendicular to $\boldsymbol{E}$. In such cases, we can generally write

$$\delta\alpha_{ij} = C_{ij}\sigma(\phi) + D_{ij}\sigma^2(\phi), \quad 0 \le \phi \le 2\pi, \tag{7.7}$$

with which intensity anomalies may be analyzed. The coefficients C_{ij} and D_{ij} are tabulated for instance in the book of Wilson, Decius and Cross [59]. Scattered intensities can be formulated with the classical radiation theory, as shown in the next section, although no such serious analyses have so far been reported in the literature. Scott published two review articles [60, 61], summarizing Raman results from various structural phase transitions.

7.2 Rayleigh and Brillouin Scatterings

Light scattering from some liquids shows evidence for induced dielectric fluctuations, where the Rayleigh scattering is due to random fluctuations in the density, and the Brillouin scattering occurs in conjunction with an acoustic excitation in the liquid. Similar light scattering can take place in a transparent dielectric crystal as well, where the Brillouin scattering is caused by an acoustic excitation at a wavevector $\boldsymbol{Q}$ in the lattice. If involved in the structural change, a specific wavevector $\boldsymbol{Q}_c$ accompanies a softening frequency ϖ, which can be detected with a specific scattering geometry $\Delta \boldsymbol{K} = \boldsymbol{Q} \parallel \boldsymbol{Q}_c$. The phonon wavelength is comparable with that of visible light in dielectric crystals, so that the scatterings can always take place with anisotropic phonon excitations $\pm\boldsymbol{Q}$. Nevertheless, we look for a specific $\boldsymbol{Q}_c$ together with a softening frequency ϖ in light scattering experiments when T_c is approached. In practice, however, being dominated by elastic Rayleigh scattering, the critical anomalies may not be revealed in the Brillouin lines, as in the case of Raman scattering.

Macroscopically, light scattering occurs primarily with dielectric fluctuations $\delta\chi(\boldsymbol{r}, t)$ as induced in a crystal by the electric field $\boldsymbol{E}_o \exp i(\boldsymbol{K}_o.\boldsymbol{r} - \omega_o t)$ of light, whereas the phonon excitation originates from electro-acoustic nature of dielectric crystals. Taking a space-time point $(\boldsymbol{r}_o, t_o)$ of interaction, the induced polarization per volume of a crystal can be expressed as

$$\delta \boldsymbol{P}(\boldsymbol{r}_o, t_o) = \varepsilon_o \delta\chi(\boldsymbol{r}_o, t_o)\boldsymbol{E}_o \exp i(\boldsymbol{K}_o.\boldsymbol{r}_o - \omega_o t_o), \tag{7.8}$$

which then reradiate scattering waves. Here, $\delta\chi$ represents an induced variation in the dielectric susceptibility at ω_o, which should be signified by a softening frequency ϖ if $\boldsymbol{Q} \parallel \boldsymbol{Q}_c$. If this is the case, we can write

$$\delta\chi \propto \sigma(\boldsymbol{r}_o, t).$$

Similar to X-ray diffraction, the electric field of the scattered light at a distant point $\boldsymbol{r}$ at time t can be expressed as

$$\boldsymbol{E}(\boldsymbol{r}, t) \approx r^{-1} \exp i(\boldsymbol{K}.\boldsymbol{r} - \omega t) \iint \varepsilon_o \delta\chi(\boldsymbol{r}_o, t_o)\boldsymbol{E}_o \times \exp i\{(\boldsymbol{K}_o - \boldsymbol{K}).\boldsymbol{r}_o - (\omega_o - \omega)t_o\} \mathrm{d}^3\boldsymbol{r}_o \mathrm{d}t_o. \tag{7.9}$$

Clearly, *elastic* scattering, called Rayleigh scattering, can take place in all directions, as (7.9) is maximum when $|\boldsymbol{K}| = |\boldsymbol{K}_o|$ and $\omega = \omega_o$. Denoting the scattering angle as φ from the relation $\boldsymbol{K}_o.\boldsymbol{r}_o = (2\pi r_o/\lambda)\cos\varphi$, we have $(\boldsymbol{K}_o - \boldsymbol{K}).\boldsymbol{r}_o = (2\pi r_o/\lambda)\cos^2(\frac{1}{2}\varphi)$ for elastic scattering. In this case, the scattered intensity $I_R(\boldsymbol{K}) \propto E^*(\boldsymbol{r}, t).\boldsymbol{E}(\boldsymbol{r}, t)$ can be calculated from (7.9), for which Rayleigh has obtained the formula

$$\frac{I_R(\boldsymbol{K})}{I_o(\boldsymbol{K}_o)} \propto \left(\frac{2\pi^2 v^2}{r^2\lambda^4}\right) \delta\chi^2(1 + \cos^2\varphi) \quad \text{for} \quad r_o \ll \lambda, \tag{7.10}$$

where the two terms on the right are contributed by $\boldsymbol{E}_\perp$ and $\boldsymbol{E}_\parallel$ components, respectively, referring to the plane of $(\boldsymbol{K}, \boldsymbol{K}_o)$. Although occurring in all directions, when measured particularly at the right angle $\varphi = \frac{1}{2}\pi$, Rayleigh scattering is plane polarized with intensity proportional to λ^{-4}.

The Brillouin scattering is inelastic, where the photon energy changes from $\hbar\omega_o$ to $\hbar\omega$, and the corresponding wavevector changes from $\boldsymbol{K}_o$ to $\boldsymbol{K}$, and the differences $\Delta\omega(\boldsymbol{Q}) = \mp(\omega_o - \omega)$ and $\boldsymbol{K}_o - \boldsymbol{K} = \pm\boldsymbol{Q}$ originate from interactions with the lattice mode. It is noted that such an inelastic change for scattered waves can be attributed essentially to electroacoustic properties of the crystal, so that the acoustic excitation at $\pm\boldsymbol{Q}$ and $\mp\Delta\omega$ should be interpreted as caused by the "stress field" associated with $\boldsymbol{E}_\mathrm{o}$ at $(\boldsymbol{r}_\mathrm{o}, t_\mathrm{o})$. Denoted here by double signs, these changes are independent of each other, so that two scatterings are observed at $\omega = \omega_\mathrm{o} \pm \Delta\omega(\boldsymbol{Q})$ with frequency shifts $\pm\Delta\omega(\boldsymbol{Q})$. The intensities of these Brillouin lines are given by

$$I_\mathrm{B}(\boldsymbol{K}, \pm\boldsymbol{Q}) \propto r^{-2}\boldsymbol{E}_\mathrm{o}^2 \int\int \delta\chi^*(\boldsymbol{r}'_\mathrm{o}, t'_\mathrm{o})\delta\chi(\boldsymbol{r}_\mathrm{o}, t_\mathrm{o}) \times \exp[\pm i\boldsymbol{Q}.(\boldsymbol{r}'_\mathrm{o} - \boldsymbol{r}_\mathrm{o}) - (\omega_\mathrm{o} \pm \Delta\omega)(t'_\mathrm{o} - t_\mathrm{o})]\mathrm{d}^3(\boldsymbol{r}'_\mathrm{o} - \boldsymbol{r}_\mathrm{o})\mathrm{d}(t'_\mathrm{o} - t_\mathrm{o}).$$

Here, it is noted that the double integral is the Fourier transform of the binary correlations $\delta\chi^*(\boldsymbol{r}'_\mathrm{o}, t'_\mathrm{o})\delta\chi(\boldsymbol{r}_\mathrm{o}, t_\mathrm{o})$, and the Brillouin intensities are written as

$$I_\mathrm{B}(\boldsymbol{K}, \pm\boldsymbol{Q}) \propto r^{-2}\boldsymbol{E}_\mathrm{o}^2\langle\delta\chi^*(\boldsymbol{K}, \pm\boldsymbol{Q})\delta\chi(\boldsymbol{K}, \pm\boldsymbol{Q})\rangle_t = r^{-2}\boldsymbol{E}_\mathrm{o}^2\tau^{-1}\int_\mathrm{o}^\tau \delta\chi^*(\boldsymbol{K}, \pm\boldsymbol{Q})\delta\chi(\boldsymbol{K}, \pm\boldsymbol{Q})\mathrm{d}t. \quad (7.11)$$

Similar to Raman spectra, these two Brillouin lines at $\mp\Delta\omega(\boldsymbol{Q})$ are referred to as Stokes- and anti-Stokes lines, respectively.

Although (7.8) expresses an electric polarization induced by the electric field of light at $(\boldsymbol{r}_\mathrm{o}, t_\mathrm{o})$, the Brillouin scattering in dielectric crystals is caused by a stress field due to electro-elastic excitations. We consider an effective electric field $\boldsymbol{E}_\mathrm{o}$, representing the complex nature of the crystal. Further, for scattering experiments, we use linearly polarized light that is composed of two oppositely circulating component fields. Therefore, we can write the following equations for these circular polarizations $\delta\boldsymbol{P}_\pm(\boldsymbol{r}_\mathrm{o}, t)$:

$$\left\{\frac{\mathrm{d}^2}{\mathrm{d}t^2} + \gamma\frac{\mathrm{d}}{\mathrm{d}t} + (\omega_\mathrm{o} \pm \varpi)^2\right\}\delta\boldsymbol{P}_\pm(\boldsymbol{r}_\mathrm{o}, t) = \varepsilon_\mathrm{o}\delta\chi_\pm(\boldsymbol{r}_\mathrm{o}, t)\left[\tfrac{1}{2}\boldsymbol{E}_\mathrm{o}\exp i(\boldsymbol{K}_\mathrm{o}.\boldsymbol{r}_\mathrm{o} \mp \omega t)\right],$$

where $\frac{1}{2}\boldsymbol{E}_\mathrm{o}\exp i\boldsymbol{K}_\mathrm{o}.\boldsymbol{r}_\mathrm{o} = \boldsymbol{E}'(\boldsymbol{r}_\mathrm{o})$ is regarded as the effective field. Letting $\delta\boldsymbol{P}_\pm(\boldsymbol{r}_\mathrm{o}, t) = \delta\boldsymbol{P}_\pm(\boldsymbol{r}_\mathrm{o})\exp(\mp i\omega t)$, we can obtain the steady-state solution that can be expressed as the susceptibility

$$\varepsilon_\mathrm{o}\delta\chi_\pm(\boldsymbol{r}_\mathrm{o}) = \delta\boldsymbol{P}_\pm(\boldsymbol{r}_\mathrm{o})/\boldsymbol{E}'(\boldsymbol{r}_\mathrm{o}) = \{-\omega^2 \mp i\omega\gamma + (\omega_\mathrm{o} \pm \varpi)^2\}^{-1}.$$

Therefore, the dielectric correlations at $\boldsymbol{r}_\mathrm{o}$ can be given by

$$\delta\chi_+{}^*(\boldsymbol{r}_\mathrm{o}, t'_\mathrm{o})\delta\chi_+{}^*(\boldsymbol{r}_\mathrm{o}, t_\mathrm{o}) + \delta\chi_-{}^*(\boldsymbol{r}_\mathrm{o}, t'_\mathrm{o})\delta\chi_-(\boldsymbol{r}_\mathrm{o}, t_\mathrm{o}) = \{|\delta\chi_+(\boldsymbol{r}_\mathrm{o})|^2 + |\delta\chi_-(\boldsymbol{r}_\mathrm{o})|^2\}\exp\{\mp i\omega(t'_\mathrm{o} - t_\mathrm{o})\}.$$

Using this result, the Brillouin intensities are given as proportional to

$$\delta\chi^*(\boldsymbol{K}, \pm\boldsymbol{Q}_\mathrm{c})\delta\chi(\boldsymbol{K}, \pm\boldsymbol{Q}) \propto \frac{2i\omega\gamma}{\{(\omega_\mathrm{o} \pm \varpi)^2 - \omega^2\}^2 + \omega^2\gamma^2} \times \int\int \exp\{\mp i\omega(t'_\mathrm{o} - t_\mathrm{o})\}\exp\{\pm\boldsymbol{Q}_\mathrm{c}.(\boldsymbol{r}'_\mathrm{o} - \boldsymbol{r}_\mathrm{o}) - (\omega_\mathrm{o} \pm \varpi)(t'_\mathrm{o} - t_\mathrm{o})\} \times \mathrm{d}^3(\boldsymbol{r}'_\mathrm{o} - \boldsymbol{r}_\mathrm{o})\mathrm{d}^3(t'_\mathrm{o} - t_\mathrm{o}). \quad (7.12)$$

Considering a one-dimensional collective mode of pseudospins along the direction of $\boldsymbol{r}'_\text{o} - \boldsymbol{r}_\text{o}$, which we call the x direction, the spatial phase $\boldsymbol{Q}_\text{c}.(\boldsymbol{r}'_\text{o} - \boldsymbol{r}_\text{o}) = \phi_\text{s}(x)$ is a continuous angle $0 \leq \phi_\text{s} \leq 2\pi$. We further note that $(\omega_\text{o} \pm \varpi) - \omega = \pm\delta\omega$ represent frequency fluctuations near the Brillouin lines, and, hence, the anomalous intensities in the critical region are expressed as

$$I_\text{B}(\boldsymbol{K}, \pm\boldsymbol{Q}_\text{c}) \propto \frac{\omega\gamma}{\{\omega_\text{o} \pm \varpi)^2\}^2 + \omega^2\gamma^2} \times \int dx(2\tau)^{-1} \int_{-\tau}^{+\tau} (\sin,\cos)(\phi_\text{s} \pm \delta\omega.t)dt,$$

where $t = t'_\text{o} - t_\text{o}$ is the time for temporal fluctuations. Redefine the phase ϕ of fluctuations by $\phi = \phi_\text{s} \pm \delta\omega.t$, where $0 \leq \phi \leq 2\pi$, we expect anomalous lineshapes characterized by $\cos\phi$ and $\sin\phi$, as in neutron inelastic scattering. Experimentally, the soft mode can be observed from $\pm\varpi$ shifting with temperature, but the critical broadening should be observable under an extreme condition for minimizing Rayleigh intensities.

Figure 7.4a shows the Brillouin spectra, accompanying the Rayleigh line at ν_R, where the Brillouin shift is measured as $\Delta\nu = \nu_\text{B} \pm \nu_\text{R}$. For a general scattering geometry at the scattering angle δ, as illustrated in Fig. 7.4b for $|\boldsymbol{K}_\text{o}| = |\boldsymbol{K}|$, the phonon vector can be expressed as

$$|\boldsymbol{Q}| = 2|\boldsymbol{K}_\text{o}| \sin\left(\tfrac{1}{2}\delta\right).$$

Writing $\Delta\omega = v|\boldsymbol{Q}|$ and $|\boldsymbol{K}_\text{o}| = n/c$, where n is the optical index of refraction, and v and c the speeds of sound and light in crystals,the Brillouin shift can

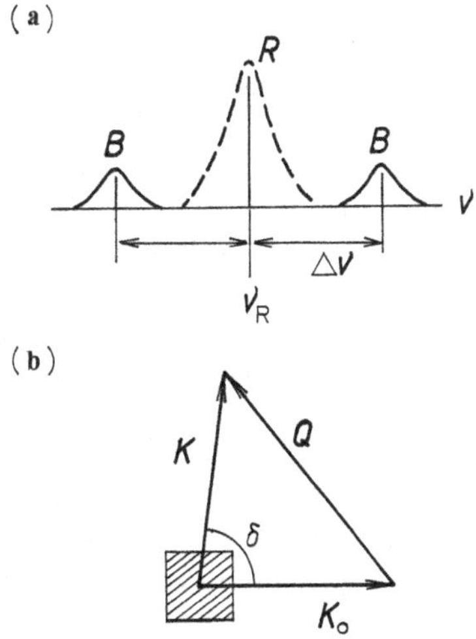

Fig. 7.4. (a) Rayleigh (R) and Brillouin (B) lines in a Brillouin spectrum. (b) A scattering geometry close to right-angle scattering.

be expressed as

$$\Delta\omega/\omega_o = \left(\frac{2nv}{c}\right)\sin\left(\tfrac{1}{2}\delta\right). \tag{7.13}$$

Here, $v/c \approx 10^{-5}$; hence, the Brillouin shift is typically of the order of 10^{-5} if considering that $\delta = \frac{1}{2}\pi$. Accordingly, with light of the wavelength $\lambda = 5000$ Å, the phonon shift in a typical transparent crystals is of the order of 10GHz. The Brillouin shift is normally measured as a frequency shift, representing the sound velocity in crystals. Therefore, it can be compared with results of ultrasound measurements to confirm the consistency.

Hikita and his collaborators [62] carried out light-scattering and ultrasonic studies on the ferroelectric phase transition in TSCC crystals, reporting that Brillouin lines showed temperature-dependent shifts in all directions of symmetric axes near 130K, where the ultrasound velocity exhibited anomalous dispersion. They consider a linear coupling between the polarization and anisotropic phonons to explain observed anomalies, which, nevertheless, appeared to be complicated by electroacoustic properties of TSCC crystals.

7.3 Dielectric Relaxation

Dielectric properties of ionic crystals cannot be discussed separately from their elastic properties because of their electroelastic coupling. Although the Brillouin scattering is basically for the acoustic response from soft modes, dielectric experiments are directly for the electric response from pseudospins in polar crystals.

The motion of dipolar ions in a crystal is slow, as characterized by the relaxation time, representing the time for returning to thermal equilibrium. If measured at a sufficiently low frequency, dipole carriers can move in near phase with an applied electric field, although the motion is significantly hindered in solid states. Being predominantly relaxational, the collective mode shows even a slower response due to heavier mass than individual dipole carriers. Therefore, dielectric studies on collective carriers should be carried out at low frequencies, constituting a major objective for experimental work in the critical region.

Dielectric experiments are normally performed in a capacitive device with an applied oscillating field $\boldsymbol{D}_o \exp(-i\omega t)$ due to a uniformly distributed charge density on the plates. It is noted that a sample crystal is deformed by dielectric displacements that occur as related to an acoustic excitation at a long wavelength. Therefore, even in the uniform field of $\boldsymbol{D}_o \exp(-i\omega t)$, the induced dielectric polarization is distributed as $\boldsymbol{P} = \boldsymbol{P}_o \exp i(\boldsymbol{q}.\boldsymbol{r} - \omega t)$, so that the internal electric field cannot be uniform, and expressed as $\boldsymbol{E} = \boldsymbol{E}_o \exp i(\boldsymbol{q}.\boldsymbol{r} - \omega t)$. In this context, the electric susceptibility defined by $\boldsymbol{P} = \varepsilon_o\chi(\omega)\boldsymbol{E}$ is signified by the redefined quantity $\chi_q(\omega) = \chi(\omega)\exp i\boldsymbol{q}.\boldsymbol{r}$ in $\boldsymbol{P}_o = \varepsilon_o\chi_q(\omega)\boldsymbol{E}_o$, although the wavevector $\boldsymbol{q}$ is insignificant for dielectric measurements at a small $|\boldsymbol{q}|$. Hence, we can deal with the effective susceptibility $\chi_q(\omega)$ defined by $\boldsymbol{P} = \varepsilon_o \chi_q \boldsymbol{E}_o \exp(-i\omega t)$, representing the response function to the uniform applied field $\boldsymbol{D} = \varepsilon_o \boldsymbol{E}_o \exp(-i\omega t)$. The equation of motion of a condensate composed of a collective pseudospin $\sigma_q(t)$ can therefore be written generally as

$$m\left(\frac{\mathrm{d}^2\sigma_q}{\mathrm{d}t^2} + \gamma\frac{\mathrm{d}\sigma_q}{\mathrm{d}t} + \varpi^2\sigma_q\right) = eE_{qo}\exp(-i\omega t), \tag{7.14}$$

where m and e are the effective mass and charge, respectively, of the condensate, and γ is the damping constant. Here, ϖ is the characteristic frequency and $m\varpi^2 = k$ represents the restoring force constant. For the steady-state solution of (7.14), we let $\sigma_q = \sigma_{qo}\exp(-i\omega t)$ and obtain the complex response function of a resonant type

$$\chi_q(\omega) = \sigma_{qo}/E_{qo} = \frac{e/m}{(-\omega^2 - i\gamma\omega + \varpi^2)}, \tag{7.15}$$

indicating dispersion in the real part and maximum absorption in the imaginary part at $\omega = \varpi$.

However, owing to a large effective mass of the condensate, the kinetic energy of fluctuations is negligible, so that with the prevailing damping effect, the equation of motion can be written as

$$\gamma\frac{\mathrm{d}\sigma_q}{\mathrm{d}t} + k\sigma_q = (e/m)E_{qo}\exp(-i\omega t). \tag{7.16}$$

The steady-state solution is given by

$$\chi_q(\omega) = \frac{e/m}{(k - i\omega\gamma)} = \frac{e/mk}{(1 - i\omega\tau)}, \tag{7.17}$$

where $\tau = \gamma/k$ is the relaxation time, and (7.17) is known as the Debye relaxation.

Such relaxation phenomena as (7.16) dominate dielectric properties of the collective pseudospins in the critical region. Corresponding to the susceptibility, the relaxational dielectric function can be expressed as

$$\varepsilon_q(\omega) = \varepsilon_o(1 + \alpha) + \frac{\varepsilon_o e/mk}{1 - i\omega\tau}, \tag{7.18}$$

where the constant α is the polarizability of the condensate.

As remarked, the wavevector $\boldsymbol{q}$ is only an implicit parameter, playing no significant role in normal dielectric measurements. However, below T_c of a phase transition, the dielectric properties are signified by collective pseudospins in unknown size, although characterized th the wavevector $\boldsymbol{q}$, changing as a function of temperature. Therefore, we retain the notation $\varepsilon_q(\omega)$ in the following discussion. At a temperature T below T_c, condesates signified by $\boldsymbol{q}$ appear to become further correlated, as observed with an oscillating field at a frequency ω, and for extreme cases at $\omega = \infty$ and $\omega = 0$, $\varepsilon_q(\omega)$ have specific values given by

$$\varepsilon_q(\infty) = \varepsilon_o(1+\alpha) \quad \text{and} \quad \varepsilon_q(0) = \varepsilon_q(\infty) + \varepsilon_o(e/mk), \tag{7.19}$$

representing uncorrelated dipoles and correlated condensates, respectively.

Although derived for oscillatory motion, the LST formula (4.10) should be revised for the relaxational function of (7.18). We notice that there are specific frequencies that characterize the relaxational $\varepsilon_q(\omega)$: first, there is a frequency ω_L for $\varepsilon_q(\omega_L) = 0$, that is, from (7.18)

$$i\omega_L\tau = \varepsilon_q(0)/\varepsilon_q(\infty),$$

and second, $\varepsilon_q(\omega)$ has a pole at $\omega = \omega_P$, namely $i\omega_P\tau = 1$. Therefore, we can write

$$\omega_L/\omega_P = \varepsilon_q(0)/\varepsilon_q(\infty), \tag{7.20}$$

analogous to the LST relation for an oscillatory mode. Applying (7.20) to a ferroelectric phase transition, where the Curie-Weiss law $\varepsilon(0) \propto |T - T_c|^{-1}$ can be used for the response to the transversal electric field E_o on both sides of T_c, the relaxation time τ is shown as temperature-dependent, i.e.

$$\tau \propto |T - T_c|^{-1}, \tag{7.21}$$

which is known as *critical slowing down.* Using notations in (7.19), the equation (7.18) can be re-expressed as

$$\varepsilon_q(\omega) - \varepsilon_q(\infty) = \frac{\varepsilon_q(0) - \varepsilon_q(\infty)}{1 - i\omega\tau} = \frac{S_q}{1 - i\omega\tau}, \tag{7.18a}$$

where $S_q = \varepsilon_q(0) - \varepsilon_q(\infty)$ is a constant for the dielectric condensate at a wavevector $\boldsymbol{q}$ that should depend on T.

For the Debye *relaxator*, the relaxation formula (7.18a) can be used for interpreting observed critical anomalies exhibited in dielectric measurements performed at various frequencies. In view of an energy loss during a cycle of variation, it is convenient to consider the frequency ω and the dielectric function $\varepsilon_q(\omega)$ are complex for the mathematical argument. Namely, we write $\varepsilon_q(\omega) = \varepsilon'_q(\omega) + i\varepsilon''_q(\omega)$ for the complex frequency $\omega = \omega' - i\omega''$. From the complex expression of (7.18a), we obtain that

$$\varepsilon'_q(\omega) = \varepsilon_q(\infty) + \frac{\varepsilon_q(0) - \varepsilon_q(\infty)}{1 + \omega^2\tau^2}$$

and

$$\varepsilon''_q(\omega) = \{\varepsilon_q(0) - \varepsilon_q(\infty)\}\frac{\omega\tau}{1 + \omega^2\tau^2}.$$

Eliminating $\omega\tau$ from these expressions, we can derive the equation

$$[\varepsilon'_q(\omega) - \tfrac{1}{2}\{\varepsilon_q(0) + \varepsilon_q(\infty)\}]^2 + \varepsilon''_q(\omega)^2 = \tfrac{1}{4}\{\varepsilon_q(0) - \varepsilon_q(\infty)\}^2, \tag{7.22}$$

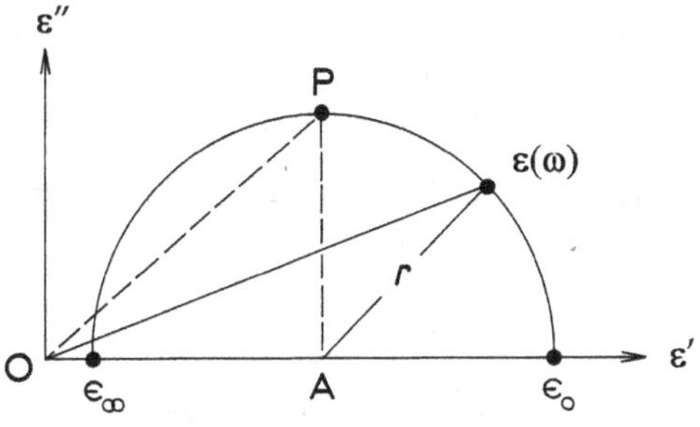

Fig. 7.5. The Cole-Cole diagram for a dielectric relaxation of the Debye type: $\epsilon(\omega) = \epsilon'(\omega) - j\epsilon''(\omega)$.

that represents a circle in the plane of ε' vs. ε'' centered at the point $A[\frac{1}{2}\{\varepsilon_q(0) + \varepsilon_q(\infty)\}, 0]$ with the radius $r = \frac{1}{2}\{\varepsilon_q(0) - \varepsilon_q(\infty)\}$. However, these dielectric components are positive, so that, as shown in Fig. 7.5, we consider a semicircle above the real axis to examine (7.22), which is known as the Cole-Cole plot. The maximum point $P(0, \varepsilon_q''(\max))$ in such a plot is determined by $\varepsilon_q'' = r$, corresponding to $\omega\tau = 1$. Therefore, we consider the frequency $\omega = i\omega_P$ on the imaginary axis, and write the relation $i\omega_P\tau = 1$, which was indeed defined for the relaxational LST relation (7.20). Hence, the imaginary part of ω at P determined on the Cole-Cole plot, i.e. ω_P'' can give direct information about the relaxation time from $\tau = 1/\omega_P''$, which should depend on $\boldsymbol{q}$ at temperature below T_c.

Needless to say, it is important to obtain supporting evidence for such collective species to exist in the critical region, as characterized by a constant τ. Unruh and his group obtained such evidence from their dielectric experiments on the ferroelectric phase transition of TSCC crystals. They observed a series of Cole-Cole semicircles plotted with various frequencies, indicating that these are related to the effective mass and relaxation time that are functions of temperatures. In such dielectric experiments, it is important to realize that the semicircular curve of $\varepsilon(\omega)$ in the range $0 < \omega < \infty$ signifies a cluster of collective pseudospins observed at various frequencies ω, where the constants $\varepsilon_q(0)$ and $\varepsilon_q(\infty)$ can be interpreted as reflecting collective behaviors of condensates that are observed at $\omega = 0$ and $\omega = \infty$, respectively. We can consider the relaxational mode for slow collective motion of dipolar pseudospins in the critical region, whereas the oscillational mode represents generally uncoupled pseudospins in the lattice. Although attributed to masses of dipolar carriers, the image of collective modes is not at all clear dynamically. Nevertheless, these different modes are considered for clustered dipoles under an applied oscillatory electric field, as discussed already in Chapter 4.

In the critical region, binary modes of fluctuations σ_P and σ_A are present, both being signified by $\boldsymbol{q}$. However, in a dielectric observation these polar modes are pinned at $\phi = 0$ and $\frac{1}{2}\pi$, respectively, where the electric fields are considered as $E_o \cos \omega t$ and $E_o \sin \omega t$. Therefore, we can write relaxatinal equations for binary fluctuations in phase and amplitude modes, i.e

$$\gamma\frac{d\sigma_P}{dt}+k\sigma_P=\frac{e}{m}E_o\cos\omega t \quad \text{and} \quad \gamma\frac{d\sigma_A}{dt}+k\sigma_A=\frac{e}{m}E_o\sin\omega t.$$

Using complex notations, $\sigma_q=\sigma_P+i\sigma_A=\sigma_o\exp i\phi=\sigma_{qo}\exp(-i\omega t)$, where $\sigma_{qo}=\sigma_o\exp iqx$, these equations are combined as

$$\gamma\frac{d\sigma_q}{dt}+k\sigma_q=\frac{e}{m}E_{qo}\exp(-i\omega t).$$

In any case, the steady-state solution is expressed as

$$\sigma_q=\sigma_{qo}\exp(-i\omega t) \quad \text{where} \quad \sigma_{qo}=\frac{e}{m}\frac{E_{qo}}{k-i\omega\gamma}=\frac{eE_{qo}/mk}{1-i\omega\tau},$$

as in (7.17), but these relaxational modes are indintinguishable in dielectric measurements.

On the other hand, as T_c is approached from above, the structural change of the lattice is signified by the soft lattice mode with the characteristic frequency ϖ, for which the equation of motion is written generally as

$$\frac{d^2u_q}{dt^2}+\gamma\frac{du_q}{dt}+\varpi^2u_q=\frac{e}{m}E_{qo}\exp(-i\omega t).$$

In the critical region, assuming that these displacements in the lattice and pseudospins are coupled as $u_q\propto\sigma_q$, the oscillational energy is dissipated via interactions with the relaxational σ_q, in addition to direct damping γ into the lattice. In this context, the constant γ can be replaced by $\gamma+\delta(d\sigma_q/dt)$, where δ is the coupling between the oscillator and the relaxator, for which we have already derived the following susceptibility formula

$$\chi_q(\omega)=\left[\varpi^2-\omega^2-i\omega\gamma-\delta\frac{e}{m}E_{qo}\frac{\omega\tau}{1+i\omega\tau}\right]^{-1},$$

where the factor $\delta(e/m)E_o$ was replaced by δ^2 in (4.25) for brevity. In Section 4.6, we showed that $\chi''_q(\omega)$ consists of a resonant absorption at $\omega=\varpi$ and a relaxational absorption at $\omega=0$, provided that $\gamma\ll\delta^2\tau$ and $\varpi\tau\gg 1$. In practice however, the separation of these modes may not so simple because the resonant mode is often overdamped, appearing as if relaxational.

In practical crystals with defects, we cannot rule out a possible contribution to $\chi''_q(\omega)$ due to the oscillatory behavior of pinned condensates at very low frequencies, as discussed in Section 5.2, so that the absorption in the vicinity of zero frequency, which is the central peak, is not easily interpretable.

7.4 Dielectric Spectra in the Ferroelectric Phase Transition of TSCC

Among many dielectric experiments on structural phase transitions, the ferroelectric phase transition in TSCC crystals has been most extensively studied.

Kozlov and his group [63] found in their experiments with a submillimeter-wave technique that typical soft-mode spectra in TSCC were underdamped to temperatures close to T_c, while Deguchi and his coworkers [64] observed a relaxational mode at uhf frequencies in the critical region, showing the critical relaxation characterized by a slowing-down behavior. Sawada and Horioka [38] observed dielectric responses of resonant and relaxational types at microwave frequencies that coexisted at temperatures near T_c (= 163K). Petzelt, Kozlov and Volkov [65] attempted to generalize the *oscillator-relaxator* model to cover many ferroelectric systems including TSCC, however it was unsuccessful with only submillimeter-frequency data. Their results, as shown in Fig. 7.6, indicate clear evidence for an underdamped soft-mode to about 50GHz, which was, in fact, the low-frequency limit of their backward-wave oscillator, and not sufficiently close to the critical region. As demonstrated by Sawada and his co-worker, more significant information about the critical state are expected at lower frequencies, whereas the central peak problem needs to be properly delineated in terms of the intrinsic mechanism. These authors obtained dielectric anomalies in $\varepsilon'(\omega)$ from TSCC, as shown in Fig. 4.6b, exhibiting relaxational

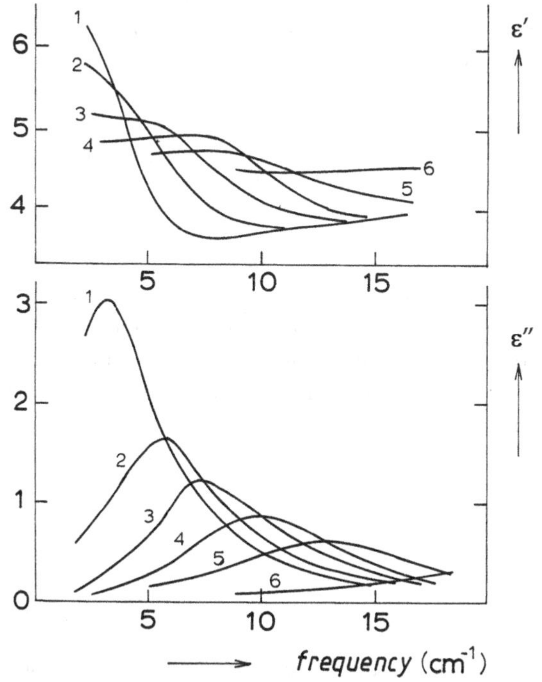

Fig. 7.6. Real and imaginary parts of the complex dielectric function $\epsilon(\omega)$ in TSCC measured at submillimeter-wave frequencies. (From A. A. Kozlov, J. P. Scott, G. E. Feldkamp and J. Petzelt, Phys. Rev. B28, 255 (1983).)

behavior at high microwave frequencies 35 and 24GHz, which were analyzed with the oscillator-relaxator formula (4.25).

Pawlaczyk and Unruh [66] carried out dielectric measurements on TSCC in a wide low-frequency range between 6GHz and 0.1kHz in the critical region of the ferroelectric phase transition. Confirming the earlier results of Decuchi et al., their relaxation studies were performed on high-quality samples characterized by low values of the internal bias field ($E_b \sim 15$ V/cm), where the results were subjected to the Cole-Cole analysis. Figures 7.7a and 7.7b summarize their observation of dielectric behaviors above and below T_c, respectively, with added microwave results from [38] for comparison. It is noticed that the plots above T_c are hardly considered as semi-circular, but resemble a circlular curve in the low-frequency region. The deviation from a semicircle in the high-frequency side should be attributed partly to the oscillatory nature of dielectric responses. On the other hand, the plots in Fig. 7.7b consist of two semi-circles in different size, indicated as mode I and mode II, showing a marked difference from plots in Fig. 7.7a. The transition temperature was determined as $T_c = 130.7$K from the changing over in the ε' vs. ε'' plots observed in these high-quality crystals. Clearly, these discrete dielectric modes I and II are signified by different wavevectors, indicating that clusters exist with differences in size and densities depending on temperature. Such a variation can be speculated by the model for clusters stabilized in a simple cnoidal potential as discussed in Section 5.9, although the details are yet to be worked out.

In Fig. 7.7b, it is appreciable that the low-frequency limit of mode I coincides with the high-frequency limit of mode II within experimental accuracy, i.e.

$$\varepsilon_{\mathrm{I}}(0) = \varepsilon_{\mathrm{II}}(\infty), \tag{7.23}$$

which allows us to interpret that these modes represent dielectric responses from stepwise clustering in succession. Whereas mode I may be regarded as due to uncorrelated pseudospins, these responses represent collective clusters in finite size characterized by different constants, $(S_{\mathrm{I}}, \tau_{\mathrm{I}})$ and $(S_{\mathrm{II}}, \tau_{\mathrm{II}})$. We can therefore write

$$\varepsilon(\omega) = \varepsilon_{\mathrm{I}}(\omega) + \varepsilon_{\mathrm{II}}(\omega) = \frac{S_{\mathrm{I}}}{(1 - i\omega\tau_{\mathrm{I}})} + \frac{S_{\mathrm{II}}}{(1 - i\omega\tau_{\mathrm{II}})} + \text{const.}, \tag{7.24}$$

where

$$S_{\mathrm{I}} = \varepsilon_{\mathrm{I}}(0) - \varepsilon_{\mathrm{I}}(\infty), \quad S_{\mathrm{II}} = \varepsilon_{\mathrm{II}}(0) - \varepsilon_{\mathrm{II}}(\infty) \quad \text{and} \quad \varepsilon_{\mathrm{I}}(0) = \varepsilon_{\mathrm{II}}(\infty).$$

Equation (7.23) was in fact introduced by Petersson [67] for dealing with lattice defects, although it may be applied to double relaxations in the present case, assuming the coexistence of independent clusters of correlated pseudospins in the critical region. Nevertheless, the effect of a pinning potential originating from lattice defects cannot be ignored, and dynamic pinning fluctuations were discussed in Section 5.2. For mode II, it may be more appropriate

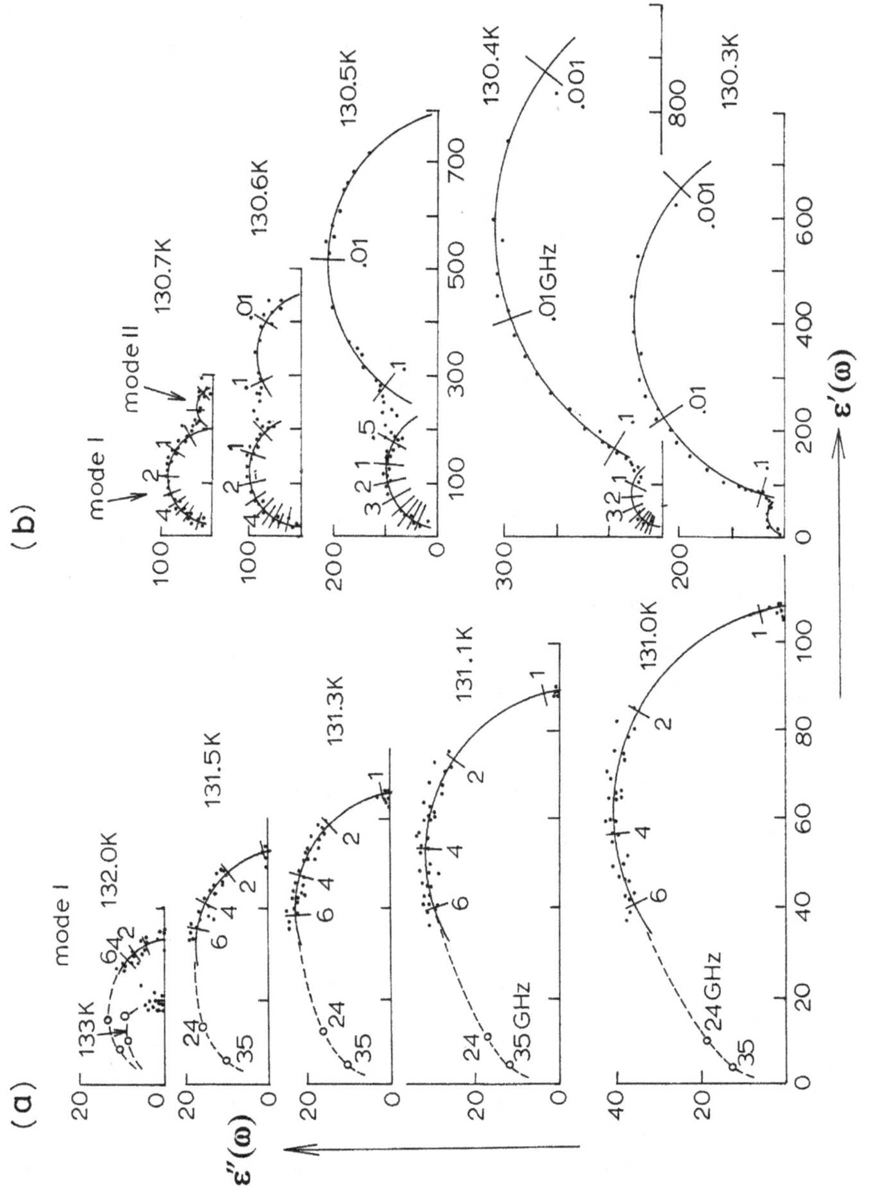

Fig. 7.7. Cole-Cole plots of $\epsilon(\omega)$ in the critical region of the ferroelectric phase transition in TSCC. (From Cz. Pawlaczyk, H.-G. Unruh and J. Petzelt, Phys. Stat. Sol. (b) 136,435 (1986); M. Fujimoto, Cz. Pawlaczyk and H.-G. Unruh, Phys. Mag. 60, 919 (1989).)

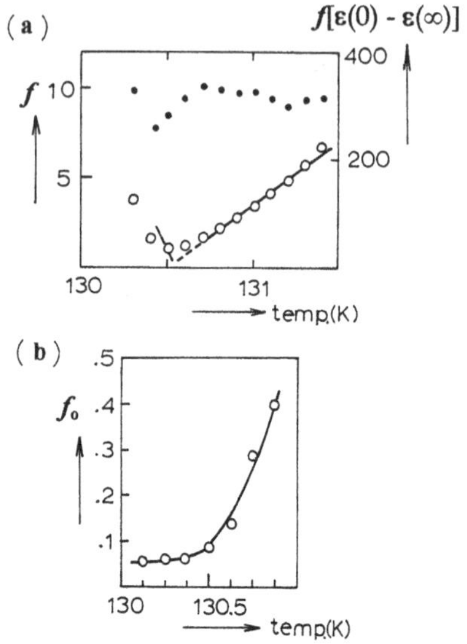

Fig. 7.8. (a) Critical slowing-down exhibited by the "relaxation mode I" in TSCC. (From Cz. Pawlaczyk, H.-G. Unruh and J. Petzelt, Phys. Stat. Sol. (b)136, 435 (1986).) (b) Pinning frequencies f_o of the phase mode vs. temperature in TSCC, determined from anomalous line-broadening of an Mn^{2+} line. (From M. Fujimoto, Cz. Pawlaczyk and H.-G. Unruh, Phil. Mag. 60, 919 (1989).)

to write the relaxation equation as

$$\frac{1}{\tau_I}\frac{d\sigma_{II}}{dt} + \omega_o^2\sigma_{II} = \frac{e}{m}E_o\exp(-i\omega t),$$

where $\omega_o^2 = k/m$, and we can consider that the effective relaxation time for Mode II is given by

$$\tau_{II} = 1/(\omega_o^2\tau_I). \tag{7.25}$$

Using measured values of τ_I and τ_{II}, the value of ω_o can be estimated from using (7.24). Summarizing such analysis of the dielectric data of Fig. 7.7b, Fig. 7.8 shows a plot of the pinning frequency ω_o against the temperature below T_c, indicating an interesting curve with the terminal frequency $f_o = \omega_o/2\pi \sim 50\text{Hz}$, which is an acceptable value for pinned domain wall in TSCC [44].

8

The Spin-Hamiltonian and Magnetic Resonance Spectroscopy

In this chapter, principles of magnetic resonance are outlined for those readers who are not familiar with experimental practice in this field of spectroscopy. In my opinion, spin-Hamiltonian parameters are generally complicated for non-specialists, and, hence, the basic concepts are reviewed to minimum necessity for the problem of phase transitions. Although usable only in nonconducting and nonmagnetic materials, the magnetic resonance provides a unique method for studying ordering processes in three-dimensional crystals. Nevertheless, those who are already familiar with the language of magnetic resonance can skip this chapter to proceed to the next one for modulated crystals.

8.1 Introduction

Active groups responsible for a structural transformation in crystals cannot be identified only from energy-momentum exchanges in inelastic neutron- and light-scattering experiments. Although inferable from the chemical composition in simple cases, active groups for structural changes can be identified logically from diffuse X-ray diffraction results. On the other hand, nuclear and paramagnetic probes embedded in a crystal can be utilized for sampling condensates, providing information about a structural change if the probe is associated with the active group. In a modulated system, the spectra exhibit anomalous lineshapes originating from the modulated structure. Sampling by magnetic resonance probes offers a unique technique for obtaining information about the local structural change in three dimensions, whereas other methods can deal basically with dynamical aspects of condensates. Being complementary to scattering and dielectric experiments, magnetic resonance sampling is indispensable for investigating the nature of condensates during a structural change.

Primarily, the magnetic resonance can be applied only to nonmagnetic and nonconducting crystals, where probes with spins higher than $\frac{1}{2}$ are useful for detecting a distortional change in the local potential. In addition, when the Larmor frequency ω_L is comparable with the characteristic frequency of critical fluctuations of the order of $10^{10} \sim 10^{11}$Hz, the spectra exhibit anomalies reflecting the nature of fluctuations. For conventional magnetic resonance experiments, the Larmor frequency ω_L is in the range of $5 \sim 35$GHz for paramagnetic probes, whereas of $1 \sim 20$MHz for nuclear probes. Critical fluctuations are generally slow in the timescale $t_o = 2\pi/\omega_L$, so that their spatial profile can be obtained from the observed lineshape. Moreover, comparative studies at different microwave frequencies can be carried out for further information about the slow dynamics of condensates. In spite of these advantages, however, the chemical compatibility of magnetic probes with a given crystal structure is a significant factor in such experiments. Therefore, it is a usual practice to perform experiments with as many usable probes as possible to obtain reliable results by comparative studies. Although it is not always possible to find a suitable probe, the method is a useful one with a limited choice of probes.

With diffuse X-ray data, magnetic resonance of a suitable probe provides a direct method for identifying active species for a structural change. Sampling a modulated structure by magnetic probes, one can determine the basic features of pseudospin condensate in the form $\sigma = \sigma_o f(\phi)$, where σ_o and ϕ are the amplitude and phase of modulation, respectively. In nuclear magnetic resonance, we can further obtain evidence for equilibrium between pseudospins and phonons in terms of the spin-lattice relaxation time T_1.

Magnetic resonance is a well-established method of investigation for many problems in condensed matter, although properties of paramagnetic ions in crystals are a little too sophisticated to analyze, when expressed in terms of *spin-Hamiltonian* parameters. However, the three-dimensional tensor character of these parameters is very informative for structural studies, and, therefore, we discuss the relevant principles prior to Chapter 9 where the modulation effect is described by modulated spin-Hamiltonian parameters.

8.2 Principles of Magnetic Resonance and Relaxation

We consider nuclear and electronic magnetic moments that are carried by atomic nuclei and paramagnetic ions, respectively, in crystals. We assume that these probes are situated at fixed lattice sites in crystals. Although diffusive migration through the lattice cannot be ruled out in some case, these microscopic magnetic moments are considered to be stationary at their lattice sites, while freely rotatable. In the presence of a uniform magnetic field $\boldsymbol{B}_o$, the torque $\boldsymbol{\mu} \times \boldsymbol{B}_o$ is responsible for changing the direction of magnetic moments $\boldsymbol{\mu}$, as described by the equation of motion

$$\frac{\mathrm{d}\boldsymbol{L}}{\mathrm{d}t} = \boldsymbol{\mu} \times \boldsymbol{B}_o, \qquad (8.1)$$

where $\boldsymbol{L} = \gamma^{-1}\boldsymbol{\mu}$ is the angular momentum and γ is the gyromagnetic ratio. Considering that $\boldsymbol{B}_o$ is parallel to the z axis, the steady-state solution of (8.1)

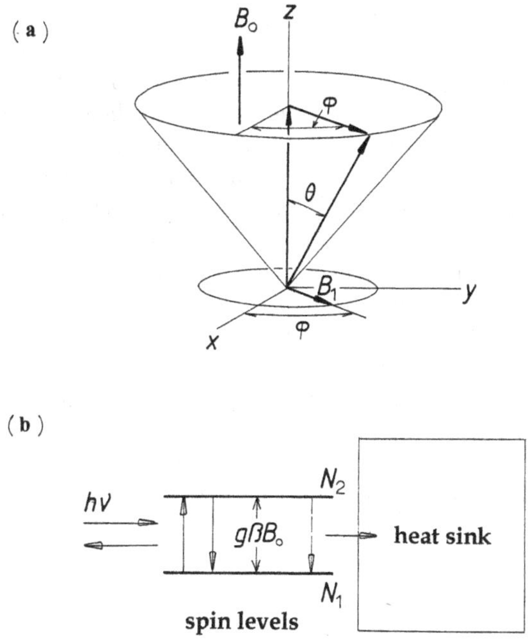

Fig. 8.1. (a) Magnetic resonance of a classical spin $\boldsymbol{S}$. (b) Magnetic resonance of a quantum-mechanical spin $S = 1/2$.

is given by

$$\mu_z = \mu \cos \theta = \text{constant} \tag{8.2a}$$

and

$$\mu_\perp = \mu \sin \theta \exp i\varphi, \quad \text{where} \quad \varphi = \gamma B_o t. \tag{8.2b}$$

Here, θ and φ are polar and azimuthal angles of $\boldsymbol{\mu}$ with regard to the z axis, as illustrated in Fig. 8.1a. Equations (8.2a) and (8.2b) describe the motion of $\boldsymbol{\mu}$ in precession around $\boldsymbol{B}_o$ at the angular frequency $\omega_L = \gamma B_o$, known as the Larmor frequency.

Magnetic resonance occurs when a circularly rotating magnetic field $B_\perp = B_1 \cos(\omega t + \varphi_o)$, where φ_o is a phase constant, is applied in perpendicular direction to the z axis. The energy associated with the rotating field B_1 is absorbed by $\boldsymbol{\mu}$, if $B_\perp$ and $\mu_\perp$ are in phase, and the condition

$$\omega = \omega_L = \gamma B_o, \tag{8.3}$$

is fulfilled. Called the *magnetic resonance condition*, (8.3) signifies that the precession of $\boldsymbol{\mu}$ changes by increasing $\mu_\perp$ and decreasing μ_z. Such a resonance phenomenon can be described conveniently in the frame of reference that is rotating at the frequency ω_L around the z axis, where the resonance is characterized by the polar angle θ increased by $\Delta\theta$ in the zx meridian plane.

Although describing an isolated magnetic moment in the above, the resonant condition in a macroscopic material must be revised for a system of a large number of moments that are in precession with random phases. In a uniform crystal, all unit cells are identical and, hence, all microscopic moments located at lattice points behave exactly in the same way with respect to the applied uniform field $\boldsymbol{B}_o$, however the phases are at random without the rotating field $\boldsymbol{B}_1$. The macroscopic magnetization $\boldsymbol{M}$ at resonance should be associated with distributed microscopic moments $\boldsymbol{\mu}$ in the crystal, but dynamically their precession is all synchronized by the applied $\boldsymbol{B}_1$. It is noted that the motion of $\boldsymbol{\mu}$ is determined by external magnetic fields, although considered as independent of the crystalline environment.

We can, therefore, assume that $M_z = \text{const.}$ and $\langle M_\perp \rangle = 0$ in the absence of $\boldsymbol{B}_1$, whereas in the presence of a rotating field $\boldsymbol{B}_1$ at resonance $\omega = \omega_L$, these $\boldsymbol{\mu}$ are forced to be synchronized by $\boldsymbol{B}_1$. Nevertheless, such a synchronous motion will return to individual motion in random phases in a characteristic time T_2 after $\boldsymbol{B}_1$ is removed. Bloch [68] proposed a relaxational process for $M_\perp$, and wrote the equations

$$\mathrm{d}M_\perp/\mathrm{d}t = -\frac{M_\perp}{T_2}, \quad \text{and} \quad M_z = M_o = \chi B_o, \tag{8.4}$$

where χ is the susceptibility of the system of $\boldsymbol{\mu}$, and such a relaxation time T_2 is called the *spin-spin relaxation time*, for which in a simple spin system, magnetic dipolar interactions among these $\boldsymbol{\mu}$ are responsible. Here, M_o represents the value of the magnetization in thermal equilibrium. It is noted that the magnitude of dipolar interactions ΔB should be less than B_1, in order for such a description to be valid; that is

$$\frac{1}{T_2} = \gamma \Delta B \ll \gamma B_1, \tag{8.5}$$

is the necessary condition for the Bloch precession, and is generally referred to as the condition for a *slow passage.* This means that the characteristic frequency of phase fluctuations $1/T_2$ should be lower than the driving frequency ω_L of B_1, sharing the same principle as slow critical fluctuations at ϖ that is observable within a timescale t_o if $\varpi t_o \leq 1$. In fact, the value of ΔB is of the order of 10G in typical diluted paramagnetic systems, giving rise to $T_2 \sim 10^{-8}$s, whereas for a typical system of nuclear moments $\Delta B \sim 1$G and so $T_2 \sim 10^{-4}$s. Hence, the condition (8.7) is generally met for many electronic systems exhibiting well-resolved spectra, as well as most nuclear systems.

It is also noted that M_z is not constant during the process for reaching M_o in thermal equilibrium. For this thermal process, he wrote another relaxational equation

$$\frac{\mathrm{d}M_z}{\mathrm{d}t} = -\frac{M_z - M_o}{T_1}, \tag{8.6}$$

where T_1 is called the *spin-lattice relaxation time.*

Taking these two relaxation mechanisms into account, Bloch expressed the macroscopic equation of motion as

$$\frac{\mathrm{d}M_\pm}{\mathrm{d}t} \pm iB_\mathrm{o}M_\pm + \frac{M_\pm}{\mathrm{T}_2} = -i\gamma B_1 M_z \exp(\mp i\omega t)$$

and (8.7)

$$\frac{\mathrm{d}M_z}{\mathrm{d}t} + \frac{M_z - M_\mathrm{o}}{\mathrm{T}_1} = \tfrac{1}{2} i\gamma B_\mathrm{o}\{M_+ \exp(-i\omega t) + M_- \exp(i\omega t).$$

In these equations (8.7), known as the Bloch equations, the transversal $M_\pm$ represent the rotating components of $\boldsymbol{M}$, being synchronized with the rotating fields $B_1 \exp(\mp i\omega t)$.

The Bloch equations have a steady-state solution. Under a slow-passage condition, the dynamical process can be described by such a solution at least approximately. In a steady state signified by $\mathrm{d}M_\mathrm{z}/\mathrm{d}t = 0$, M_z accompanies transversal rotating components $M_\pm$, which are expressed in the laboratory frame of reference as

$$M_\pm = \gamma B_1 M_z \exp(\mp i\omega t)/(\mp\omega + \gamma B_\mathrm{o} + i/\mathrm{T}_2).$$

Using this result in the solutions (8.6), we obtain

$$\frac{M_z}{M_\mathrm{o}} = \frac{1 + (\omega - \omega_\mathrm{L})^2\mathrm{T}_2^2}{1 + (\omega - \omega_\mathrm{L})^2\mathrm{T}_2^2 + \gamma^2 B_1^2 \mathrm{T}_1\mathrm{T}_2} \tag{8.8a}$$

and

$$\frac{M_\pm}{M_\mathrm{o}} = \frac{\{(\omega - \omega_\mathrm{L})\mathrm{T}_2 + i\}\gamma B_1 \mathrm{T}_2 \exp(\mp i\omega t)}{1 + (\omega - \omega_\mathrm{L})^2\mathrm{T}_2^2 + \gamma^2 B_1^2 \mathrm{T}_1\mathrm{T}_2}. \tag{8.8b}$$

If $(\gamma B_1 \mathrm{T}_1\mathrm{T}_2)^2 \ll 1$, from (8.8a) we have $M_\mathrm{z} \approx M_\mathrm{o}$ at $\omega = \omega_\mathrm{L}$, in which case the tilting angle θ of $\boldsymbol{M}$ from the z axis can be calculated from

$$\tan\theta = \frac{M_\pm}{M_\mathrm{o}} \approx \frac{\gamma B_1 \mathrm{T}_2}{1 + (\omega - \omega_\mathrm{L})^2\mathrm{T}_2^2}.$$

The high-frequency susceptibility is defined by writing

$$M_\pm = \chi(\omega) B_1 \exp(\mp i\omega t) \quad \text{and} \quad M_\mathrm{o} = \chi_\mathrm{o} B_\mathrm{o},$$

and

$$\frac{\chi(\omega)}{\chi_\mathrm{o}} = \frac{\gamma B_1 \mathrm{T}_2\{(\omega - \omega_\mathrm{L})\mathrm{T}_2 + i\}}{1 + (\omega - \omega_\mathrm{L})^2\mathrm{T}_2^2 + (\gamma B_1 \mathrm{T}_1\mathrm{T}_2)^2}.$$

The real and imaginary parts of this complex susceptibility are therefore expressed as

$$\frac{\chi'(\omega)}{\chi_\mathrm{o}} = \frac{\omega_\mathrm{L}(\omega - \omega_\mathrm{L})}{(\omega - \omega_\mathrm{L})^2 + \delta\omega^2 + \gamma^2 B_1^2 \mathrm{T}_1 \delta\omega}$$

and

$$\frac{\chi''(\omega)}{\chi_\mathrm{o}} = \frac{\omega_\mathrm{L}\delta\omega}{(\omega - \omega_\mathrm{L})^2 + \delta\omega^2 + \gamma^2 B_1^2 \mathrm{T}_1 \delta\omega},$$

respectively, where $\delta\omega = 1/T_2$. The function $\chi''(\omega)$ becomes maximum at $\omega = \omega_L$, where

$$\chi''(\omega_L)/\chi_o \approx \omega_L/\delta\omega, \quad \text{if} \quad \gamma^2 B_1^2 T_1 \ll \delta\omega. \tag{8.9}$$

It is noted that the value of $\chi''(\omega_L)$ is substantially larger than χ_o if $\delta\omega \ll \omega_L$, giving a sharp resonance at $\omega = \omega_L$.

Microscopically, the magnetic resonance can be interpreted by quantum mechanics where a system of ionic or nuclear magnetic moments absorbs radiation quanta from the applied high-frequency field. Considering the angular momentum $\boldsymbol{L} = \hbar \boldsymbol{l}$ as specified by the quantum number $\boldsymbol{l}$, the magnetic moment is given by $\boldsymbol{\mu} = \gamma\hbar\boldsymbol{l}$ and the energy levels for the precession in a static field $\boldsymbol{B}_o$ are discrete as $\varepsilon_m = -\beta B_o m$, where $\beta = \gamma\hbar$. These energies are specified by the magnetic quantum number, $m = l, l-1, \ldots, -l$, and separated by an equal difference $\varepsilon_m - \varepsilon_{m-1} = \beta B_o$. Being proportional to B_o, such an energy difference can be adjusted exactly to the radiation energy $\hbar\omega$, so that the resonance condition is expressed by $\hbar\omega = \beta B_o$, namely $\omega = \gamma B_o = \omega_L$. Here, unlike in classical theory, the quantum number l can take either integers or half-integers, and the magnetic resonance condition can be generally written as

$$\hbar\omega = g\beta_o B_o, \tag{8.10}$$

where the gyromagnetic ratio γ for a quantum number l is replaced by $g\beta_o$. Here, β_o is called the Bohr *magneton* and g is the Lande factor, which is equal to 2 for $l = \frac{1}{2}$, referring to an electron in an s-state. Half-integral angular momenta arise from orbital electrons with intrinsic spins, which are coupled in ionic states, whose magnetic moments are expressed by the Landé factor g.

The resonance condition (8.10) was obtained in the above microscopic argument, where such relaxation processes as in the Bloch theory are ignored. First, as described by T_2, the resonant frequency ω_L cannot be sharply defined due to interactions among magnetic moments in the system. The effect of an oscillating field B_1 must be interpreted as causing transitions between the discrete energy levels, as described in terms of transition probabilities induced by B_1 in quantum theory. Taking the spin-spin interactions into consideration, induced transitions are distributed as described by the probability $w_{m,m+1}d\omega$ between ω and $\omega + d\omega$, for which

$$w_{m,m+1} = \tfrac{1}{2}\pi\hbar^{-2}B_1^2|\mu_{m,m+1}|^2 f(\omega),$$

where $\mu_{m,m+1}$ represents an off-diagonal matrix element of $\boldsymbol{\mu}$ between states m and m + 1, and $f(\omega) = T_2^{-1}$ is called the shape function normalized as $\int f(\omega)d\omega = 1$. Under conventional experimental conditions, the frequency ω is in the ranges of radio and microwaves, so that the transitions are dominated by those induced by B_1, and the probability for spontaneous transition is negligible.

Second, a macroscopic system of a large number of magnetic moments in a uniform field B_o is considered to be in equilibrium with the applied radiation energy of an oscillating field B_1. Figure 8.1b illustrates a simple case of $\mathrm{m} = \pm\frac{1}{2}$, where the two energy levels $\varepsilon_{\pm 1/2}$ are populated by the Boltzmann statistics, namely $N_{\pm 1/2} = N_o \exp(-\varepsilon_{\pm 1/2}/\mathrm{k_B}T)$. Assume that $\varepsilon_{+1/2} < \varepsilon_{-1/2}$, and $N_{+1/2} > N_{-1/2}$, which allows absorption of radiation quanta $\hbar\omega$ when $\omega = \omega_L$. The average "pumpimg" rate per cycle is expressed as

$$\begin{aligned}\left\langle \frac{\mathrm{d}W}{\mathrm{d}t} \right\rangle_t &= w_{+-}(\hbar\omega_L)(N_{+\frac{1}{2}} - N_{-\frac{1}{2}}) \\ &= w_{+-}(\hbar\omega_L)N_{+\frac{1}{2}}\{1 - \exp(-\hbar\omega_L/\mathrm{k_B}T)\} \\ &\approx w_{+-}(\hbar\omega_L)N_{+\frac{1}{2}}(\hbar\omega_L/\mathrm{k_B}T) \\ &= N_{+\frac{1}{2}}(\pi\omega_L^2 B_1^2/\mathrm{k_B}T)|\mu_{+-}|^2 f(\omega_L),\end{aligned}$$

which is equivalent to the macroscopic expression

$$\left\langle \frac{\mathrm{d}W}{\mathrm{d}t} \right\rangle_t = \tfrac{1}{2}\omega_L \chi''(\omega_L) B_1^2,$$

and hence

$$\chi''(\omega_L) = N_{+\frac{1}{2}} \pi\omega_L |\mu_{+-}|^2 f(\omega_L)/\mathrm{k_B}T. \tag{8.11}$$

The result indicates that the magnetic resonance absorption can be appreciable at lower temperatures due to a higher population difference.

Although interpretable simply as described in the above quantum theory, the observation of magnetic resonance in a system of magnetic moments must be interpreted macroscopically, where the magnetization M is the basic quantity to be described with two relaxation times $\mathrm{T_1}$ and $\mathrm{T_2}$. At resonance, the system must be in equilibrium with radiation quanta from the high-frequency field B_1 at a given temperature T. The process can be described by the rate equation for the population difference $\mathrm{n} = N_{+1/2} - N_{-1/2}$:

$$\frac{\mathrm{d}(\mathrm{n} - \mathrm{n_o})}{\mathrm{d}t} = -\frac{(\mathrm{n} - \mathrm{n_o})}{\mathrm{T_1}},$$

where $\mathrm{n_o}$ is the value of n with no radiation at T, and is identical to (8.5). On the other hand, the relaxation process described by $\mathrm{T_2}$ of (8.4) can be interpreted as the time for microscopic moments in Larmor precessions in random phases at T to become in phase with the applied B_1 after being switched on or to return to random when switched off.

8.3 Magnetic Resonance Spectrometers

Magneticmagnetic resonance spectrometers resonance is signified by a maximum absorption of radiation energy of the oscillating field B_1 at $\omega = \gamma B_o$ in

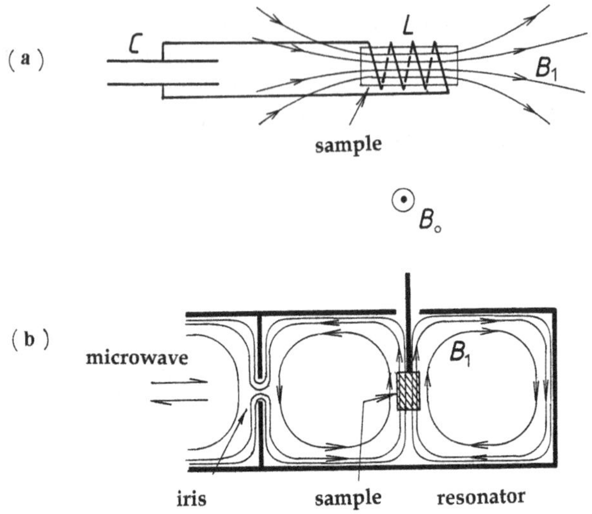

Fig. 8.2. (a) A radio-frequency resonator consisting of a capacitor C and an inductor L with a sample crystal. B_1 represents magnetic rf lines. (b) A microwave resonator. A sample can be placed at a position (e.g. the center of the resonator box) where the microwave field B_1 has maximum amplitude. The iris is adjustable for a desired value of the quality factor Q.

a uniform magnetic field B_o. Therefore, for a system of magnetic moments of known γ, the resonance experiments can be performed either scanning ω at a constant B_o, or scanning B_o at a fixed ω. Although theoretically a trivial matter, the latter arrangement is more practical than the former and therefore used in most spectrometers with conventional laboratory magnets that produce a uniform field up to about 15kG. With such a magnet, the frequency is in the range $1 \sim 100$MHz for nuclear resonance, whereas microwaves in the frequency range $5 \sim 40$GHz are conventional for paramagnetic resonance experiments.

Further, for experiments on samples in small size, it is practical to observe magnetic resonance at a fixed position in a resonator where B_1 is at maximum strength. At radio frequencies, a sample is placed in a resonator of an inductance L_o connected with a capacitor C, where the tuning frequency is given by $\omega = (L_oC)^{-1/2}$, as illustrated in Fig. 8.2a. At microwave frequencies, on the other hand, a sample can be placed inside a cavity resonator at a maximum B_1 of the standing wave, as shown in Fig. 8.2b. At such a position, the sample occupies a finite partial volume V_s of the inductor, so that the resonant frequency depends generally on the sample volume.

When a sample is placed in such a resonator, the inductance can be expressed as

$$L = (1+\chi)L_o = (1+\chi')L_o - i\chi''L_o,$$

where the value of the susceptibility $\chi = \chi' - i\chi''$ depends on the sample volume. The impedance of a "loaded" resonator can therefore be expressed by

$$Z = R + i\omega L + 1/i\omega C = R + \omega\chi'' L_o + i\omega(1+\chi')L_o - i/\omega C,$$

where R is the effective resistance and $\omega\chi'' L_o$ represents the energy loss due to magnetic resonance. The resonant frequency can approximately be determined by letting the imaginary part zero, i.e. $\omega_o^2 = 1/(1+\chi')L_o C$. Assuming that the field B_1 is practically uniform over the sample volume, the susceptibility can be expressed as $(V_s/V_m)\chi$, where V_m is the effective volume for maximum B_1. The volume ratio $\alpha = V_s/V_m$ is called the *filling factor*, which is less than 1 in practical resonators.

In practical applications, the energy loss in a resonator can be described in terms of the *quality factor* Q that is defined by

$$1/Q = (\text{total energy loss/energy stored}) \text{ per cycle},$$

which is compared with the factor Q_o of the unloaded resonator. The electromagnetic energy stored per cycle is given by $\omega(\frac{1}{2}\mu_o B_1{}^2 V_m)/Q_o$, and the energy loss due to magnetic resonance is $\frac{1}{2}\omega_L\mu_o\chi''(\omega_L)B_1 2V_s$. Therefore, the energy loss at resonance $\omega = \omega_L$ is expressed by a fractional change in the quality factor:

$$\frac{\Delta Q}{Q_o} = \frac{Q_o - Q}{Q_o} = \alpha\chi''(\omega_L),$$

being proportional to the imaginary part of $\chi(\omega_L)$. In practical observation, $\Delta Q/Q_o$ is deterimed from $\chi''(B_o)$ by impedance measurements at a constant ω_L, while B_o is scanned around ω_L/γ. Figure 8.3 shows a block diagram of a typical magnetic resonance spectrometer consisting of an *impedance bridge*, where the impedance $Z(B)$ of a sample resonator is compared with a reference impedance Z_o. In the vicinity of a magnetic resonance, the sample impedance can be expressed by

$$Z(\omega) = R + \omega_L\chi'' L_o + i\omega_L(1+\chi')L_o\left(\frac{\omega}{\omega_L} - \frac{\omega_L}{\omega}\right).$$

In such a bridge, reflected waves from $Z(B)$ and Z_o can be balanced either in phase or out of phase, so that the detector signal is related to χ'' or χ', respectively, namely for $B_o = \omega_L/\gamma \pm \Delta B_o$, we have the detector signal that is proportional either to

$$R + \omega_L L_o\chi''(\Delta B_o) \quad \text{or} \quad 2L_o\{1+\chi'(\Delta B_o)\}\gamma B_o.$$

Although independently measurable, these χ'' and χ' are related by the Kramers-Krönig formula, and so only one of these is sufficient for magnetic resonance to be detected.

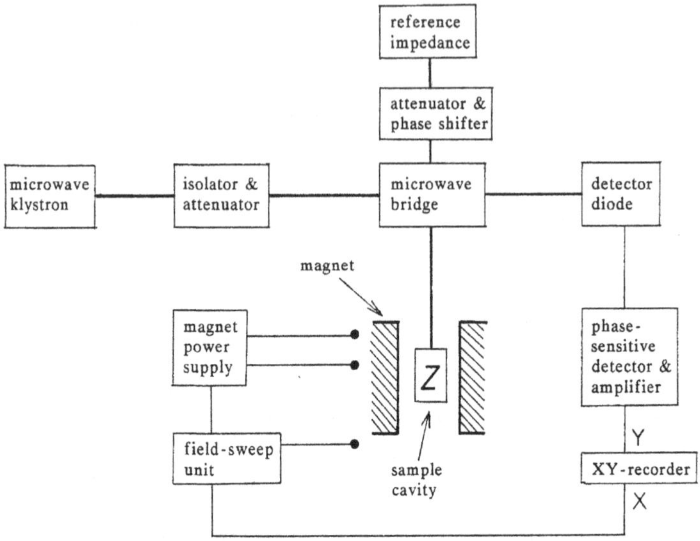

Fig. 8.3. A block diagram of a microwave bridge spectrometer.

8.4 The Crystalline Potential

Magnetic probes for studying structural phase transitions should preferably be of a spin larger than $\frac{1}{2}$, although requiring a very complex spectral analysis, depending on the magnitude of the applied field B_o. As will be explained in Chapter 9, the effect of lattice modulation can be discussed in the first-order accuracy, offering sufficient information in many applications. We could carry out spectral analysis in full, which may, however, be unnecessary for structural studies. In the following sections, the method for spin-Hamiltonians in normal crystals is outlined, prior to discussing modulated crystals by means of magnetic resonance parameters.

In normal crystals, the electronic state of a paramagnetic probe with unpaired spins can be described primarily by the total orbital and spin angular momenta, $\boldsymbol{L}$ and $\boldsymbol{S}$, which are perturbed by the surroundings at the probe site represented by a *crystalline potential.* Known as the Russell-Saunders scheme in atomic spectroscopy, such a description of ions in crystals is adequate for "transition elements" commonly used as magnetic resonance probes.

Orbiting around a nucleus at a lattice site, electrons in the ion are considered as perturbed by a crystalline potential that is expressed by power series of coordinates (x, y, z) with respect to the local lattice symmetry. In an orthorhombic crystal, for example, the crystalline potential can be expressed in the lowest order as a quadratic form

$$V(x, y, z) = Ax^2 + By^2 + Cz^2,$$

where the relation $A + B + C = 0$ holds for the coefficients A, B and C, because any static potential V must satisfy the Laplace equation $\Delta V = 0$. A uniaxial case of tetragonal or trigonal symmetry can be specified as $A = B$, and, hence, $C = -2A$, for which the potential is given by

$$V(x, y, z) = A(x^2 + y^2 - 2z^2) = A(r^2 - 3z^2),$$

where $r^2 = x^2 + y^2 + z^2$, being uniaxial along the z direction.

Another example is the quartic potential

$$V(x, y, z) = D(x^4 + y^4 + z^4),$$

representing cubic symmetry in the lowest order. For cubic symmetry, we can write a quardratic potential $A(x^2 + y^2 + z^2)$, which is just equal to Ar^2, and hence, it does not represent a crystalline potential. Only a deviation from spherical symmetry is essential for the crystalline potential; hence, a cubic potential should be quartic in the lowest order.

Further, to be consistent with the crystal symmetry, x, y and z should be taken as parallel to symmetry axes a, b and c of the crystal, if only one probe can be accommodated in a unit cell. On the other hand, if there are two or more crystallographically equivalent sites for a probe, these x, y and z for the crystalline potential at the site must be chosen as related consistently to the symmetry of the probe site. The concept of a crystalline potential was founded by Bethe's work in 1929, which has since been refined for applications to many solid-state problems. For magnetic problems, Abragam and Pryce [69] have laid the foundation for the method of spin Hamiltonians, providing the basic formulation for magnetic resonance spectroscopy.

8.5 The Zeeman Energy and the g Tensor

Reflecting local symmetry that is orthorhombic in most cases, the crystalline potential can often be signified by a unique axis, allowing a simplified analysis. In such a case, the orbital angular momentum $\boldsymbol{L}$ can be considered as quantized along the unique axis, resulting in the perturbed ionic energy that is split into several levels. In a quadratic potential where the constant A is greater than $k_B T$, the ground state is well separated from upper levels, and hence the magnetic moment can be described as related to the spin angular momentum $\boldsymbol{S}$ only, resulting in a doubly-degenerate ground state. In such a case, the orbital momentum $\boldsymbol{L}$ is said *quenched* by the crystalline potential, and such a spin-degenerated ground state is called the Kramers doublet.

For a magnetic probe of the doublet ground level, we have to further consider the role played by the spin-orbit coupling $\lambda \boldsymbol{L}.\boldsymbol{S}$, where λ is a parameter of the order of $100 \sim 800\ \text{cm}^{-1}$ for iron-group elements, which gives rise to a significant interaction between $\boldsymbol{S}$ and the crystal field. Although typically smaller than the crystalline field splitting, the ground state ψ_o separated from an excited state ψ_ε is perturbed by $\lambda \boldsymbol{L}.\boldsymbol{S}$ in the second-order that is signified by the off-diagonal elements $\int \psi_\varepsilon{}^* L_\perp \psi_o \mathrm{d}v$, resulting in the perturbed wavefunction ψ' expressed by

$$\psi' = \psi_o + \sum\nolimits_\varepsilon \frac{\lambda}{\Delta\varepsilon}\left(\int \psi_o{}^* L_\perp \psi_\varepsilon \mathrm{d}v\right)\psi_\varepsilon,$$

where $\Delta\varepsilon$ is the energy gap between the excited and ground states. In this case, the Zeeman energy of the ion in an applied uniform field $\boldsymbol{B}_o$ is given effectively by

$$\mathsf{H}_Z = -\boldsymbol{\mu}_e.\boldsymbol{B}_o,$$

where

$$\boldsymbol{\mu}_e = -\beta \int \psi'^*(g_e \boldsymbol{S})\psi' \mathrm{d}v,$$

$\beta = e\hbar/2m_e c = -0.927 \times 0^{-20}$ emu known as the Bohr magneton, and $g_e = 2.0023$ the Landé factor of a free electron. For such a magnetic moment modified by the spin-orbit coupling in a crystalline potential, we first consider a case where B_o is applied parallel to the z axis of the crystalline potential, that is $B_o \parallel z$. In this case, the perturbation originates from the off-diagonal element of $L_\perp$, i.e. $\int \psi_\varepsilon{}^* L_\perp \psi_o \mathrm{d}v = 2\exp(i\Delta\varepsilon t/\hbar)$, and only the component μ_{ez} can be stationary. The effective g_z factor in the Zeeman energy can therefore be expressed as

$$\mathsf{H}_Z = \beta g_z S_z B_o \quad \text{where} \quad g_z = g_e\left(1 + \frac{2\lambda}{\Delta\varepsilon}\right). \tag{8.12a}$$

For $\boldsymbol{B}_o$ applied parallel to x or y axis, we consider that $L_\perp = L_y + iL_z$ or $L_z + iL_x$, respectively, for the perpturbing spin-orbit coupling, and we obtain similar expressions of Zeeman energies, namely

$$\mathsf{H}_Z = \beta g_x S_x B_o \quad \text{or} \quad = \beta g_y S_y B_o,$$

where

$$g_x = g_y = g_e\left(1 - \frac{\lambda}{\Delta\varepsilon}\right). \tag{8.12b}$$

If $\boldsymbol{B}_o$ is applied in a general direction $\boldsymbol{n}$, the effective g factor behaves as a tensor quantity $\mathbf{g}$, and the Zeeman energy can be written as

$$\mathsf{H}_Z = \beta \boldsymbol{S}.\mathbf{g}.\boldsymbol{B}_o. \tag{8.13a}$$

In (8.13a), the vectors $\boldsymbol{S}$ and $\boldsymbol{B}_o$ can be considered mathematically as row and column matrices, respectively, in order to accommodate $\mathbf{g}$ of a 3×3 matrix in the product form. To facilitate such a matrix calculation, it is convenient to write (8.13a) as

$$\mathsf{H}_Z = \beta\langle \boldsymbol{S}|\mathbf{g}|\boldsymbol{B}_o\rangle, \tag{8.13b}$$

which will be used in the following discussion. The matrix product as in (8.13b) can be considered as a scalar product of row and column vectors. For instance, the product (8.13b) may be interpreted either that the effective magnetic moment $-\beta\langle \boldsymbol{S}|\mathbf{g}$ is in precession around $|\boldsymbol{B}_\mathrm{o}\rangle$, or that the spin $\langle \boldsymbol{S}|$ is in precession around the effective magnetic field $\mathbf{g}|\boldsymbol{B}_\mathrm{o}\rangle$. Experimentally, the former is convenient, because $\boldsymbol{B}_\mathrm{o}$ provides a convenient reference direction that is fixed in the laboratory frame. On the other hand, the crystal field axes are also a convenient crystallographic reference, in which the direction $\boldsymbol{n}$ of the static field $\boldsymbol{B}_\mathrm{o}$ is expressed as $\boldsymbol{B}_\mathrm{o} = B_\mathrm{o}\boldsymbol{n}$ and the effective field to quantize $\boldsymbol{S}$ can be defined as $\boldsymbol{B}' = \mathbf{g}.\boldsymbol{B}_\mathrm{o}$. Using the relation

$$\langle \boldsymbol{B}'|\boldsymbol{B}'\rangle = B_\mathrm{o}^2\langle \boldsymbol{n}|\hat{\mathbf{g}}\mathbf{g}|\boldsymbol{n}\rangle,$$

we can define

$$B' = g_\mathrm{n}B_\mathrm{o} \quad \text{where} \quad g_\mathrm{n}^2 = \langle \boldsymbol{n}|\hat{\mathbf{g}}\mathbf{g}|\boldsymbol{n}\rangle, \tag{8.14}$$

where $\hat{\mathbf{g}}$ expresses a "transposed matrix" of $\mathbf{g}$. Physically $\mathbf{g}$ is a symmetrical tensor, so that $\hat{\mathbf{g}}\mathbf{g}$ is identical to $\mathbf{g}^2$. The Zeeman energy in a field $B_\mathrm{o}|\boldsymbol{n}\rangle$ can therefore be expressed as

$$\mathsf{H}_\mathrm{Z} = \beta g_\mathrm{n} S_\mathrm{n} B_\mathrm{o}, \tag{8.15}$$

implying that the magnetic moment is effectively given by $\beta g_\mathrm{n} S_\mathrm{n}$. Equation (8.15) is a convenient formula for magnetic resonance, in which the spin $\boldsymbol{S}$ is quantized along the direction $\boldsymbol{n}$, while the g_n-factor is modified as related to the axes of the squared tensor $\mathbf{g}^2$ of (8.14). It is significant that the symmetry axes of the crystalline potential can be determined by the principal axes of $\mathbf{g}^2$, which can be obtained by a coordinate transformation to the principal form from the experimentally determined quadratic $\langle \boldsymbol{n}|\mathbf{g}^2|\boldsymbol{n}\rangle$.

In a usual magnetic resonance practice, spectra are recorded for various directions of $\boldsymbol{B}_\mathrm{o}$, when a sample crystal is rotated around a crystallographic axis (in practice, $\boldsymbol{B}_\mathrm{o}$ is rotated in a laboratory frame where a sample is fixed), so that a set of three elliptical angular variations of g_n^2 showing a sinusoidal curve can be obtained in symmetry planes independently. Fitting these variations to the equation of an ellipse $g_\mathrm{ii}{}^2 n_\mathrm{i}{}^2 + 2g_\mathrm{ij}{}^2 n_\mathrm{i} n_\mathrm{j} + g_\mathrm{jj}{}^2 n_\mathrm{j}{}^2 = g_\mathrm{e}{}^2$, we can evaluate three on-axis elements $g_\mathrm{ii}{}^2$ and three off-axis elements g_ij for the 3×3 symmetrical tensor $\mathbf{g}^2$, which can then be diagonalized numerically to determine the principal axes, X, Y and Z, with zero off-diagonal elements; namely in the principal form, the variation of $g_\mathrm{n}{}^2$ can be expressed as

$$g_\mathrm{n}^2 = g_\mathrm{X}^2 n_\mathrm{X}^2 + g_\mathrm{Y}^2 n_\mathrm{Y}^2 + g_\mathrm{Z}^2 n_\mathrm{Z}^2 \quad \text{and} \quad n_\mathrm{X}^2 + n_\mathrm{Y}^2 + n_\mathrm{Z}^2 = 1,$$

where $(n_\mathrm{X}, n_\mathrm{Y}, n_\mathrm{Z})$ represents the direction cosines of $\boldsymbol{n}$ with respect to X, Y and Z. The principal values g_X, g_Y and g_Z of $\mathbf{g}$ should all be positive, which are directly determined from $g_\mathrm{X}{}^2$, $g_\mathrm{Y}{}^2$ and $g_\mathrm{Z}{}^2$.

A significant feature of the $\mathbf{g}$ tensor is

$$\text{trace}\,\mathbf{g} = g_\mathrm{X} + g_\mathrm{Y} + g_\mathrm{Z} = 3g_\mathrm{e}, \tag{8.16a}$$

as confirmed from (8.12a) and (8.12b). It is noted that deviations from the free-electron value $g_e = 2.0023$ are significant measure of the symmetry in the crystalline potential, and so we determine experimentally the tensor $\Delta\mathbf{g} = \mathbf{g} - g_e\mathbf{e}$, where $\mathbf{e}$ is the unit tensor ($e_{ij} = \delta_{ij}$). By this definition, it is clear that

$$\text{trace}\,\Delta\mathbf{g} = 0. \tag{8.16b}$$

As will be discussed later, these equations (8.16a) and (8.16b) are important formula for a three-dimensional analysis of observed Zeeman terms.

8.6 The Fine Structure

In Section 8.5, we discussed the spin-orbit coupling that perturbs the ionic ground state in Zeeman levels, where the g factor is modified by the crystalline potential. Due to quenching by the crystal potential, the orbital momentum becomes insignificant in the first order; however, the anisotropic g shift arises from $\lambda\boldsymbol{L}.\boldsymbol{S}$ in the second order, reflecting the symmetry of the crystalline potential.

In addition, the spin-orbit coupling will deform the charge cloud in the second order, which is particularly significant for ions in non-S states. Classically, such a charge deformation represents an electric quadrupole moment induced in the crystalline potential, which is quantum mechanically calculated from the second-order perturbation of the spin-orbit coupling. The quadrupole energy of an ion induced in the crystalline potential is generally referred to as the *fine structure* in magnetic resonance spectroscopy.

The spin-orbit coupling in a crystalline potential can generally be written as

$$\mathbf{H}_{LS} = \lambda(L_X S_X + L_Y S_Y + L_Z S_Z) \tag{8.17}$$

with respect to the principal axes X, Y and Z, where the coupling constant λ is assumed to remain isotropic as in a unperturbed ion. In the crystal field, the orbital angular momentum $\boldsymbol{L}$ "quenched" in the first order and, hence, the second-order energy of $\mathbf{H}_{LS}$ can be calculated as

$$E_{LS}^{(2)} = \frac{\lambda^2}{\Delta\varepsilon}\sum_{ij} S_i S_j \left\{\left(\int \psi_o{}^* L_i \psi_\varepsilon \mathrm{d}v\right)\left(\int \psi_\varepsilon{}^* L_j \psi_o \mathrm{d}v\right)\right\},$$

that is determined by non-vanishing off-diagonal elements of $\boldsymbol{L}$ between the ground state and the excited state separated by $\Delta\varepsilon$. It is noted that $E_{LS}{}^{(2)}$ is a quadratic form with respect to the spin components S_X, S_Y and S_Z and, therefore, expressed as

$$E_{LS}^{(2)} = \sum_{ij} S_i D_{ij} S_j = \langle \boldsymbol{S}|\mathbf{D}|\boldsymbol{S}\rangle, \tag{8.18a}$$

where

$$D_{ij} = \frac{\lambda^2}{\Delta\varepsilon}\left(\int \psi_o{}^* L_i \psi_\varepsilon \mathrm{d}v\right)\left(\int \psi_\varepsilon{}^* L_j \psi_o \mathrm{d}v\right) \tag{8.18b}$$

are elements of the fine-structure tensor **D**.

In (8.18b), we notice that $D_{ij} = D_{ji}$ by definition and, hence, **D** is a symmetrical tensor that is further characterized as traceless. We have

$$\operatorname{trace}\mathbf{D} = \sum_i D_{ii} = (\lambda^2/\Delta\varepsilon)\sum_i \langle 0|L_i{}^*|\varepsilon\rangle\langle\varepsilon|L_i|0\rangle = (\lambda^2/\Delta\varepsilon)\sum_i \langle 0|L_i{}^* L_i|0\rangle,$$

which can be verified as vanishing, regardless of the value of $\boldsymbol{L}$; hence

$$\operatorname{trace}\mathbf{D} = 0. \tag{8.19}$$

Transforming a **D** tensor to the principal form, we can express (8.19) as

$$D_X + D_Y + D_Z = 0, \tag{8.19a}$$

which is held for the diagonal elements. Therefore, in a uniaxial crystal field, we have specific relations $D_X = D_Y$ and $D_Z = -2D_X$, and the fine-structure energy can be expressed by

$$\begin{aligned}\langle \boldsymbol{S}|\mathbf{D}|\boldsymbol{S}\rangle &= D_X(S_X^2 + S_Y^2) + D_Z S_Z^2 = -\tfrac{1}{2}D_Z(S^2 - S_Z^2) + D_Z S_Z^2 \\ &= \tfrac{1}{2}D_Z\{3S_Z^2 - S(S+1)\},\end{aligned} \tag{8.20}$$

where $S^2 = S_X^2 + S_Y^2 + S_Z^2$. The fine structure is therefore determined effectively by the spin S and the component S_Z, which however vanishes if $S = S_Z = \frac{1}{2}$. The principal directions signify the symmetry of an ionic charge cloud deformed by the crystalline potential. Being consistent with the classical argument, a charge cloud represented by (8.20) can be interpreted as non-spherically deformed and is expressed in terms of an electric quardrupole moment in the second-order approximation.

Similar to the $\mathbf{g}^2$ tensor, (8.18a) is generally proportional to a quadratic form $\langle n|\mathbf{D}|n\rangle$ with respect to the direction $|n\rangle$ of the applied field $\boldsymbol{B}_o$. For example, when $\boldsymbol{B}_o$ is on the plane XZ, the magnitude D_n changes along an elliptic locus against the direction $|n\rangle$, if the crystal is rotated around the b axis in a uniform field $\boldsymbol{B}_o$, indicating an elliptically deformed charge on this plane, as shown in Fig. 8.4a. The charge cloud is deviated either positively or negatively from otherwise circular along the X and Z axes, being expressed by an electric quadrupole moment. In three-dimensional measurements, such elliptic loci can be obtained on three symmetry planes, representing projections of the ellipsoidal charge. In Fig. 8.4b, the elliptical variation on the ab plane is illustrated.

For magnetic resonance of a paramagnetic ion, we have so far considered a strong applied field $\boldsymbol{B}_o$, along which the electronic spin $\boldsymbol{S}$ is quantized, representing the local symmetry at the site of a magnetic probe. However, in practice, the tensor $\Delta\mathbf{g}$ and **D** may show different symmetries, depending on the strength of an applied field $\boldsymbol{B}_o$. Therefore, we write more generally

$$\mathbf{H} = \mathbf{H}_Z + \mathbf{H}_F = \beta\langle \boldsymbol{S}|\mathbf{g}|\boldsymbol{B}_o\rangle + \langle \boldsymbol{S}'|\mathbf{D}|\boldsymbol{S}'\rangle. \tag{8.21}$$

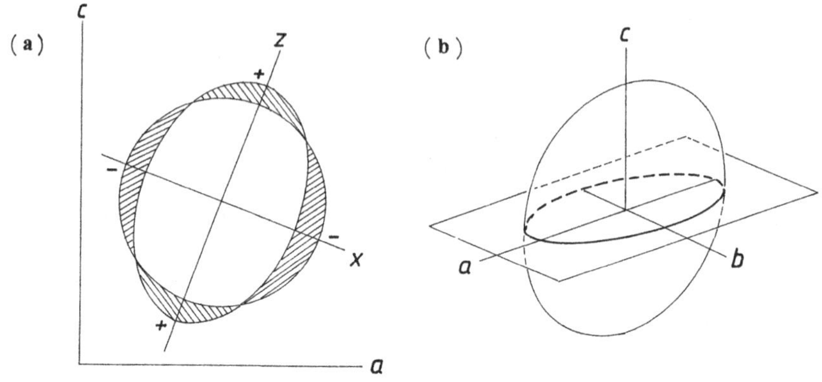

Fig. 8.4. (a) An ellipsoidal deformation of an electronic charge due to a crystalline potential, resulting in the fine structure in the spin-Hamiltonian. (b) An elliptic cross section between the ellipsoid of the bilinear $\langle n|\mathbf{D}|n\rangle$ and a plane of observation, where the applied field $\boldsymbol{B}_\mathrm{o}$ is rotated.

If the Zeeman term $\mathbf{H}_\mathrm{Z}$ is larger than the fine-structure energy $\mathbf{H}_\mathrm{F}$, the spin $\boldsymbol{S}$ is in precession around the effective field $|\boldsymbol{B}'\rangle = \mathbf{g}|\boldsymbol{B}_\mathrm{o}\rangle = g_\mathrm{n}B_\mathrm{o}|\boldsymbol{n}\rangle$, whereas the spin vector $|\boldsymbol{S}'\rangle$ in $\mathbf{H}_\mathrm{F}$ is quantized effectively along $B_\mathrm{o}|\boldsymbol{n}\rangle$ and, hence, expressed as $|\boldsymbol{S}'\rangle = |\boldsymbol{S}\rangle/g_\mathrm{n} = (S_\mathrm{n}/g_\mathrm{n})\mathbf{g}|\boldsymbol{n}\rangle$. Accordingly, (8.21) is expressed effectively as

$$\mathbf{H} = \beta g_\mathrm{n} S_\mathrm{n} B_\mathrm{o} + D'_\mathrm{n} S_\mathrm{n}^2, \tag{8.22}$$

where

$$D'_\mathrm{n} = \langle \boldsymbol{n}|\mathbf{gDg}|\boldsymbol{n}\rangle / g_\mathrm{n}^2. \tag{8.23}$$

However, basically both $\Delta\mathbf{g}$ and $\mathbf{D}$ tensors are originated from the spin-orbit coupling mechanism in a crystalline potential, and hence characterized as *coaxial*, if $\mathbf{H}_\mathrm{F} > \mathbf{H}_\mathrm{Z}$, in particular. Accordingly, we write $\mathbf{D}' = \mathbf{gDg}/{g_\mathrm{n}}^2 = \mathbf{g}^2\mathbf{D}/g_\mathrm{n}^2$ in (8.23), for which the principal axes X, Y and Z are often considered as common for $\mathbf{g}$ and $\mathbf{D}$ tensors. Referring to these common axes, (8.23) can be expressed as

$$g_\mathrm{n}^2 D_\mathrm{n} = g_\mathrm{X}^2 D_\mathrm{X} + g_\mathrm{Y}^2 D_\mathrm{Y} + g_\mathrm{Z}^2 D_\mathrm{Z}. \tag{8.24}$$

Eigenvalues of the spin-Hamiltonian (8.22) are determined by the magnetic quantum number M_S for the spin state S_n:

$$E(M_\mathrm{S}) = g_\mathrm{n}\beta B_\mathrm{o} M_\mathrm{S} + D'_\mathrm{n} M_\mathrm{S}^2, \tag{8.25}$$

where D'_n is such an effective fine-structure parameter as defined above for the direction $|n\rangle$, and the magnetic resonance occurs when the selection rule $\Delta M_\mathrm{S} = \pm 1$ is fulfilled between these energy levels; namely the resonance conditions are given by

$$\hbar\omega = E(M_\mathrm{S}+1) - E(M_\mathrm{S}) = g_\mathrm{n}\beta B_\mathrm{o} + D'_\mathrm{n}(2M_\mathrm{S}+1), \tag{8.26}$$

showing the basic resonance line $\hbar\omega = g_n\beta B_o$ between spin levels $M_S = \pm\frac{1}{2}$, plus a ladder of lines that are equally spaced by D'_n.

On the other hand, if $\mathbf{H}_F \gg \mathbf{H}_Z$, the spin $\boldsymbol{S}$ can conveniently be quantized along the direction for the largest principal value of $\mathbf{D}$, and $\mathbf{H}_Z$ is considered as a perturbation. In this case, the first-order calculation is generally not accurate, and the perturbation calculation should be carried out to a higher order. Generally it is too complicated to extract structural information from $\mathbf{g}$ and $\mathbf{D}$ calculated in high-order from observed spectra. Experimentally, it is preferable to select probes for easier interpretation in such a case; therefore, we do not continue our spectroscopic discussion beyond this point, leaving it to general references for magnetic resonance spectroscopy. Nevertheless, such a low-field analysis is required in some applications, e.g. Zeeman studies of a nuclear quadrupole resonance and of triplet states in molecular probes.

8.7 Hyperfine Interactions and Forbidden Transitions

For magnetic resonance spectra, the interaction between the electronic magnetic moment $\boldsymbol{\mu}_e$ and nuclear magnetic moments $\boldsymbol{\mu}_n$ located within the electronic orbital yields often useful structural information. Called the *hyperfine interaction*, such an interaction arises from the quantum-mechanical "contact" mechanism, known as the Fermi interaction, as well as the classical "dipole-dipole" interaction when the electron is orbiting at a finite distance $\boldsymbol{r}$ from the nucleus. Combining these two mechanisms, the hyperfine interaction can be expressed as

$$\mathbf{H}_{HF} = (8\pi/3)|\psi(0)|^2\boldsymbol{\mu}_e.\boldsymbol{\mu}_n + \{\boldsymbol{\mu}_e.\boldsymbol{\mu}_n/r^{-3} - 3(\boldsymbol{\mu}_e.\boldsymbol{r})(\boldsymbol{\mu}.\boldsymbol{r}_n)/r^{-5}\}, \qquad (8.27)$$

where the first Fermi interaction term is essential for an s-electron that has a finite density $|\psi(0)|^2$ at the nucleus, and the second dipolar term represents a dipole-dipole interaction between $\boldsymbol{\mu}_e$ and $\boldsymbol{\mu}_n$ at a finite distance $\boldsymbol{r}$. Using direction cosines (l, m, n) for the vector $\boldsymbol{r}$ with respect to the fixed position of $\boldsymbol{\mu}_n$, the hyperfine interaction energy can be written as

$$\mathbf{H}_{HF} = (8\pi/3)|\psi(0)|^2\langle\boldsymbol{\mu}_e|\boldsymbol{\mu}_n\rangle + \langle\boldsymbol{\mu}_e|\mathbf{A}_d|\boldsymbol{\mu}_n\rangle,$$

where

$$\mathbf{A}_d = r^{-3}\begin{pmatrix} 1-3l^2 & -3lm & -3nl \\ -3m & 1-3m^2 & -3mn, \\ -3ln & -3nm & 1-3n^2 \end{pmatrix}$$

which is a traceless tensor, i.e. trace $\mathbf{A}_d = 0$, because of the relation $l^2+m^2+n^2 = 1$. Writing the Fermi term as $\mathbf{A}_f = (8\pi/9)|\psi(0)|^2\mathbf{E}$, where $\mathbf{E} = (\delta_{ij})$, (8.27) can be expressed as

$$\mathbf{H}_{HF} = \langle\boldsymbol{\mu}_e|\mathbf{A}|\boldsymbol{\mu}_n\rangle \quad \text{where} \quad \mathbf{A} = \mathbf{A}_f + \mathbf{A}_d$$

and

$$\operatorname{trace}\mathbf{A} = (8\pi/3)|\psi(0)|^2. \qquad (8.28)$$

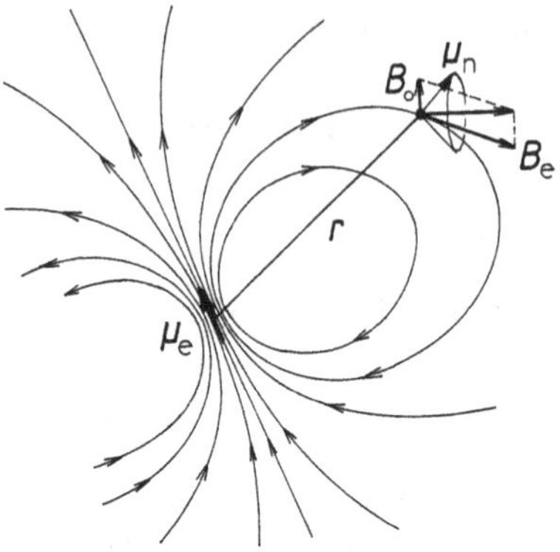

Fig. 8.5. Magnetic dipole interaction between an electronic moment $\boldsymbol{\mu}_e$ and a nuclear moment $\boldsymbol{\mu}_n$ at a distance r. The effective magnetic field at $\boldsymbol{\mu}_n$ is composed of the field of the electron $\boldsymbol{B}_e$ and an applied field $\boldsymbol{B}_o$.

Because of $\mathsf{H}_{HF} \ll \mathsf{H}_Z$ in usual cases, we consider that $\boldsymbol{\mu}_e = -g_n\beta\boldsymbol{S}$ is primarily quantized in the direction of an applied field $\boldsymbol{B}_o = B_o|\boldsymbol{n}\rangle$, where the Bohr magneton β is expressed as positive. Therefore, the hyperfine energy in the strong field approximation can be expressed as

$$\mathsf{H}_{HF} = -\beta\gamma\hbar\langle\boldsymbol{S}|\mathbf{g.A}|\boldsymbol{I}\rangle,$$

allowing one to interpret that the nuclear moment $\gamma\hbar|\boldsymbol{I}\rangle$ is in precession around the magnetic field $\langle\boldsymbol{B}_e| = \beta\langle\boldsymbol{S}|\mathbf{gA}$ that originates from the orbiting electron at the position of the nucleus. In fact, the applied field $\langle\boldsymbol{B}_o|$ also contribute to such a field for nuclear precession, although the magnitude B_o may be significantly small as compared with the hyperfine field $\langle\boldsymbol{B}_e|$ in some cases. Generally, for the nuclear motion, the effective field is given by the vector sum $\langle\boldsymbol{B}_{hf}| = \langle\boldsymbol{B}_e| + \langle\boldsymbol{B}_o|$, as shown in Fig. 8.5. In this case, the hyperfine energy can be expressed in the first-order approximation as

$$E^{(1)}_{HF} = -\gamma\hbar\langle\boldsymbol{B}_{hf}|\boldsymbol{I}\rangle = -\gamma\hbar B_{hf}m',$$

where

$$B_{hf} = B_o + \beta M\langle\boldsymbol{n}|\hat{\mathbf{g}}\hat{\mathbf{A}}\mathbf{Ag}|\boldsymbol{n}\rangle^{1/2}$$

and m' is the nuclear spin quantum number with respect to the effective hyperfine field $\langle\boldsymbol{B}_{hf}|$.

If $B_{hf} \approx B_e$, the hyperfine energy is written as

$$E^{(1)}_{HP} = -K_nMm', \tag{8.29}$$

where

$$g_n^2\beta^2K_n^2 = \langle\boldsymbol{n}|\hat{\mathbf{g}}\hat{\mathbf{A}}\mathbf{Ag}|\boldsymbol{n}\rangle.$$

Here, the coefficient K_n is in units of energy, where the factor $\gamma\hbar$ is included in the definition of the tensor $\mathbf{A}$. For convenience in magnetic resonance for scanning the external field, it is desirable to express K_n in units of magnetic field, for which K_n (energy) is replaced by K_n (field)/$g_n\beta$, resulting in the expression

$$K_n = \langle \boldsymbol{n} | \hat{\mathbf{g}}\hat{\mathbf{A}}\mathbf{A}\mathbf{g} | \boldsymbol{n} \rangle^{1/2} \quad \text{(in field unit)}, \tag{8.30}$$

which is usually called the hyperfine splitting.

Unlike $\mathbf{g}$ and $\mathbf{D}$ tensors characterized by the crystalline field, the symmetry of the $\mathbf{A}$ tensor are generally determined by the nuclear location with respect to the electron. Therefore, unless the nucleus is at the center of the crystal field, the tensor $\mathbf{A}$ is related to symmetry axes at the nuclear site, which is different from $\mathbf{g}$ and $\mathbf{D}$. Nevertheless, in practical analysis, we determine the tensor $\mathbf{g}^2$ and the hyperfine splitting tensor $\hat{\mathbf{g}}\hat{\mathbf{A}}\mathbf{A}\mathbf{g} = \mathbf{g}^2\mathbf{A}^2$ from observed spectra, and then proceed to numerical calculation of the squared tensor $\mathbf{A}^2$ to obtain the tensor $\mathbf{A}$. Examples of practical tensor analysis are shown in later discussions.

In the above description, the electronic and nuclear spins are in precession primarily at frequencies, $\omega_\varepsilon = g_n\beta B_o/\hbar$ and $\omega_n = \gamma B_{hf}$, independently around $\boldsymbol{B}_o$ and $\boldsymbol{B}_{hf}$, respectively. These frequencies are very different in a given field $\boldsymbol{B}_o$, so that these magnetic resonances can be independently observed, as indicated by the selection rules

$$\Delta M = \pm 1, \quad \Delta m' = 0 \quad \text{and} \quad \Delta M = 0, \quad \Delta m' = \pm 1,$$

for electronic and nuclear transitions, respectively.

For a coupled electron-nuclear system, the spin-Hamiltonian is expressed as

$$\mathbf{H} = g_n\beta S_n B_o + S_n^2 \langle \boldsymbol{n} | \mathbf{D}' | \boldsymbol{n} \rangle + S_n \langle \boldsymbol{n} | \mathbf{A} | \boldsymbol{I} \rangle - \gamma \boldsymbol{I}.\boldsymbol{B}_o.$$

In magnetic resonance experiments, the direction $|\boldsymbol{n}\rangle$ of $\boldsymbol{B}_o$ can often be referred to the principal directions X, Y and Z of the crystal field, for which the tensors $\mathbf{g}^2$ and $\mathbf{D}'$ are diagonalized.

Expressing $|\boldsymbol{n}\rangle$ by polar and azimuthal angles θ and φ, $\mathbf{H}$ can be written as

$$\begin{aligned}\mathbf{H} = {} & g_n\beta B_o S_Z + \tfrac{1}{2}D'\{S_n^2 - S(S+1)/3\}(3\cos^2\theta - 1) \\ & + D'(S_Z S_X + S_X S_Z)\cos\theta\sin\theta + \tfrac{1}{4}D'(S_+^2 + S_-^2)\sin^2\theta(K_n S_Z - \gamma B_o)I_n',\end{aligned} \tag{8.31}$$

in which the nuclear spin I_n' is quantized along the combined field $\boldsymbol{B}_{hf}$ composed of the electronic spin field $\boldsymbol{B}_e$ and the external field $\boldsymbol{B}_o$. It is significant that in (8.31) the eigenvalue of $\mathbf{H}$ is determined only by S_n if $\theta = 0$ and $\frac{1}{2}\pi$; namely when $\boldsymbol{B}_o$ is parallel to the principal direction either Z or X; in all other directions, **H** depends on other components of $\boldsymbol{S}_n$ as well. When $\boldsymbol{B}_o \parallel Z$ and X, the eigenvalues of **H** can be expressed in the first-order accuracy as

$$E_Z(M, m') = g_Z\beta B_o M + 2D'M^2 + (K_Z M - \gamma B_o)m' \quad \text{for n} \parallel Z$$

and

$$E_X(M, m') = g_X\beta B_o M - D'M^2 + (K_X M - \gamma B_o)m' \quad \text{for n} \parallel X,$$

for which the electronic magnetic transitions between $M = +\frac{1}{2}$ and $-\frac{1}{2}$, i.e. $\Delta M = \pm 1$, are allowed, accompanying hyperfine energies proportional to m', where m' is the nuclear spin quantum number in the effective hyperfine field B_{hf}.

In contrast, for $0 < \theta < \frac{1}{2}\pi$, energy eigenvalues E_n are significantly contributed by the second-order perturbations due to cross products of the terms in D' and K_n, consisting of $(S_Z S_+)(S_- I'_+)$, $(S_Z S_-)(S_+ I'_-)$ and so forth, where these coefficients are of the order of ($D'K_n$/Zeeman energy). According to Abragam and Bleaney [70], such terms can be reduced to

$$(3D'K_n \sin 2\theta / 4g_n\beta B_o)(I'_{n+} + I'_{n-})\{S_Z^2 - S(S+1)/3\},$$

if $D' < g_n\beta B_o$, which includes operators connecting different hyperfine levels, e.g. by mixing a nuclear state $|m'\rangle$ with $|m' \pm 1\rangle$ with the amounts of order of $3D' \sin 2\theta / 4g_n\beta B_o$, resulting in so-called *forbidden* transitions ($\Delta M = \pm 1, \Delta m' = \pm 1$). The intensities of these forbidden transitions relative to the allowed transitions ($\Delta M = \pm 1, \Delta m' = 0$) are given by the square of this mixing constant. Because of the factor $\sin^2(2\theta)$ in the relative intensities, such forbidden lines vanish at $\theta = 0$ and $\frac{1}{2}\pi$, while they are maximum at $\theta = \frac{1}{4}\pi$. As will be discussed in Chapter 9 for Mn^{2+}spectra from BCCD crystals, this result provides a useful method for determining principal directions, while such forbidden lines are easily identified by diminishing intensities when the direction is close to the principal axes.

Forbidden nuclear transitions are not only given by $\Delta m' = \pm 1$ but also by $\Delta m' = \pm 2$, as related to other cross-products, such as $D'S_+{}^2(K_n S_- I'_+)^2$ and $D'S_-^2(K_n S_+ I'_-)^2$, in the second-order perturbation. Forbidden lines for $\Delta m' = \pm 2$ occur in the range $0 < \theta < \frac{1}{2}\pi$ with an angular dependence different from $\Delta m' = \pm 1$, exhibiting very complex spectra. However, the forbidden lines for $\Delta m' = \pm 1$ are identifiable in the vicinity of the principal axes, whereas those lines for $\Delta m' = \pm 2$ are not easily identified. All forbidden lines appear in comparable intensities in the vicinity of $\theta = \frac{1}{4}\pi$ depending on the strength of B_o, making the spectral analysis almost impossible.

Although fully discussed by Bleaney and his group, the complete analysis is not always necessary for studying modulation effects, as will be discussed in Chapter 9. In a modulated crystal, principal directions of magnetic probes are spatially distributed, providing useful information about the lattice modulation. The hyperfine tensor **A** is also useful for studying modulation schemes, as characterized by a change in the distinctive direction. Normally, a hyperfine tensor **A** from an off-center nucleus in the *ligand*, known also as a *superfine* interaction, can be used for analyzing the lattice modulation.

9

Magnetic Resonance Sampling and Nuclear Spin Relaxation Studies on Modulated Crystals

9.1 Paramagnetic Probes in a Modulated Crystal

In normal crystals characterized by translational symmetry, magnetic impurity probes distributed among lattice sites are all identical, exhibiting the same spectra, whereas in a modulated phase, unit cells are not exactly identical, resulting in inhomogeneously broadened magnetic resonance lines as related to the type of lattice modulation.

Doped with a small quantity of impurity ions, such probes in a crystal can be assumed as randomly distributed among lattice sites, thereby showing a weak *diluted* paramagnetism. For studies of phase transitions, these probes should be associated with active groups in the crystal, otherwise remaining as insensitive to the structural change. Furthermore, we are aware of properties of a doped crystal that are only weakly modified, where such impurities allow us to study the intrinsic dynamics of pseudospins. The electronic wavefunction in outer orbits extends to the surroundings, hence serving as a more sensitive probe than a nuclear spin at a lattice site. On the other hand, nuclear spins larger than $\frac{1}{2}$ can also interact with the surroundings through the quadrupole energy, which should also be informative about the local structural change.

In practice, ions of transition-group elements are commonly used for doping crystals, serving as useful probes, if showing well-resolved magnetic resonance spectra. However, it is realized that such probes accompany charge defects in the lattice if the impurity charge is different from the ionic charge at the lattice site, making the situation more complex than necessary. Irradiated crystals can also be used for that purpose, because damaged constituents, namely free radicals, are paramagnetic. In any case, we have to consider the physical compatibility of probes with the properties of undoped crystals.

In a nonmagnetic crystal, the spin $\boldsymbol{S}$ of a probe ion is coupled with the oribital moment $\boldsymbol{L}$, which determines the spin orientation in the crystal, although $\boldsymbol{L}$ is quenched by the lattice in the first order. In a modulated lattice, the spin motion is further modified by the pseudospin vector $\boldsymbol{\sigma}(\phi)$ and, hence, the spin is denoted by $\boldsymbol{S}'$ that is quantized along a direction different from $\boldsymbol{S}$

in the unmodulated crystal. We assume that the local quantization is modified by a transforming matrix **a** that is written as

$$S' = \mathbf{a}.S, \tag{9.1}$$

where $\mathbf{a} = \mathbf{1} + \langle\sigma|\mathbf{e}$ expresses that the local distortion tensor $\mathbf{e} = (\mathrm{e_{ij}})$ is modulated with the site-dependent order variable $\langle\sigma|$. The tensor **e** represents the basic strains for distorting the crystal structure, whereas the unit tensor **1** given by the Kronecker delta δ_{ij} keeps the unit cell undistorted.

A displacive binary phase transition is characterized by specific displacements e and $-e$ of the pseudospin, that is $\boldsymbol{\sigma} = \pm\sigma_o e$. The unit vector e specifies the inversion of $\boldsymbol{\sigma}$ in a crystal undergoing a structural change, where the unit cell can generally contain even number of pseudospin sites. On the other hand, if the unit cell has only one pseudospin site, e should occur along a symmetry axis of the crystal.

Representing local deformation, the tensor elements $\mathrm{e_{ij}}$ are generally defined as composed of symmetric and antisymmetric components: that is,

$$(\mathrm{e_{ij}}) = \tfrac{1}{2}(\mathrm{e_{ij}} + \mathrm{e_{ji}}) + \tfrac{1}{2}(\mathrm{e_{ij}} - \mathrm{e_{ji}}),$$

where the first and second terms on the right side represent local directional and orientational changes, respectively, in the cell structure. In a displacive crystals, these occur exclusively, namely either one represents the basic mode of distortion, and the corresponding strains in the unit cell are significant for the structural change.

If the symmetric components of $\mathrm{e_{ij}}$ vanish,

$$\mathrm{e_{ij}} = -\mathrm{e_{ji}} \tag{9.2}$$

indicates that the tensor **e** is antisymmetric, and the crystal is locally strained by a pure rotational twist as determined by the second term, for which the off-diagonal elements $\mathrm{e_{ij}}$ are essential. On the other hand, if symmetrical, i.e. $\mathrm{e_{ij}} = \mathrm{e_{ji}}$, the second term vanishes, and the deformation is characterized by the diagonal elements $\mathrm{e_{11}}$, $\mathrm{e_{22}}$ and $\mathrm{e_{33}}$ in the principal form. In this case, the deformation is characterized by a volume change, for which the expansion coefficient α can be defined by

$$\alpha = 1 + (\mathrm{e_{11}} + \mathrm{e_{22}} + \mathrm{e_{33}}).$$

The condition

$$\mathrm{trace}\,\mathbf{e} = 0 \tag{9.3}$$

signifies the specific case characterized by no volume change.

9.2 The spin-Hamiltonian in Modulated Crystals

For sampling pseudospin condensates in a modulated crystal, we can write the spin-Hamiltonian **H'** of a paramagnetic probe as

$$\mathbf{H}' = -\beta\langle S'|\mathbf{g}|B_o\rangle + \langle S'|\mathbf{D}|S'\rangle + \langle S'|\mathbf{A}|I\rangle,$$

where $\langle \boldsymbol{S}'|$ represents the modified spin vector by a modulated structure, giving $\mathbf{H}'$ different from $\mathbf{H}$ in the unmodulated crystal.

Assuming (9.1), $\mathbf{H}'$ can be separated into two terms as

$$\mathbf{H}' = \mathbf{H} + \mathbf{H}_1,$$

where

$$\mathbf{H} = -\beta\langle \boldsymbol{S}|\mathbf{g}|\boldsymbol{B}_\mathrm{o}\rangle + \langle \boldsymbol{S}|\mathbf{D}|\boldsymbol{S}\rangle + \langle \boldsymbol{S}|\mathbf{A}|\boldsymbol{I}\rangle$$

expresses the unmodulated spin-Hamiltonian, whereas

$$\mathbf{H}_1 = -\sigma\beta\langle \boldsymbol{S}|\hat{\boldsymbol{\varepsilon}}.\mathbf{g}|\boldsymbol{B}_\mathrm{o}\rangle \tag{9.4a}$$

$$+\sigma\langle \boldsymbol{S}|\hat{\boldsymbol{\varepsilon}}.\mathbf{D} + \mathbf{D}.\boldsymbol{\varepsilon}|\boldsymbol{S}\rangle + \sigma^2\langle \boldsymbol{S}|\hat{\boldsymbol{\varepsilon}}.\mathbf{D}.\boldsymbol{\varepsilon}|\boldsymbol{S}\rangle \tag{9.4b}$$

$$+\sigma\langle \boldsymbol{S}|\hat{\boldsymbol{\varepsilon}}.\mathbf{A}|\boldsymbol{I}\rangle \tag{9.4c}$$

represents effects of modulation. In these expressions, we set $\boldsymbol{\varepsilon} = e.\mathbf{e}$ for local strains at a probe site, while $\mathbf{D}$ should be written as $\mathbf{D}'$ according to the definition given in Section 8.6. However, we use the same $\mathbf{D}$ to reduce the number of notations, whereas for observable coefficients in experiments, we do not need to distinguish $\mathbf{D}'$ from $\mathbf{D}$. In normal crystals, the electronic spin $\langle \boldsymbol{S}|$ can be quantized along the external field $\langle \boldsymbol{B}_\mathrm{o}|$, while the nuclear spin $\langle \boldsymbol{I}|$ is quantized effectively in the field of fine structure. For practical applications, it is convenient to use the quantum numbers M and m referring to these fields.

The equations (9.4a), (9.4b) and (9.4c) express modulation effects, for which $\mathbf{H}_1$ consists of terms proportional to σ and σ^2, whose coefficients do not vanish in general because of the strain tensor $\boldsymbol{\varepsilon}$. In the following, these modulation terms are discussed individually for magnetic resonance spectra observed in the critical regions.

9.2.1 The g Tensor Anomaly

In a strong applied field $\boldsymbol{B}_\mathrm{o} = B_\mathrm{o}\boldsymbol{n}$, the Zeeman energy of a paramagnetic ion can be expressed in terms of the stationary spin component $\boldsymbol{S}_\mathrm{n}$ with respect to $\boldsymbol{B}_\mathrm{o}$. If the crystal becomes modulated, the effective magnetic moment $\beta\langle \boldsymbol{S}|\mathbf{g}'$ can be considered as quantized along $\boldsymbol{B}_\mathrm{o}$, and hence its steady component can be expressed as $g'_\mathrm{n}\beta\langle \boldsymbol{S}_\mathrm{n}| = g'_\mathrm{n}\beta M\langle \boldsymbol{n}|$. Here, g'_n represents the effective g-factor in the modulated crystal. In the first-order approximation, the Zeeman energy can therefore be expressed as

$$E'_\mathrm{Z}{}^{(1)} = E_\mathrm{Z}{}^{(1)} + E_{1\mathrm{Z}}{}^{(1)} = -g'_\mathrm{n}\beta B_\mathrm{o}M,$$

where the modulated g'_n is determined by

$$g'_\mathrm{n}{}^2 = \langle n|\mathbf{g}'^2|n\rangle = \langle n|\hat{\mathbf{a}}.\mathbf{g}^2.\mathbf{a}|n\rangle,$$

where $\mathbf{a} = \mathbf{1} + \langle\boldsymbol{\sigma}|\boldsymbol{\varepsilon}$. Therefore, the effect of modulation is calculated from

$$g_n'^2 = g_n^2 + \sigma\langle n|\hat{\boldsymbol{\varepsilon}}.\mathbf{g}^2 + \mathbf{g}^2.\boldsymbol{\varepsilon}|n\rangle + \sigma^2\langle n|\hat{\boldsymbol{\varepsilon}}.\mathbf{g}^2.\boldsymbol{\varepsilon}|n\rangle, \quad \text{where} \quad g_n^2 = \langle n|\mathbf{g}^2|n\rangle,$$

Here, the two additional terms proportional to σ and σ^2 are generally nonzero, if the tensor $\boldsymbol{\varepsilon}$ has asymmetric off-diagonal elements.

The threshold of binary ordering can be identified in magnetic resonance lines that *split* into two components at T_c as related to inversion symmetry of σ; that is,

$$g_n'(\pm\sigma_o)^2 = g_n(0)^2 \pm \sigma_o\langle n|\hat{\boldsymbol{\varepsilon}}.\mathbf{g}^2 + \mathbf{g}^2.\boldsymbol{\varepsilon}|n\rangle + \sigma_o^2\langle n|\hat{\boldsymbol{\varepsilon}}.\mathbf{g}^2.\boldsymbol{\varepsilon}|n\rangle,$$

and, hence, we can write

$$g_n'(+\sigma)^2 - g_n'(-\sigma)^2 = 2\sigma(\phi)\langle n|\hat{\boldsymbol{\varepsilon}}.\mathbf{g}^2 + \mathbf{g}^2.\boldsymbol{\varepsilon}|n\rangle$$

in the critical region, where the pseudospin $\sigma(\phi)$ is a function of the phase ϕ, fluctuating between $+\sigma$ and $-\sigma$ in phase- and in amplitude-modes. Therefore, such distributed g are responsible for the g_n-anomaly that is expressed by

$$\Delta g_n = g_n'(+\sigma) - g_n'(-\sigma),$$

being spread around the center

$$g_n = \tfrac{1}{2}\{g_n'(+\sigma) + g_n'(-\sigma)\}.$$

Accordingly, we can write the following expressions for the fluctuating cos and sin modes:

$$\Delta g_n = c_n \cos\phi \quad \text{and} \quad c_n \sin\phi, \tag{9.5}$$

where

$$c_n = \sigma_o\langle n|\hat{\boldsymbol{\varepsilon}}.\mathbf{g}^2 + \mathbf{g}^2.\boldsymbol{\varepsilon}|n\rangle / g_n. \tag{9.6}$$

It is noted that the anomalies Δg_n signified by the factor c_n is proportional to σ_o, whereas the term of σ^2 is cancelled out, allowing one to identify the binary transition in terms of inversion symmetry.

For a one-dimensional modulation along an x direction, the phase ϕ represents propagation of σ, i.e. $\phi = \Delta q.x - \Delta\omega.t + \phi_o$, although the equations (9.5) suggest that these pseudospin modes σ are pinned by the anharmonic quartic potential, and the phase is distributed in the range $0 \le \phi \le 2\pi$. In magnetic resonance experiments, such sampling results of σ should be interpreted as time averages of $\cos\phi$ and $\sin\phi$ over the timescale $t_o = 2\pi/\omega_L$, which do not vanish if $\Delta\omega.t_o \le 1$. Letting the spatial part of the phase as $\phi_s = \Delta q.x + \phi_o$ for brevity,

$$\begin{aligned}\langle\cos\phi\rangle_t &= (1/t_o)\int_0^{t_o} \cos(\phi_s - \Delta\omega.t)dt \\ &= \frac{\sin(\Delta\omega.t_o)}{(\Delta\omega.t_o)}\cos\{\phi_s - \Delta\omega(t + \tfrac{1}{2}t_o)\}\end{aligned}$$

and

$$\langle \sin \phi \rangle_t = (1/t_o) \int_o^{t_o} \sin(\phi_s - \Delta\omega.t)dt$$
$$= \left[\frac{\sin(\Delta\omega.t_o)}{\Delta\omega.t_o} \right] \sin\{\phi_s - \Delta\omega(t + \tfrac{1}{2}t_o)\}.$$

Here, the mathematical formula $\lim_{\Delta\omega t \to 0}\{\sin(\Delta\omega.t_o)/(\Delta\omega.t_o)\} = 1$ suggests that the observed amplitude is reduced by the factor in the square brackets, that is numerically close to 1 if $2\pi/\Delta\omega \leq t_o$. In this case, the observed g-anomaly due to the sinusoidal variation can be interpreted as proportional to the fluctuating σ with the reduced amplitude $\sigma_o \sin(\Delta\omega.t_o)/(\Delta\omega.t_o)$ and phase $\phi_s - \Delta\omega(t - \frac{1}{2}t_o)$.

The factor c_n given by (9.6) must be nonzero to obtain such a **g** anomaly. It is noted that for a pure rotational **e**, c_n vanishes when the direction $|n\rangle$ is parallel to one of the principal **g** tensor axes, but is nonzero in all other directions. For a pure dilatational **e**, on the other hand, such anomalies are expected only along the axes. Furthermore, the anomaly depends on the timescale t_o of observation and, hence, $\Delta\omega$ for critical fluctuations can be estimated if comparing such Δg_n observed at different Larmor frequencies.

The Zeeman energies of magnetic probes are thus distributed in the critical region, resulting in broad resonance lines in magnetic resonance experiments. The two modes of fluctuations, as characterized by $\cos\phi$ and $\sin\phi$, are independent under critical conditions. Experimentally, it is significant that binary fluctuations can be explicitly in the cos-mode, leading to domain formation on lowering the temperature. On the other hand, the sin-mode vanishes with long-range order.

In magnetic resonance, the anomaly Δg_n arises from fluctuations in the resonance frequency at a constant resonance field B_o, i.e.

$$\hbar\Delta\omega_n = \beta\Delta g_n B_o.$$

Writing $\nu_n = c_n\sigma_o\beta B_o/h$,

$$\Delta\nu_c = \nu_n \cos\phi \quad \text{and} \quad \Delta\nu_s = \nu_n \sin\phi,$$

for the cos and sin modes, respectively.

These formula are directly applicable to magnetic resonance spectra by scanning frequency at a constant $\boldsymbol{B}_o$. Theoretically, the absorption intensity is expressed as a function of distributed ϕ in the form $f(\phi)d\phi$, whereas, experimentally, the resonance is observed with varying $\Delta\nu_c$ and $\Delta\nu_s$ in this case. Therefore, for these fluctuating modes, we can write

$$f_c(\phi)d\phi = \frac{F_c(\Delta\nu_c)d\Delta\nu_c}{|d\Delta\nu_c/d\phi|} = \frac{F_c(\Delta\nu_c)d\Delta\nu_c}{|\nu_n \sin\phi|} = \frac{F_c(\Delta\nu_c)d\Delta\nu_c}{(\nu_n^2 - \Delta\nu_c^2)^{1/2}}$$

and

$$f_s(\phi)d\phi = \frac{F_s(\Delta\nu_s)d\Delta\nu_s}{|d\Delta\nu_s/d\phi|} = \frac{F_s(\Delta\nu_s)d\Delta\nu_s}{|\nu_n \cos\phi|} = \frac{F_s(\Delta\nu_s)d\Delta\nu_s}{|\Delta\nu_s|}.$$

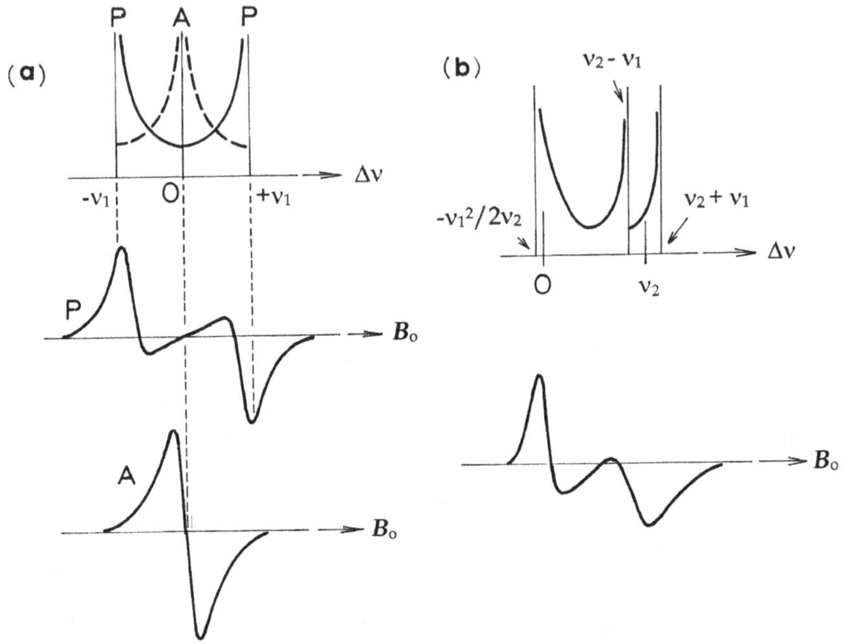

Fig. 9.1. (a) The first curve shows density distributions of anomalous magnetic resonance frequencies $\Delta\nu \propto \sigma$ in the phase mode (P) and in the amplitude mode (A). The second and third ones are the corresponding magnetic resonance lineshapes of P and A modes, respectively, as observed with magnetic field scanning. (b) Magnetic resonance anomaly for $\Delta\nu = a\sigma + b\sigma^2$. The top curve shows a density distribution in frequency scan, and the bottom the corresponding resonance line by magnetic-field scan.

In the cos mode, the density of distributed intensities has two edge singularities at $\Delta\nu_c = \pm\nu_n$ corresponding to $\phi = 0$ and π, whereas in the sin-mode the density is characterized by a singularity at $\Delta\nu_s = 0$ for $\phi = \frac{1}{2}\pi$.

In Chapter 6 we discussed the intensity anomalies in neutron inelastic scattering illustrated in Fig. 6.6. In magnetic resonance experiments, the static field B_o is modulated additionally at a low frequency for recording spectra, so that the spectrum is displayed as proportional to the derivative $d\chi''/dB_o$. Such anomalous spectra detectable with a conventional spectrometer are also shown in Fig. 9.1a.

A binary structural phase transition is signified by anomalies arising from fluctuations in cos- and sin-modes. It is significant that such anomalies are anisotropic, as indicated by Δg_n depending on the direction of $\boldsymbol{B}_o$. The binary splitting Δg_n signifies fluctuations between σ and $-\sigma$, in a symmetry plane where the mirror symmetry is violated, for which the experimental detail will be discussed in Section 9.3. In practice, however, such g_n-anomalies are too small to measure in sufficient accuracy, whereas similar anomalies in hyperfine and fine-structure splittings are larger for accurate analyses.

9.2.2 The Hyperfine Structure Anomaly

The magnetic hyperfine Hamiltonian in a modulated crystal can be written as

$$\mathbf{H}_{\mathrm{HF}} = \beta\gamma\langle \boldsymbol{S}'_{\mathrm{n}}|\mathbf{A}|\boldsymbol{I}\rangle + \gamma\langle \boldsymbol{B}_{\mathrm{o}}|\boldsymbol{I}\rangle,$$

which expresses the nuclear spin motion in the combined field of the electronic spin and the external field $\boldsymbol{B}_{\mathrm{o}}$. Usually, the first term prevails in the hyperfine structure with a negligible contribution of the second term; hence, we consider only the first one for spectral analysis, where the tensor $\mathbf{A}$ can be redefined including the factor γ:

$$\mathbf{H}_{\mathrm{HF}} = \beta\langle \boldsymbol{S}_{\mathrm{n}}|\mathbf{g}.\mathbf{A}|\boldsymbol{I}\rangle,$$

and eigenvalues of $\mathbf{H}_{\mathrm{HF}}$ are given in the first order by

$$E_{\mathrm{HF}}{}^{(1)} = K_{\mathrm{n}}Mm,$$

where

$$K_{\mathrm{n}}^2 = \beta^2\langle n|\mathbf{g}^2.\mathbf{A}^2|n\rangle.$$

Here, the *hyperfine splitting factor* K_{n} is expressed in energy units, whereas for magnetic resonance experiments at a constant microwave frequency, it is expressed in field units for convenience; K_{n} in energy units divided by $g_{\mathrm{n}}\beta$. Replacing K_{n} by $g_{\mathrm{n}}\beta K_{\mathrm{n}}$ in the above expression, the first-order hyperfine energy can be written as

$$E_{\mathrm{HF}}{}^{(1)} = g_{\mathrm{n}}\beta K_{\mathrm{n}}Mm, \quad \text{where} \quad K_{\mathrm{n}}^2 = \langle n|\mathbf{g}^2.\mathbf{A}^2|n\rangle/g_{\mathrm{n}}^2.$$

Defining further $\mathbf{C} = (\mathbf{g}/g_{\mathrm{n}}).\mathbf{A}$ for brevity, the tensor $\mathbf{C}$ represents a directly measurable splitting from observed magnetic resonance spectra.

The modulated hyperfine splitting can then be expressed similar to the modulated $\mathbf{g}'$; that is,

$$\begin{aligned} K_{\mathrm{n}}'^2 &= \langle n|\hat{\mathbf{a}}.\mathbf{C}^2.\mathbf{a}|n\rangle \\ &= \langle n|\mathbf{C}^2|n\rangle + \sigma\langle n|\hat{\boldsymbol{\varepsilon}}.\mathbf{C}^2 + \mathbf{C}^2.\boldsymbol{\varepsilon}|n\rangle + \sigma^2\langle n|\hat{\boldsymbol{\varepsilon}}.\mathbf{C}^2\boldsymbol{\varepsilon}|n\rangle. \end{aligned}$$

In the critical region, we can write

$$K_{\mathrm{n}}'^2 - K_{\mathrm{n}}^2 = 2\sigma\langle n|\hat{\boldsymbol{\varepsilon}}.\mathbf{C}^2 + \mathbf{C}^2.\boldsymbol{\varepsilon}|n\rangle, \tag{9.7a}$$

for a binary splitting that is proportional to σ. Therefore, hyperfine anomalies in a binary phase transition can be related directly to fluctuations in cos and sin modes. As discussed for g anomalies, the sin-mode gives a featureless single line, whereas the cos-mode signifies distributed lines between two edges with a separation

$$\Delta K_n = K'_n - K_n = (\sigma_o/\underline{K}_n)\langle n|\hat{\boldsymbol{\varepsilon}}.\mathbf{C}^2 + \mathbf{C}^2\boldsymbol{\varepsilon}|n\rangle, \quad \underline{K}_n = \tfrac{1}{2}(K'_n + K_n). \quad (9.7b)$$

It is interesting that $\Delta K_n = 0$ when $\boldsymbol{B}_o$ is applied parallel to the axis of an antisymmetric tensor $\boldsymbol{\varepsilon}$, which may correspond to the axis for librational fluctuations of active groups, as will be discussed in the next section for the VO^{2+} spectra in BCCD crystals.

9.2.3 The Fine-Structure Anomaly

The modulated fine structure shows an anomaly of different type from Δg_n and ΔK_n. In the critical region, the former is observed in a quardratic form with respect to the direction $\boldsymbol{n}$, whereas in the latter, the square tensors $\mathbf{g}^2$ and $\mathbf{C}^2$ are involved in line splitting below the transition temperature.

In the strong-field approximation, the fine-structure Hamiltonian in a normal crystal can be written as $\langle \boldsymbol{S}_n|\mathbf{D}'|\boldsymbol{S}_n\rangle$, where $\langle \boldsymbol{S}_n| = S_n\langle \boldsymbol{n}|$. On the other hand, in a modulated crystal, the spin $\boldsymbol{S}$ of a probe is modified as $|\boldsymbol{S}'_n\rangle = \mathbf{a}|\boldsymbol{S}_n\rangle$ and, hence, the Hamiltonian is modified as

$$\mathbf{H}'_F = \langle \boldsymbol{S}'_n|\mathbf{D}'|\boldsymbol{S}'_n\rangle = \langle \boldsymbol{S}_n|\hat{\mathbf{a}}.\mathbf{D}'.\mathbf{a}|\boldsymbol{S}_n\rangle.$$

As previously mentioned, we use in the following the unprimed notation $\mathbf{D}$ for the fine structure tensor and write

$$\mathbf{H}'_F = \mathbf{H}_F + \sigma\langle \boldsymbol{S}_n|\hat{\boldsymbol{\varepsilon}}.\mathbf{D} + \mathbf{D}.\boldsymbol{\varepsilon}|\boldsymbol{S}_n\rangle + \sigma^2\langle \boldsymbol{S}_n|\hat{\boldsymbol{\varepsilon}}.\mathbf{D}.\boldsymbol{\varepsilon}|\boldsymbol{S}_n\rangle. \quad (9.8)$$

Normally, these fine-structure terms are greater than the hyperfine term, giving a dominant contribution to spectra of probes of $S > \frac{1}{2}$, where the modulation effects are described by those two terms proportional to σ and σ^2 in (9.8). Therefore, the first-order eigenvalues of $\mathbf{H}'_F$ are given by

$$E'_F{}^{(1)} = E_F{}^{(1)} + (a_n\sigma + b_n\sigma^2)M^2, \quad (9.9a)$$

where

$$a_n = \langle n|\hat{\boldsymbol{\varepsilon}}.\mathbf{D} + \mathbf{D}.\boldsymbol{\varepsilon}|n\rangle \quad \text{and} \quad b_n = \langle n|\hat{\boldsymbol{\varepsilon}}.\mathbf{D}.\boldsymbol{\varepsilon}|n\rangle.$$

Setting hyperfine interactions aside, the magnetic resonance condition in this case can be written as

$$h\nu = E'_n(M+1) - E'_n(M) = g'_n\beta B_o + (D_n + \Delta D_n)(2M+1), \quad (9.9b)$$

where

$$\Delta D_n = a_n\sigma + b_n\sigma^2,$$

indicating that the fine-structure anomaly is enhanced in transition lines at larger $\pm M$ as multiplied by the factor $2M+1$. Equation (9.9c) can be applied to any directions of $\boldsymbol{B}_o$, whereas for binary splitting we consider

$$\Delta D_n = 2a_n\sigma$$

just as discussed for Δg_n and ΔK_n, identifying the mirror plane geometrically by the shape of anomalies. In all directions other than in the mirror plane, transition anomalies in the magnetic resonance parameters are characterized by σ and σ^2.

In the latter case, the anomaly observed at $\pm M$ can be expressed for a cos mode as

$$\Delta\nu_n = \nu_1(n)\cos\phi + \nu_2(n)\cos^2\phi, \tag{9.10a}$$

where

$$\nu_1(n) = (2M+1)\sigma_o a_n/h \quad \text{and} \quad \nu_2(n) = (2M+1)\sigma_o b_n/h.$$

In this case, the intensity distribution is described by the density of resonance frequencies

$$\frac{1}{(d\phi/d\Delta\nu_n)} = \left[\left\{\frac{\nu_1^2}{2\nu_2} + \Delta\nu_n\right\}\{\nu_1^2 - (\nu_2 - \Delta\nu_n)^2\}\right]^{-1}, \tag{9.10b}$$

which is characterized by three singularities at $\nu_2 \pm \nu_1$ and $-\nu_1^2/2\nu_2$ as illustrated in Fig. 9.1b for $\nu_2 > \nu_1$. If, on the other hand, $\nu_2 < \nu_1$, the lineshape is similar to that in Fig. 9.1a.

Emerging at the threshold of the critical region, cos- and sin-modes are modified by local fields E of long-range correlations on decreasing temperature. For one-dimensional correlations, the sin mode will be stabilized by shifting the phase by $\frac{1}{2}\pi$ to join the cos mode, as discussed in Section 5.2, while such a cos-mode turns out to be an elliptical mode cnϕ', where ϕ' is temperature dependent while in the range $0 \le \phi' \le 2\pi$. Although the elliptical variation can be observed as distributed resonances between the edge frequencies, the anomalies are most appreciable in the sinusoidal variation at temperatures very close to T_c.

9.3 Structural Phase Transitions in TSCC and BCCD Crystals as Studied by Paramagnetic Resonance Spectra

The nature of structural phase transitions cannot be sufficiently analyzed by only one anomalous spectrum at T_c. Signified by the outset of collective pseudospin modes, transition anomalies should arise from a three-dimensional structural change in the lattice. Related to the loss of a symmetry element, the structural deformation should be explicit in magnetic resonance anomalies. However, the fine structure of spins larger than $\frac{1}{2}$ are often too complex to extract information of the collective pseudospin σ, unless allowed transitions are all identified in every direction of the applied field $\langle B_o|$. On the other hand, for $S = \frac{1}{2}$, it is necessary for the hyperfine tensor $\mathbf{A}$ to be anisotropic, in order to obtain structural information. Phase transitions in TSCC and *betaine calcium chloride dehydrate* (BCCD) crystals were thoroughly investigated systems by magnetic resonance spectra, where the modulated structures in the critical regions were analyzed in light of the condensate model. In TSCC crystals, the spectra show anomalies at the ferroelectric phase transition at 130K, giving a visual example for growing domains, in which the long-range field along the b axis was identified. On the other hand, BCCD crystals exhibit successive structural changes under ambient pressure, among which the modulated phase is characterized by the absence of long-range order. In this section, we summarize these results as examples for the modulated spin-Hamiltonian, providing a useful guideline for investigating transition processes in other systems.

9.3.1 The Ferroelectric Phase Transition in TSCC Crystals

(i) The Active Complex

TSCC crystals can be doped with a variety of paramagnetic impurities, such as Mn^{2+}, Cr^{3+}, Fe^{3+} and VO^{2+}, substituting for the Ca^{2+} ion at the center of a double pyramidal complex coordinated by six sarcosine molecules in near-tetrahedral symmetry, as illustrated in Fig. 3.4 [26]. Unlike divalent $Mn^{2+}$and VO^{2+}, trivalent impurities Cr^{3+} and Fe^{3+} accompany charge defects that are associated with one of the carboxyl groups of sarcosine molecules lying in the mirror plane, on which two long-range internal fields in opposite directions are related with reflection symmetry below T_c. On the other hand, the dipolar axes of VO^{2+} ions are associated with one of the six ligand sarcocines with an equal probability, giving an internal electric field at the Ca^{2+} site in the polar phase of TSCC crystals.

(ii) Breaking Mirror Symmetry at the Ferroelectric Phase Transition

Mn^{2+} impurities in TSCC at room temperature show relatively simple spectra under the magnetic resonance condition, where the spin-Hamiltonian consists of the Zeeman energy, the fine structure-energy, and the hyperfine energy with a ^{55}Mn nucleus ($I = 5/2$) in decreasing order of magnitudes. The $g_n$factor and the hyperfine structure due to ^{55}Mn ($I = 5/2$) are isotropic and, hence, not suitable for studying the structural change, whereas the fine-structure tensor is very anisotropic and, thus, useful for detecting the symmetry violation in the lattice at and below T_c. A simple resonance condition (8.26) can be applied in this case to determine D_n as a function of the direction $|n\rangle$ of B_o.

Figure 9.2 shows a representative Mn^{2+} spectrum from TSCC for $B_o \parallel a$. The angular dependences of D_n in three symmetry planes, ab, bc and ca, as observed at a temperature above T_c, are shown by solid curves in Fig. 9.3. On lowering the temperature below T_c however, the lines in the ab and bc planes split into two, as shown by dotted lines in Fig. 9.3, whereas in the mirror plane ca, the spectra remain virtually unchanged. The splitting in the bc plane is large among others, particularly in the direction 45° between the b and c, as indicated by 1-2, where the transition was observed with the

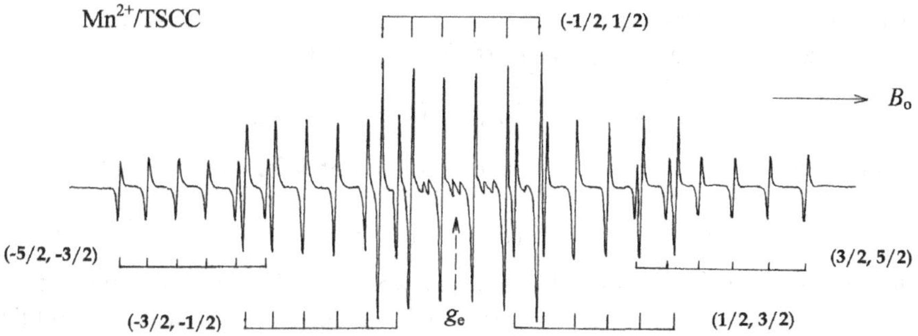

Fig. 9.2. A representative Mn^{2+} spectrum in TSCC, characterized by a well-resolved fine structure of five allowed transitions $M \to M+1$ with six hyperfine lines $\Delta m = 0$ on each.

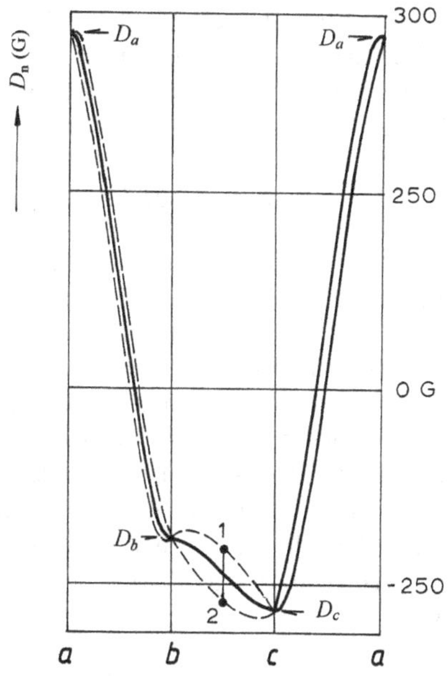

Fig. 9.3. Angular variations of the fine-structure splitting in Mn^{2+} spectra in TSCC below T_c. The anomaly was largest between the b and c axes, as indicated by $1-2$.

largest separation, whereas the splitting in the ab plane is small, and no such splitting was recognized in the mirror plane ac. The splitting in the bc and ab planes signifies two opposite condensates, which can be considered as "seeds" of domains related by the mirror reflection on the b plane, being consistent with the symmetry change at T_c. Noting that the splitting in the bc plane is sensitive to such a binary relation between $+\sigma_o$ and $-\sigma_o$, the temperature dependence was studied, particularly at 45° in the bc plane. Writing

$$\Delta D_{n+} = +a_n\sigma_o + b_n\sigma_o^2 \quad \text{and} \quad \Delta D_{n-} = -a_n\sigma_o + b_n\sigma_o^2,$$

the observed splitting given by $\Delta D_{n+} - \Delta D_{n-}$ can be expressed as entirely proportional to σ, as the terms of σ^2 are canceled [21]. Therefore,

$$\Delta D_{\rm n} = (\Delta D_{n+}) - (\Delta D_{n-}) = 2a_{\rm n}\sigma \tag{9.11}$$

represents critical fluctuations between condensates for $\pm\sigma$ if σ can be written as

$$\sigma = \sigma_{\rm o} \cos\phi \quad \text{and} \quad \sigma_{\rm o} \sin\phi$$

for the cos and sin modes of fluctuations, respectively, where $0 \le \phi \le 2\pi$.

Below $T_{\rm c}$, the sinusoidal variation of σ changes to elliptical, as modified by the internal field $E_{\rm int}$ of long-range correlations, although at the threshold, the sin mode can be stabilized by a weak external field E, converting it to a cos mode. The noncritical region is dominated by the phase mode modified by $E_{\rm int}$, which is expressed as elliptic $\sigma_{\rm o}{\rm sn}(\mu\phi/2^{1/2})$, where $0 \le \phi \le 2\pi$; $\sigma_{\rm o} \propto \kappa/(1+\kappa^2)^{1/2}$ and $\mu = 2^{1/2}/(1+\kappa^2)^{1/2}$. Here, the phase ϕ is not the same as in the critical region, varying as a function of temperature. Nevertheless, the timescale $t_{\rm o}$ of magnetic resonance, i.e. $t_{\rm o} = 2\pi/\omega_{\rm L}$, plays an essential role for sampling, and the observed splitting (9.11) should be expressed by the time average

$$\Delta D_{\rm n} = 2a_{\rm n}\sigma_{\rm o}\langle {\rm sn}(\mu\phi/2^{1/2})\rangle_{\rm t},$$

where

$$\langle {\rm sn}(\mu\phi/2^{1/2})\rangle_{\rm t} = t_{\rm o}^{-1}\int_0^{t_{\rm o}} {\rm sn}\{\mu(\Delta q x - \Delta\omega t)\}{\rm d}t. \tag{9.12}$$

The value of (9.12) is determined by the average phase $\bar\phi = \Delta qx - \Delta\omega.t_{\rm o}$ of distributed pseudospins, which does not vanish if $\Delta\omega.t_{\rm o} \le 1$. For such a phase mode, the magnetic resonance frequencies are distributed as $\nu_{\rm n} = A_{\rm n}\sigma(\bar\phi)$ in the range $0 \le \bar\phi \le 2\pi$, whose spatial distribution can be evaluated in terms of the distributed off-axis angle θ of a classical vector $\boldsymbol{\sigma}$ defined by $\sin\theta = {\rm sn}(\mu\bar\phi/2^{1/2})$. The broadened magnetic resonances can, therefore, be expressed as $\nu_{\rm n} = \nu_{1n}\sin\theta$, where $A_{\rm n}\sigma_{\rm o} = \nu_{1n}$ is called the edge frequency. By definition, we have

$$(\mu/2^{1/2}){\rm d}\bar\phi/{\rm d}\theta = (1-\kappa^2\sin^2\theta)^{-1/2},$$

Writing $\nu_{1n} - \nu_{\rm n} = \Delta\nu_{\rm n}$,

$$(\mu/2^{1/2})({\rm d}\bar\phi/{\rm d}\Delta\nu_{\rm n}) = 1/[\nu_{1\rm n}(1-\sin^2\theta)^{1/2}(1-\kappa^2\sin^2\theta)^{1/2}],$$

or

$${\rm d}\bar\phi/{\rm d}\Delta\nu_{\rm n} = \lambda\nu_{1\rm n}/[2^{1/2}(\nu_{1\rm n}^2 - \Delta\nu_{\rm n}^2)^{1/2}\{(\nu_{1\rm n}/\kappa)^2 - \Delta\nu_{\rm n}^2\}^{1/2}].$$

Therefore, in the limit $\kappa \to 1$, we obtain the density distribution given by

$$({\rm d}\bar\phi/{\rm d}\Delta\nu_{\rm n})_{\kappa\to 1} = \lambda\nu_{1\rm n}/2(\nu_{1\rm n}^2 - \Delta\nu_{\rm n}^2),$$

which is characterized by two edges $+\nu_{1n}$ and $-\nu_{1n}$. The distribution between the edges is less than the sinusoidal case, diminishing as $\kappa \to 1$, and the spectrum shows no appreciable difference from two independent lines in this limit.

For a very small κ in the critical region characterized by no appreciable long-range field, observed fine-structure anomalies ΔD_n show the symmetry-breaking fluctuations that can be evaluated with the integral

$$\langle \cos \bar{\phi} \rangle_t = t_o^{-1} \int_0^{t_o} (\cos(\Delta q.x - \Delta\omega.t + \phi_o) dt = \alpha(t_o) \cos(\Delta q.x - \Delta\omega.t + \phi_o),$$

where

$$\alpha(t_o) = \sin(\Delta\omega.t_o)/(\Delta\omega.t_o) \quad \text{and} \quad \underline{t} = t + \tfrac{1}{2} t_o.$$

Applying this to the transition $M = 3/2 \to 5/2$, for example, the edge frequency is expressed as

$$\nu_{1n}(t_o) = 4\alpha(t_o)\sigma_o a_n / h, \tag{9.13}$$

indicating that the edge separation depends on the timescale t_o of experiments [21]. At standard microwave frequencies, 9.2 and 35GHz, the edge separation of critical fluctuations is proportional to the factor $\alpha(t_o)$ in (9.13), which is determined by ω_L. Accordingly, the higher is Larmor frequency, the wider the edge separation. At these frequencies, critical anomalies in the Mn^{2+} spectra exhibited different separations in different shape as illustrated in Fig. 9.4. The characteristic frequency $\Delta\omega$ in the critical region is comparable with measuring Larmor frequencies ω_L, where the observed anomaly at 35GHz was characterized by a wider edge separation than at 9.2GHz.

The critical fluctuations are sinusoidal in the limit $\kappa \to 0$, and Fig. 9.5 shows examples of simulated spectra at small values of κ, which were numerically fitted to observed ones, allowing to estimate values of κ at temperatures close to T_c. Figure 9.6 demonstrates that the polar sin-mode (marked "a") in the critical region can be converted to a cos mode (marked p and p′) by applying a weak electric field E externally. We can therefore postulate that the sin mode diminishes spontaneously with increasing long-range field E_{int} below T_c.

(iii) The Long-Range Weiss Field in Ferroelectric TSCC.

The presence of an internal field of long-range order was not explicit in the Mn^{2+} spectra in the noncritical region. In VO^{2+}-doped crystals, the ferroelectric phase transition cannot be critical due to the electric field of dipolar impurity ions, thus, the transition is first order. Therefore, it is not surprising to see the absence of critical anomalies in the VO^{2+} spectra.

The ordering progresses with decreasing temperature, due essentially to the long-range dipolar field $\boldsymbol{E}_{dip}$ that behaves as if applied externally. In the presence of an applied field $\boldsymbol{F}$, the total internal field $\boldsymbol{F} + \boldsymbol{E}_{dip}$ should be considered to act on the pseudospin mode $\boldsymbol{\sigma}$ at a temperature below T_c and, therefore,

$$\lambda\boldsymbol{\sigma} = \boldsymbol{F} + \boldsymbol{E}_{dip}.$$

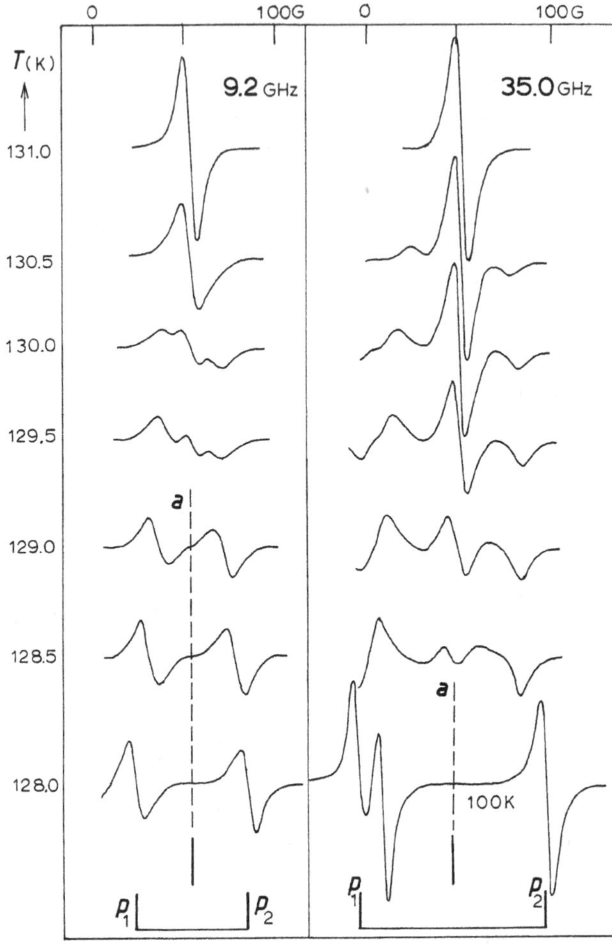

Fig. 9.4. Critical anomalies below T_c of the ferroelectric phase transition in TSCC observed from the separation between the hyperfin lines in the bc plane, i.e., $B_o \parallel (90°, 45°, 45°)$. At 9.2 and 35GHz, these anomalies showed different patterns.

For binary order in the ferroelectric phase of TSCC, this relation can be written for two domains as

$$\pm\lambda\sigma = F \pm E_{\text{dip}}.$$

For sampling the variables $\pm\sigma$ with dipolar VO^{2+} impurities, we can utilize the electric dipole moment μ of the VO^{2+} molecule, which is associated with $\pm\sigma$ at random positions. The dipole μ can be oriented by these internal fields $F \pm E_{\text{dip}}$ at energies $-\mu(F \pm E_{\text{dip}})$ with the Boltzmann probabilities $\exp\{\mu(F \pm E_{\text{dip}})/k_B T\}$, respectively. In TSCC crystals, the VO^{2+} spectra showed splitting into two lines below T_c with the intensity ratio varying with

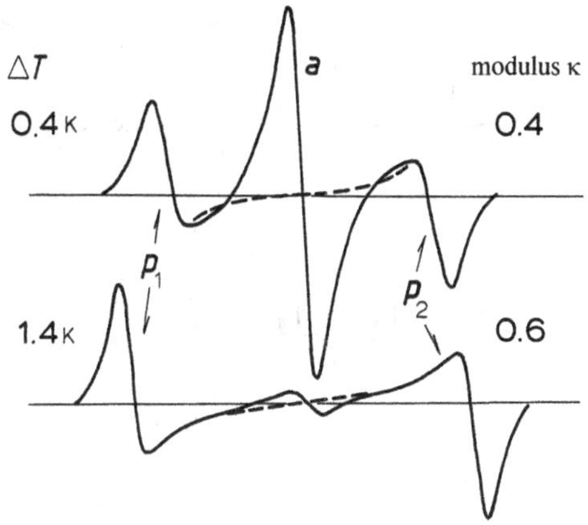

Fig. 9.5. Anomalous lineshapes numerically simulated by the equation $\sigma = \sigma_o \lambda \mathrm{sn}(\mu\phi/2^{1/2})$ with $\kappa = 0.4$ and 0.6. These lines could be fitted to observed ones at temperatures $\Delta T = 0.4$ and 1.4K, respectively.

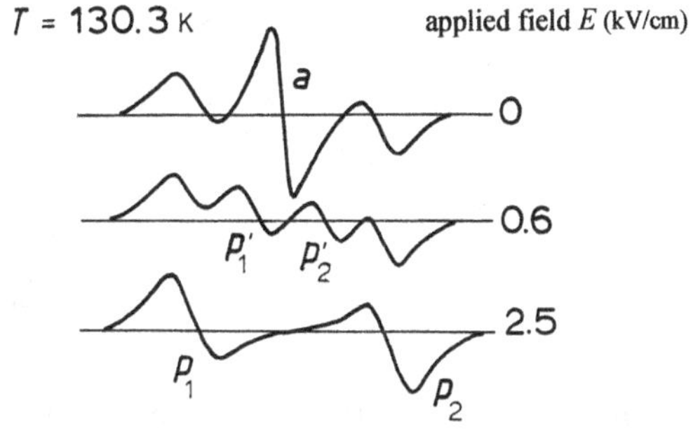

Fig. 9.6. Effects of a weak applied electric field on an observed anomalous line at 130.3K, showing that the central amplitude mode (marked a) splits into two (marked p_1' and p_2') with applying a weak electric field $E = 0.6\mathrm{kV/cm}$, joining into the outside components p_1 and p_2 in phase mode on increasing E to 2.5kV/cm.

temperature, as shown in Fig. 9.7a. We can, thus, interpret the observed intensity ratio as determined by E_{dip}. Under the noncritical condition, this assumption is consistent with the statistical theory of binary order, as discussed in Chapter 2. In this analysis, the intensity ratio $r_{\mathrm{n}} = I_-/I_+$ between these lines of VO^{2+} spectra below T_{c} is given by

$$r_{\mathrm{n}} = \exp \frac{\mu(F + E_{\mathrm{dip}})}{\mathrm{k_B}T} \Big/ \exp \frac{\mu(F - E_{\mathrm{dip}})}{\mathrm{k_B}T} = \exp \frac{2\mu E_{\mathrm{dip}}}{\mathrm{k_B}T},$$

therefore,

$$|E_{\mathrm{dip}}| \propto |T \ln r_{\mathrm{n}}|. \tag{9.14}$$

Using (9.14), the temperature dependence of E_{dip} can be evaluated experimentally from r_{n} for any direction of $\boldsymbol{B}_{\mathrm{o}}$ as a function of temperature, and

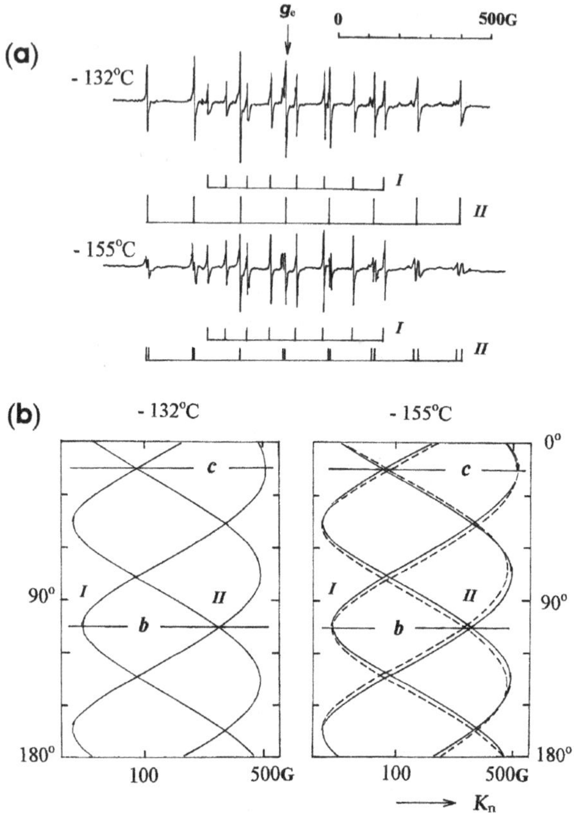

Fig. 9.7. (a) Representative VO^{2+} spectra for $B_o \parallel b$ in paraelectric and ferroelectric phases of TSCC. The spectrum is composed of types I and II in different symmetries, and the spectrum II splits below T_c into two. (b) Angular variation of the ^{51}V-hyperfine splitting K_n for B_o in the bc plane of TSCC.

Fig. 9.8 shows the results from TSCC. Using VO^{2+} probes and NH_3^+ radicals, the presence of E_{int} was also confirmed in other systems, for example, ferroelectric *triglycine sulphate* (TGS) crystals [71] and the *ferrielectric* phase of $(NH_4)_2SO_4$ crystals [72, 73], respectively. For binary ferroelectric systems, we can generally write $|E_{int}| \propto (T_c - T)^{1/2}$ in the mean-field approximation and, hence, the internal field related to long-range correlations can be regarded as a valid concept substantiated by resonance experiments [74].

In the following, we review the published magnetic resonance spectra of VO^{2+} impurities. The ground state of the unpaired $(3d)^1$-electron in a VO^{2+} ion is an S-state ($L = 0$), and the g_n factor is anisotropic in the uniaxial structure, and there is no fine structure because of the spin $S = \frac{1}{2}$. On the other hand, the hyperfine interaction with the ^{51}V nucleus ($I = 5/2$) is very anisotropic, reflecting uniaxial symmetry of the diatomic molecule. The mag-

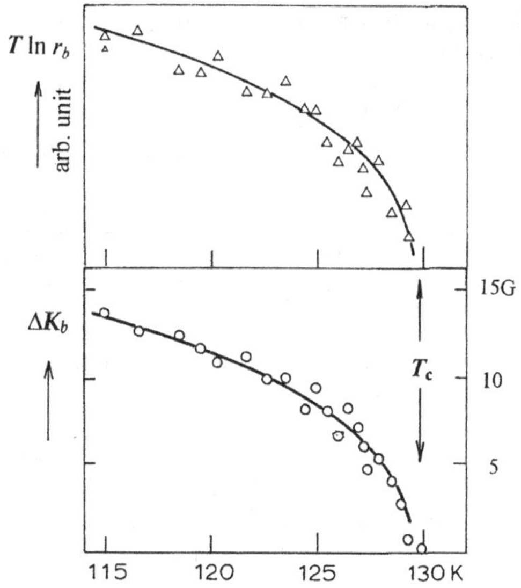

Fig. 9.8. Temperature dependences of $T \ln r$ and ΔK_b below T_c in VO^{2+}-doped TSCC.

netic resonance condition can therefore be expressed as

$$\hbar\omega = g_n\beta B_o + g_n\beta K_n M m,$$

or in terms of resonance fields $B_n = \hbar\omega/g_n\beta$

$$B_n(M) = B_o + K_n M m.$$

Figure 9.7a shows representative spectra of VO^{2+} at temperatures above and below T_c, where all allowed transitions for $\Delta M = \pm 1$ and $\Delta m = 0$ are indicated by the "stick" diagram. In these electronic transitions, nuclear magnetic quantum numbers $m = -5/2, -3/2, \ldots, 5/2$ are unchanged, but the hyperfine separations are unequal due to the second-order contributions. The value of K_n can be determined from the separations between adjacent hyperfine lines, where the second-order contributions proportional to $(K_n m/B_o)^2$ are canceled between the lines for $+m$ and $-m$.

In the spectrum at $-155°C$, additional lines appeared as related to progressing binary order. Figure 9.7b shows the anisotropic hyperfine splitting K_n observed from a crystal rotated in the bc plane, in which the splitting is plotted against angles of rotation in two kinds of VO^{2+} spectra marked I and II showing different symmetries. The overall pattern (of the angular dependence) is near trigonal, and the observed spectra are consistent to the molecular arrangement in Fig. 3.4a. By symmetry, we can consider that the VO^{2+} spectra of type I are equivalent to Mn^{2+} spectra, whereas no analogy was found between the VO^{2+} of type II and Mn^{2+}.

Table 9.1. ^{51}V hyperfine tensors for VO^{2+} complexes observed at 22.5°C in annealed crystals of TSCC

Spectrum	Principal values[a]	Direction cosines[b]
II	185.0	(0.624, ±0.673, ±0.396)
	70.0	(0.329, ±0.233, ∓0.915)
	68.5	(0.710, ∓0.701, ±0.072)
I	190.0	(0.622, 0, ± 0.783)
	74.6	(0, ±1, 0)
	68.1	(−0.783, 0, ±0.622)

[a] unit in gauss, [b] reference axes: orthorhombic a, b, and c.

To determine the hyperfine tensors $\boldsymbol{K}$, we require angular dependences of K_n in the other planes ac and ab as well. In Fig. 9.9 the results are summarized together with that in the bc plane for numerical calculation. Table 9.1 shows the results for the ^{51}V hyperfine tensors of spectra I and II at room temperature, which are both characterized by unique axes for the largest principal values of almost equal magnitudes. The principal values were virtually unchanged, whereas their directions changed with temperature. It is reasonable to assume that such unique directions represent the dipole axis of VO^{2+}(I) and VO^{2+}(II) at different sites.

Below T_c, these spectra showed small additional splittings ΔK_n, as seen in the spectra in Fig. 9.7a. It is realized that ΔK_n and the intensity ratio r_n are both temperature dependent, showing practically identical variations as

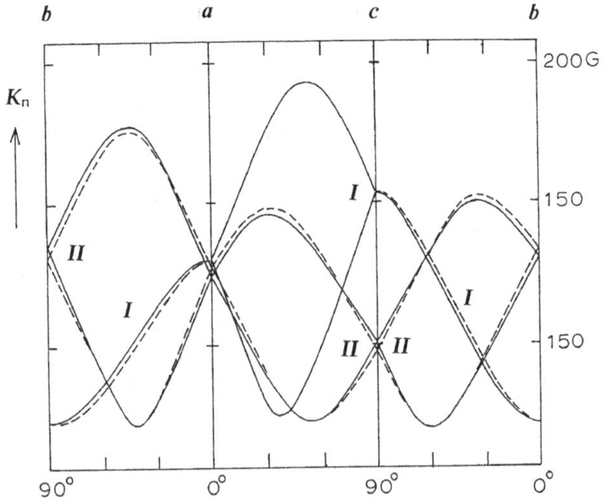

Fig. 9.9. Angular dependences of the ^{51}V hyperfine splitting K_n in VO^{2+} spectra I and II in all symmetry planes of TSCC.

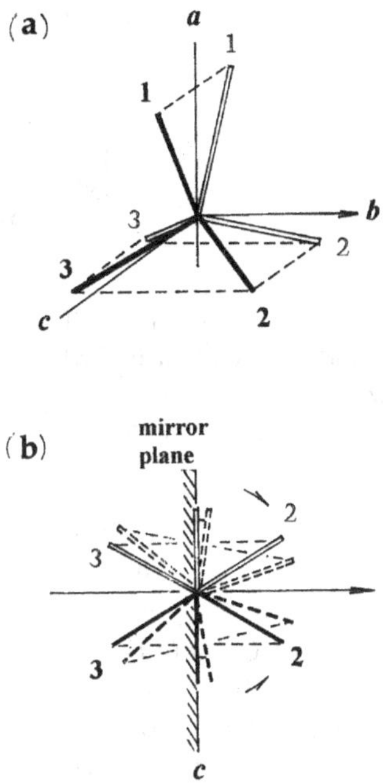

Fig. 9.10. Unique directions of the VO^{2+} complex in TSCC, as determined from VO^{2+} spectra in TSCC: (a) A view in the symmetry frame of reference above T_c. (b) Changes in the ferroelectric axes below T_c: a view with respect to the mirror plane.

proportional to $(T_c - T)^{1/2}$ within experimental errors, as shown in Fig. 9.8, which is consistent with the mean-field theory of a binary variable. Furthermore, with decreasing temperature below T_c, these unique hyperfine directions shifted by a small angle θ toward the crystallographic b direction, on which the internal field E lies, allowing one to consider that VO^{2+} dipoles are in the potential $-\mu E \cos\theta$, as illustrated in Fig. 9.10. The temperature dependence of the hyperfine anomaly ΔK_n can therefore be interpreted geometrically as proportional to $\mu \cos\theta = \lambda\sigma_o \mathrm{cn}\{\phi/(1+\kappa^2)^{1/2}\}$, while $\kappa = 1$ in the noncritical region, where the angle θ is related to local strains in the lattice. Such distributed local strains signify two opposite "microdomains" or small ordered regions during the ordering process.

9.3.2 Structural Phase Transitions in BCCD Crystals

Orthorhombic crystals of (betaine)-$CaCl_2 2H_2O$ known as BCCD, where the *betaine* is an organic amino acid $(CH_3)_3NCH_2COOH$, exhibit successive structural phase transitions under the atmospheric pressure in the range between the normal phase above 164K and the ferroelectric phase below 45K. Rother and his coworkers [75] first observed dielectric anomalies at these transitions,

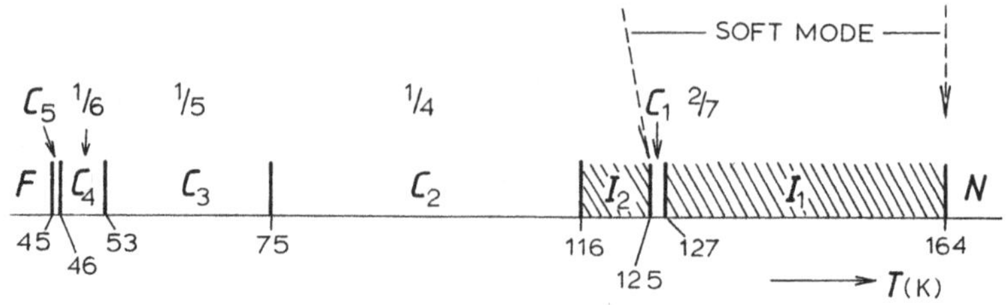

Fig. 9.11. Successive phase transitions in BCCD crystals: phases I_1 and I_2 are incommensurate, and soft modes were identified when the transition theresholds are approached from above, whereas C_1, C_2, C_3 and C_4 are commensurate, for which values of fractional parameters δ are indicated.

suggesting that a polar mechanism is involved in these sequential structural changes. Brill and Ehses [76] carried out an X-ray study on BCCD crystals and found that there are seven modulated phases between 164 and 45K, which are characterized by the wavevector $Q_c = \delta(T)c^*$, where c^* is the unit translation in the reciprocal lattice. Figure 9.11 illustrates the successive phase transitions in BCCD crystals, where values of the parameter $\delta(T)$ are indicated as suggested by the X-ray results of [76]. Using a submillimeter technique, Volkov and his collaborators [77] observed underdamped soft modes near the threshold of these incommensurate phases, which were later confirmed by Ao and Schaack [78] as of B_{2u} symmetry by their infra-red measurements. The molecular arrangement in BCCD at room temperature was determined by Brill, Schildkamp and Spilker [79], who reported the presence of appreciable librational fluctuations of Ca-(betaine)$_2$ complexes in the b plane. Figure 9.12 shows the unit-cell structure sketched after their X-ray results.

According to the X-ray results [79], the Ca-(betaine)$_2$ complex in BCCD crystals is asymmetrical with regard to two -COOH groups of ligand betaine molecules: O(1) in one of the betaines and O(2) in the other are not related by inversion. If the asymmetrical structure fluctuates between these two betaines as related by inversion in the b plane, these complexes along the b axis should be correlated with each other, resulting in collective fluctuations during the structural change to the next phase. Evidenced by VO^{2+} sampling, the active group is predominantly in slow librational motion between inversion-related structures.

Placing a pseudospin variable at each Ca^{2+} site, we can consider a chain of these variables in a BCCD crystal, which is similar to TSCC, although the correlation directions for J_d and J_c may not be exactly on the bc plane, as inferred from the diagram in Fig. 9.12. However, assuming that the correlations in the bc plane are the primary mechanism, we can simplify the correlation scheme for the normal-to-incommensurate phase transition ($N \rightarrow I_1$). Although unverified yet, out-of-plane correlations associated with J_d may be significant for subsequent transitions below phase I_1. Assuming such a model, the modulated pseudospin mode is characterized by an irrational wavevector

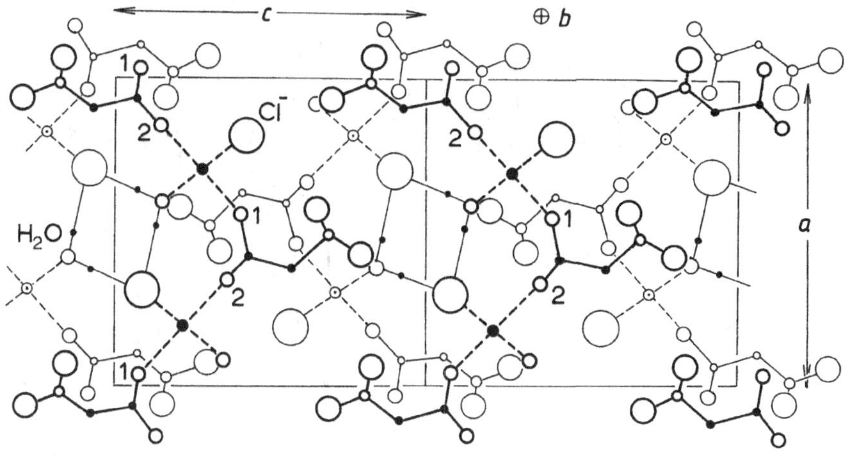

Fig. 9.12. Molecular arrangement in BCCD crystals in the *ac* plane. Betaine molecules are planar, lying in the *ac* plane. Each Ca^+ ion is coordinated by O(1) and O(2) of the ligand betaines and two Cl^- and two OH_2, where the two protons are out of the *ac* plane.

k_c determined by the relation $\cos(k_c c) = -J_d/J_c$ if $|-J_d/J_c| < 1$, which is considered as temperature-dependent. We can then write the relation

$$k_c = \{1 - \delta(T)\}c^* \quad \text{where} \quad Q_c = \delta(T)c^*.$$

(i) VO^{2+} Spectra

Diatomic VO^{2+} molecules are useful probes for studying the behavior of pseudospins in BCCD at least at the first transition from the normal phase N to the incommensurate I_1. The transition from N to I_1 was clearly observed in VO^{2+}-doped crystals; however, subsequent transitions were not identified in the spectra, which could be attributed to a lattice modification by the electric dipole moment of VO^{2+}. Nevertheless, binary fluctuations in phase I_1 were clearly identified in the VO^{2+} spectra by typical magnetic resonance anomalies. In fact, no critical region was found in VO^{2+}-doped TSCC crystals, whereas in the incommensurate phase I_1 of BCCD crystals, anomalous VO^{2+} lines were observed at all temperatures, indicating that there is no significant long-range internal electric field in this phase.

Figure 9.13 shows representative VO^{2+}spectra in the magnetic resonance at 9.2 and 35GHz, observed at temperatures above and below the critical temperature $T_i = 164$K. Considerably different lineshapes observed at these microwave frequencies can be attributed to the fluctuation frequency $\Delta\omega$ observed differently at these Larmor frequencies ω_L. It is also noticed that intensities and linewidths of eight hyperfine components exhibit an appreciable variation from one line to another even at temperatures above T_i, depending

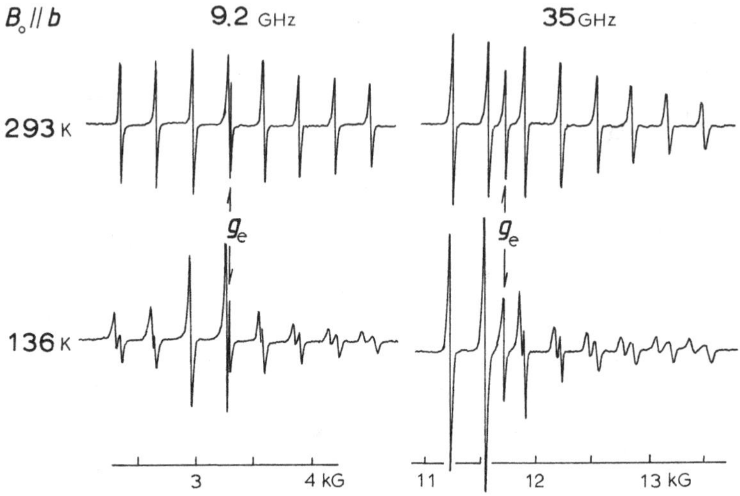

Fig. 9.13. Representative VO^{2+} spectra in BCCD for $B_o \parallel b$, observed at 9.2GHz and 35GHz in the normal and the first incommensurate phases.

on the quantum number m of the ^{51}V nucleus. Such a variation can be attributed to random librational fluctuations of betaine molecules, as discussed by Brill et al. with their X-ray results at room temperature, turning into binary fluctuations below T_i.

In the normal phase, the **g** and ^{51}**A** tensors were coaxial with well-defined principal values [80, 81]. Crystallographically, there are four independent unique directions for a VO^{2+} ion to take in the unit cell, which are symmetrical with respect to the b direction. Therefore, VO^{2+} impurities exhibit two magnetically equivalent spectra, when B_o is applied in the symmetry planes (100), (010) and (001), whereas for $B_o \parallel b$, the four complexes showed identical spectra, as shown in Fig. 9.13. Below T_i, the anomalous line is characterized by symmetrically separated edge frequencies, between which the resonance frequencies are distributed sinusoidally. Observed anomalies in BCCD were temperature dependent, as shown in Fig. 9.14, where increasing short-range correlations were recognized to some extent, but not as significant as in ferroelectric TSCC.

At the resonance fields given by $B_n(M) = B_o + K_n m$, the spectral lines are broadened as

$$\Delta B_n(m) = \Delta B_o + m\Delta K_n,$$

where $\Delta B_o = -(\Delta g_n/g_n)B_o$ and $m\Delta K_n$ is the hyperline anomaly. Therefore the edge separations at hyperfine lines are not the same but different, at different nuclear quantum number m. Replacing ΔB_o by $(\Delta g_n/g_n)B_o$, we have

$$\Delta B_n(m) = -(B_o/g_n)\Delta g_n + m\Delta K_n. \tag{9.15}$$

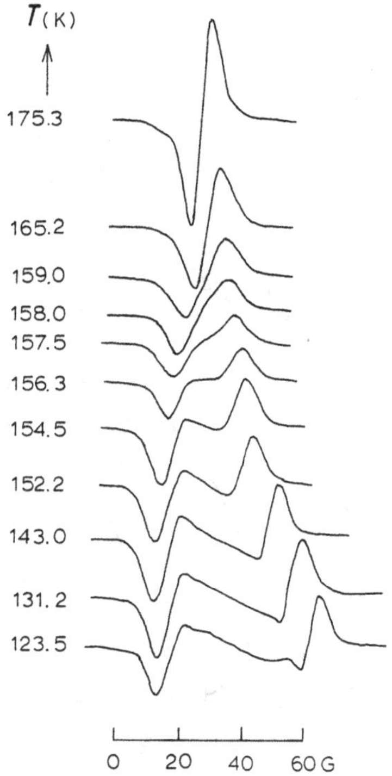

Fig. 9.14. Anomalous temperature-dependent splitting of a hyperfine line of VO^{2+} spectra in the incommensurate phase of BCCD.

Therefore, the edge separation is contributed by fluctuations Δg_n and ΔK_n. Figure 9.15 illustrated such anomalies, where the stick diagram shows how Δg_n and ΔK_n contribute to $\Delta B_n(m)$ on these hyperfine lines.

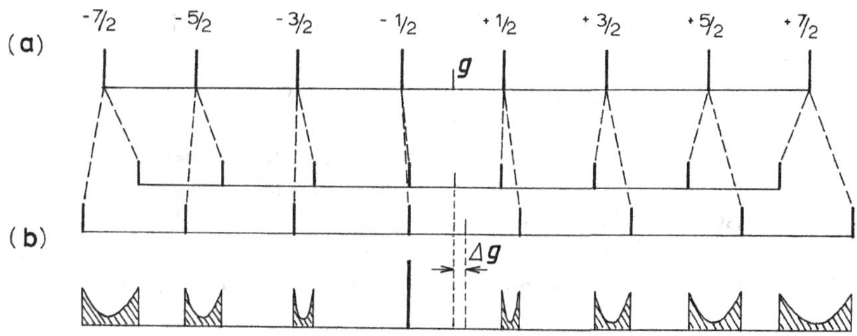

Fig. 9.15. (a) Interpretation of the anomalous hyperfine structure of VO^{2+} spectra from BCCD. Illustrated are anomalies Δg_n and ΔK_n, resulting in broadenings sketched in (b) at the bottom.

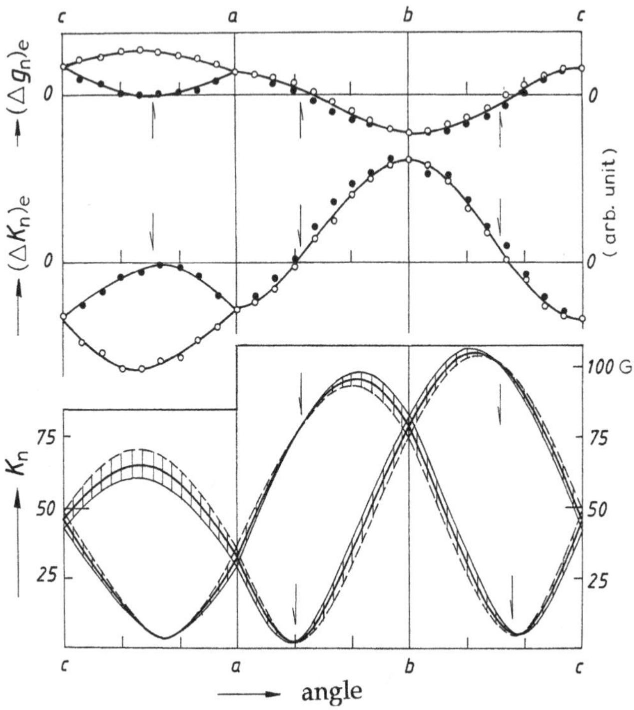

Fig. 9.16. Angular variations of K_n, $(\Delta K_n)_{edge}$ and $(\Delta g_n)_{edge}$ determined by the equation (9.14) in the symmetry planes of BCCD at 130K.

In the VO^{2+} spectra in phase I_1 of BCCD, angular variations of K_n, Δg_n and ΔK_n can be calculated from spectra analyzed as in Fig. 9.15, and plotted separately, as shown in Fig. 9.16. It is interesting to note that there are directions n for Δg_n and ΔK_n to change signs in these projections, which are consistent in three planes of rotation within experimental errors. Such a specific direction $\boldsymbol{n}$ for no anomalies can be interpreted as the librational axis in phase I_1. Table 9.2 shows the magnetic resonance data of VO^{2+} in BCCD obtained at 170K and at 9.2GHz [80, 81], in which calculated librational directions are shown.

In liquids, free VO^{2+} ions are tumbling in random fashion, from which Rogers and Pake [82] observed unequal hyperfine intensities similar to VO^{2+} spectra in BCCD. In liquids, lines were symmetrically broadened, whereas the VO^{2+} spectra in phase I_1 exhibited an anomalous lineshape. The librational axis identified in phase I_1 should be associated with the symmetry change in crystal.

Table 9.2. The g and ^{51}V hyperfine tensors of VO^{2+} in the first incommmensurate phase of BCCD at 140K.

Principal values[a]		Direction cosines[b]
g	K	
2.0315	135	$(+0.559, \mp0.581, \pm0.592)$ $(-0.559, \mp0.581, \pm0.592)$
1.9538	75	$(-0.511, \mp0.624, \mp0.591)$ $(+0.511, \mp0.624, \mp0.591)$
1.9623	22	$(-0.831, \mp0.324, \pm0.453)$ $(+0.831, \mp0.324, \pm0.453)$
Librational axes		$(+0.600, \pm0.375, \pm0.707)$ $(-0.600, \pm0.375, \pm0.707)$

[a] Unit for K in gauss. [b] Reference axes: orthorhombic a, b and c.

(ii) Transition Anomalies in Mn^{2+} Spectra

The Mn^{2+} spectra in BCCD crystals are dominated by a large fine structure in the high-field condition, whereas the ^{55}Mn hyperfine interaction is isotropic. In such a case, it is difficult to carry out a precision analysis on allowed transition lines because of a large number of forbidden lines in off-symmetry directions of the applied field. On the other hand, the principal directions of the **D** tensor can be easily identified, as characterized by the absence of forbidden lines, where the phase transition anomalies can be studied. Figure 9.17 shows a representative Mn^{2+} spectrum when the direction of B_o was close to the unique tensor axis, where allowed transitions $\Delta M = \pm 1$ are predominant, consisting of six equally spaced ^{55}Mn hyperfine lines on each. Although weak in intensities, forbidden lines appear in the central part of the spectrum, allowed transition lines from one of the impurity sites are widely spread out, whereas those from the other site are squeezed toward lower fields.

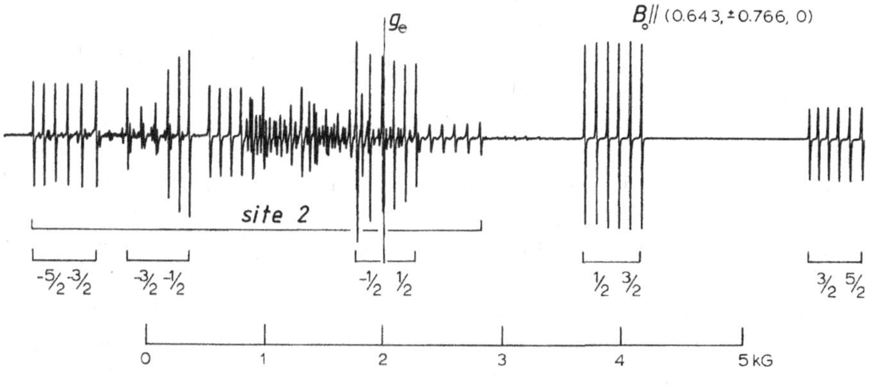

Fig. 9.17. A Mn^{2+} spectrum from BCCD at room temperature, when B_o is applied in the direction 45° from the a axis in the ab plane, showing all allowed transitions are well resolved from one impurity site, whereas lines from the other are crammed in the area of site 2.

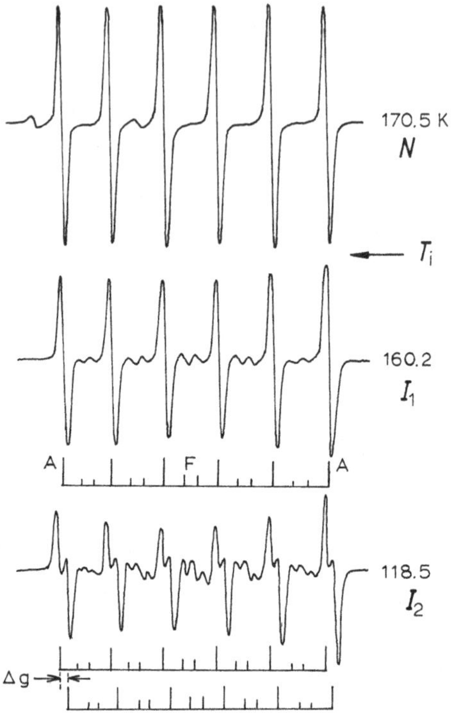

Fig. 9.18. The center group of allowed transitions $(1/2, -1/2)$ in the Mn^{2+} spectra in N, I_1 and I_2 phases of BCCD, where the commensurate phase C_1 was not distinguishable in the spetra.

In the former Mn^{2+} spectrum, those allowed lines at $m = \pm 5/2$ at the highest and lowest fields are well isolated from forbidden lines, permitting one to observe phase transition anomalies.

Also significant is that a change in the librational fluctuations at T_i can directly be detected in the central transitions $(-\frac{1}{2} \rightarrow +\frac{1}{2})$, although unrelated to the fine-structure splitting D_n. Figure 9.18 illustrates such allowed lines at the center of the spectrum observed at temperatures above and below T_i, which are compared with those in the second incommensurate phase I_2. At 160.2K, emerged forbidden lines are clearly resolved between strong allowed lines, indicating that the librational motion of the Mn^{2+} complex is slowed down below T_i, fluctuating between binary states of rotation, which is further slowed down exhibiting noticeable shifts $\pm\Delta g$ in phase I_2. As indicated by the stick diagram, the fluctuation amplitude is virtually unchanged in the transition $I_1 \rightarrow I_2$.

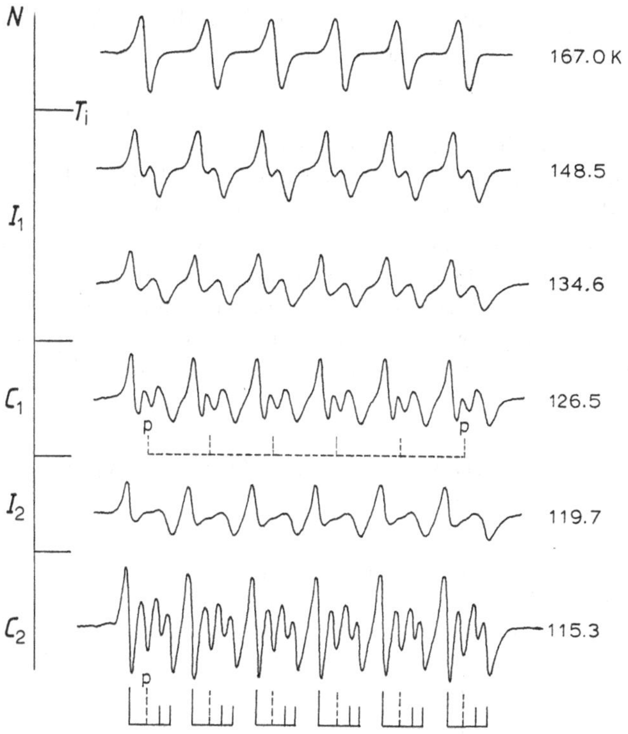

Fig. 9.19. The low-field group of allowed Mn^{2+} transitions $(-3/2, -5/2)$ in successive phases of BCCD.

(iii) Anomalies in the Low-Field Mn^{2+} Lines $(-5/2 \rightarrow -3/2)$ for $B_o \parallel c$

Although exhibiting complicated spectra, properties of Mn^{2+}-doped BCCD crystals are not much modified as in VO^{2+}-doped crystals, because all the successive phase transitions below T_i are identifiable in the Mn^{2+} spectra. These successive transitions were recognized in the electronic resonance $(-5/2 \rightarrow -3/2)$ in low fields, which is isolated from other lines in the direction $B_o \parallel c$. Figure 9.19 shows a variety of lineshapes observed in the successive phases with decreasing temperature. It is noted that six ^{55}Mn hyperfine lines in the normal phase N begin to show an anomalously broadened fine structure, exhibiting successively different shapes from one phase to another below T_i. In the incommensulate phases I_1 and I_2, the lineshape observed in this direction of B_o is similar to the simulated line in Fig. 9.1b, signifying fluctuations of the type expressed by (9.10a), arising from terms proportional to σ and σ^2, where the phase ϕ is continuously distributed in the range $0 \leq \phi \leq 2\pi$. The edge distribution in phase I_2 appear to be slightly better resolved than in I_1, although similar in these two phases. On the other hand, phases C_1 and C_2 are commensurate, where ϕ is locked in discrete values specified by $\phi = Q_c x$, where $Q_c = \delta(T)c^*$ and $x = c \times$ integer. According to Brill and Ehses [76], $\delta(C_1) = 2/7$, $\delta(C_2) = \frac{1}{4}$, and hence, these discrete phases are $\pi/7$, $2\pi/7$, $3\pi/7$, $4\pi/7$ and $5\pi/7$ in C_1, and $\pi/4$ in C_2 between edges 0 and π of symmetrical discommensuration lines, and there should be a single line (p) of asymmetrical fluctuations. In the magnetic resonance spectra so far analyzed, each hyperfine line was noticeably resolved into multiple lines, but the resolution was not sufficient to verify the theoretical consequences, for which we need to look into other directions of B_o for further studies. Nevertheless, transitions from I_1 to C_1 and from I_2 and C_2 are incommensurate-to-commensurate phase transitions, for which different types of collective mode should be respnsible. The Mn^{2+} spectra below 100K were even more complicated, suggesting that the impurity complex is locked into further restricted mechanisms.

9.4 Nuclear Quadrupole Relaxation in Incommensurate Phases

Nuclear spin probes can also be used for sampling pseudospin condensates for structural phase transitions, because a nuclear spin $I > \frac{1}{2}$ can interact with the lattice via the quadrupole energy. The quadrupole interaction can be expressed as $\mathbf{H}_Q = \langle I|\mathbf{QT}|I\rangle$, where $\mathbf{Q}$ is the nuclear quadrupole moment tensor and $\mathbf{T} = \nabla \boldsymbol{E}$ is the electric field gradient tensor at the nuclear site. In a modulated crystal, the nuclear spin in a condensate should be modified as $|I'\rangle = \mathbf{a}|I\rangle$, where $\mathbf{a} = \mathbf{1} + \sigma\boldsymbol{\varepsilon}$, and so the quadrupole energy is modulated as $\mathbf{H}'_Q = \langle I'|\mathbf{QT}|I'\rangle$, which can be written as

$$\mathbf{H}'_Q = \mathbf{H}_Q + \sigma\langle I|\hat{\boldsymbol{\varepsilon}}\mathbf{QT} + \mathbf{QT}\boldsymbol{\varepsilon}|I\rangle + \sigma^2\langle I|\hat{\boldsymbol{\varepsilon}}\mathbf{QT}\boldsymbol{\varepsilon}|I\rangle.$$

With this $\mathbf{H}'_Q$, the threshold of a binary phase transition at T_c should be described by antisymmetric fluctuations for breaking symmetry $\sigma \to -\sigma$ in a specific direction of the applied field, as well as symmetric fluctuations of a pinned σ.

In breaking inversion symmetry, the anomaly should be described by

$$\Delta\mathbf{H}_Q = \mathbf{H}'_Q(+\sigma) - \mathbf{H}'_Q(-\sigma) = 2\sigma\langle I|\hat{\boldsymbol{\varepsilon}}(\mathbf{QT} + (\mathbf{QT})\boldsymbol{\varepsilon}|I\rangle = 2\sigma\langle I|\mathbf{Q}|I\rangle(\hat{\boldsymbol{\varepsilon}}\mathbf{T} + \hat{\mathbf{T}}\boldsymbol{\varepsilon}),$$

where the nuclear quadrupole tensor $\mathbf{Q}$ is symmetrical, and therefore, separated in the last expression from the other tensors that are related to the surroundings. The gradient tensor $\mathbf{T}$ in a crystal is also symmetrical, and, hence, these product tensors of $\boldsymbol{\varepsilon}$ and $\mathbf{T}$

$$\sigma(\hat{\boldsymbol{\varepsilon}}\mathbf{T} + \mathbf{T}\hat{\boldsymbol{\varepsilon}}) = 2\sigma\bar{\boldsymbol{\varepsilon}}\mathbf{T}, \quad \text{where} \quad \bar{\boldsymbol{\varepsilon}} = \tfrac{1}{2}(\hat{\boldsymbol{\varepsilon}} + \boldsymbol{\varepsilon}),$$

represents the distorted field gradient at the nuclear site during ordering process. It is noticed that such transition anomalies can be significant only if $\boldsymbol{\varepsilon}$ is a symmetrical tensor, for which $\bar{\boldsymbol{\varepsilon}} \neq 0$. Otherwise for an antisymmetric $\boldsymbol{\varepsilon}$, and no quadrupole anomalies can be expected in this case.

In nuclear magnetic resonance experiments, the nuclear moment is quantized along the direction $|n\rangle$ of an applied field B_o, if the Zeeman energy $\mathbf{H}_Z = \gamma\hbar B_o$ is greater than $\mathbf{H}'_Q$, otherwise $|I\rangle$ should be quantized along the nuclear quadrupole axis. In a pure quadupole resonance for $I > \frac{1}{2}$, $\mathbf{H}'_Q$ is the dominant term, and radiation energies are absorbed by the nucleus via non-vanishing elements of $\langle m|I|m'\rangle$, where $m' - m = \Delta m = \pm 1$ and ± 2. The transitions $\Delta m = \pm 1$ and ± 2 occur when the r.f. magnetic field B_1 is applied perpendicular to the static field B_o, as in ordinary magnetic resonance at a high field. It is a common practice that the pure quadrupole resonance (PQR) experiments are performed in a weak applied field B_o, to observe these magnetic transitions for $\Delta m = \pm 1$ and ± 2, as identified by nuclear magnetic quantum numbers m.

Further, to detect anomalies $\Delta\mathbf{H}'_Q$, the critical fluctuations σ should be in symmetrical cos mode, for which $\omega_L \leq \Delta\omega$. As discussed for paramagnetic resonance, the sin mode of fluctuations is also detectable in principle, but is indistinguishable from the resonance line from unmodulated crystals.

In the condensate model, it is significant to deal with energy exchanges between the collective pseudospin mode and the corresponding lattice excitation. When the pseudospin mode is subjected to a nuclear quadrupole resonance experiment, the absorbed radiation energy should be exchanged with an excitation in the lattice counterpart, acting as the "primary" energy sink. It is noted that such an energy transfer should be recognized as the nuclear quadrupole relaxation through the tensor $\mathbf{T}' = \boldsymbol{\varepsilon}\mathbf{T} + \mathbf{T}\boldsymbol{\varepsilon}$, where the nuclear spin relaxation time T_1^* should be anomalous and shorter than the ordinary spin-lattice relaxation time T_1 to the surroundings. In nuclear magnetic resonance spectroscopy, it is a common practice to measure such a relaxation time at resonance conditions.

The quadrupole interaction arises from the electrostatic interaction of a charge cloud deformed by a field gradient and, hence, the product tensor $\mathbf{QT}'$ should be traceless, and $\Delta\mathbf{H}'_Q$ is a quardratic form of components of the spin vector $|I\rangle$. Expressing the spin vector by components I_z and $I_x \pm iI_y$ in the frame of reference fixed in the crystal, we can write

$$\Delta\mathbf{H}'_Q = \sigma \sum\nolimits_{\Delta m} T'_{-\Delta m} Q_{\Delta m},$$

where corresponding to transitions $\Delta m = 0$, ± 1 and ± 2 for $B_o \parallel z$, we have expressions

$$T'_o = T'_{zz}, \quad T'_{\pm 1} = T'_{xz} \pm iT'_{yz}, \quad T'_{\pm 2} = \tfrac{1}{2}(T'_{xx} - T'_{yy}) \pm iT'_{xy},$$
$$Q_0 = A(3I_z^2 - I^2), \quad Q_{\pm 1} = A(I_z I_\pm + I_\pm I_z), \quad Q_{\pm 2} = I_\pm^2$$

and

$$A = e^2 Q/4I(2I-1).$$

Here, the parameter Q is known as the *nuclear quadrupole moment.*

The nuclear spin energy contributed by the perturbation $\Delta\mathbf{H}'_Q$ in a crystal is secular in the second order, as these $Q_{\Delta m}$ multiplied by $T_{-\Delta m}$ for these Δm are time-independent. Considering that the operator $\mathbf{T}'$ provides a pathway for the change in nuclear spin energies to flow into a specific lattice excitation, such a second-order contribution of $\Delta\mathbf{H}'_Q$ should represent the internal energy exchange process in the condensate.

Although calculated for a nuclear probe located at specific space-time coordinates in principle, such an energy exchange can be considered at all lattice sites for relaxation processes, whose temporal profile is specified by the relaxation time T^*, and the spatial modulation is described by the factor σ in $\Delta\mathbf{H}'_Q$. In the critical region, such a relaxation measurement confirm the direct nuclear energy transfer process to soft modes in the condensate, which is expressed by the second-order energy $\Delta E^{(2)} = |\Delta\mathbf{H}'_Q|^2/\hbar\Delta\omega$, where $\Delta\omega$ is the fluctuation frequency of the pseudospin variable $\boldsymbol{\sigma}$. Because the quadrupole moment tensor $\mathbf{Q}$ is a nuclear property, whereas the tensor $\mathbf{T}$ represents the soft lattice mode in the condensate model, such a nuclear relaxation represents the temporal behavior of the product $\boldsymbol{\sigma}\mathbf{T}$. Accordingly, we can define the quadrupole relaxation time T_1^* by the relation $1/T_1^* = \Delta E^{(2)}/\hbar$. Signified by two fluctuation modes of $\boldsymbol{\sigma}$, such a nuclear relaxation can take place in two independent paths; hence, we can write

$$1/T_1{}^* = 1/T_{1\mathrm{A}}{}^* + 1/T_{1\mathrm{P}}{}^*, \tag{9.16}$$

where

$$1/T_{1\mathrm{A}}{}^* \propto \langle \sigma_{\mathrm{A}}^2 \sum\nolimits_{\Delta m} T'_{\Delta m}{}^* T'_{\Delta m} \exp(i\Delta\omega.t)\rangle_{\mathrm{t}} = \sigma_{\mathrm{oA}}^2 \sin^2 \underline{\phi} (\sum\nolimits_{\Delta m} J_{\Delta m})$$

and

$$1/T_{1\mathrm{P}}{}^* \propto \sigma_{\mathrm{oP}}^2 \cos^2 \underline{\phi} (\sum\nolimits_{\Delta m} J_{\Delta m}).$$

Here, the average time correlation $t_{\mathrm{o}}^{-1} \int \mathrm{T}_{\Delta m'}{}^* T_{\Delta m'} \exp(i\Delta\omega.t)dt$ over the timescale t_{o} is addreviated as $J_{\Delta m}$. It is noted that the amplitudes σ_{oA} and σ_{oP} are equal at the transition threshold, although staying nearly constant in an incommensurate phase if there is no significant long-range order. Thus, we arrive at the simple formula

$$1/T_1{}^* \propto \sigma_{\mathrm{oA}}^2 \sin^2 \underline{\phi} + \sigma_{\mathrm{oP}}^2 \cos^2 \underline{\phi}, \tag{9.17}$$

where the spatial phase $\underline{\phi}$ of corrections, $0 \le \underline{\phi} \le 2\pi$, signifies the anomalous nuclear-spin relation time in the critical region.

Blinc and his co-workers [83-86] have derived the relation (9.17) originally for analyzing ^{14}N-spin-lattice relaxation rate in the incommensurate phase of $\{N(CH_3)_4\}_2ZnCl_4$ crystals at 14°C. Figures 9.20a and 9.20b show their results of ^{14}N-resonance experiments, where observed T_1^* are plotted against distributed resonance frequencies. Figure 9.20c shows the numerical simula-

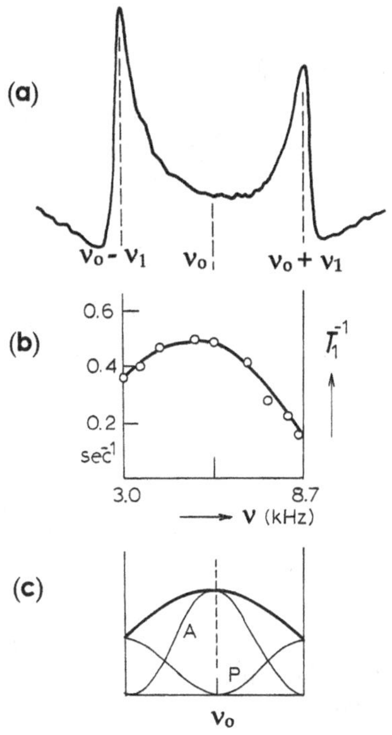

Fig. 9.20. (a) An anomalous ^{14}N-nuclear magnetic resonance line from the incommensurate phase of a $[N(CH_3)_4]_2ZnCl_4$ crystal. (From S. Zumer and R. Blinc, J. Phys. C14, 465 (1981).) (b) Observed reciprocal relaxation time distributed in the range 3.0 ∼ 8.7kHz. (c) Calculated from the amplitude mode and the phase mode with an equal proportion. The net contribution (thick line) resembles the observed T_{1-1}.

tion with appropriate values of σ_{oA} and σ_{oP} to fit to (9.16). We consider that the result signifies the energy exchange in the condensate, which was in fact proposed for the critical region and incommensurate phases. Blinc [87] published an extensive review article of nuclear relaxation studies of various systems undergoing structural phase transitions.

Although generally considered in a modulated phase, the two fluctuation modes are not always recognized, because the crystal may be modified by a local electric field of long-range order as in ferroelectric critical regions. It is interesting to see a similar effect in a CDW-state of $Rb_{0.3}MoO_3$, although in a different category. Segransan and his co-workers [88] carried out a ^{87}Rb-NMR experiment, showing that the modulated resonance line of pinned CDW condensates can be converted to free-running condensates by increasing electric current through the system. Figure 9.21 shows their results, which is similar to two coexisting modes of fluctuations in the incommensurate phase of $\{N(CH_3)_4\}_2Rb_2ZnCl_4$.

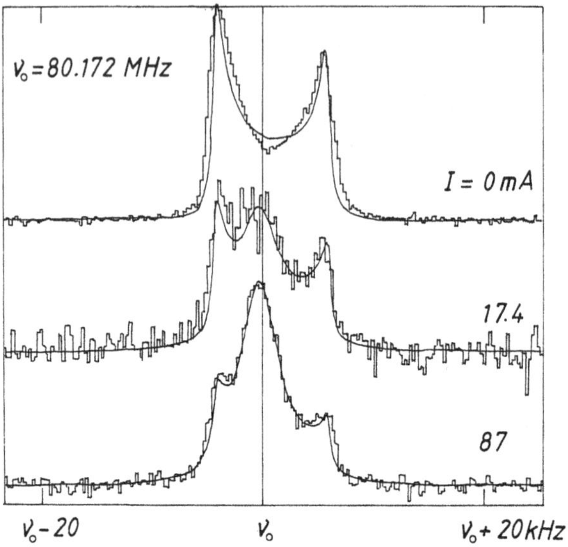

Fig. 9.21. Electric current "pinning-dipinning" phenomenon observed in a ^{87}Rb NMR of the CDW state in $Rb_{0.3}MoO_3$. The central line appeared with increasing current, as observed in the NMR spectrum at 80.17 MHz. (From P. Segransan, A. Jánossy, C. Berther, J. Mercus and P. Buraud, Phys. Rev. Lett. 56, 1654 (1986).)

10

Structural Phase Transitions in Miscellaneous Systems

In Chapter 9, transition anomalies in TSCC and BCCD crystals were discussed as examples of magnetic resonance spectra interpreted with the concept of order variable condensates. In this chapter, we look at other systems of interest in light of knowledge obtained from these examples to see how much of their transition mechanism can be elucidated with available experimental results. Whereas the principles for binary systems discussed in Part One are a useful guideline, phase transitions in crystals are of such an enormous variety that there is no established prototype other than the binary model at the present stage. In the meantime, we focus our attention on representative systems so far studied extensively, to conclude our discussions.

10.1 Cell-Doubling Transitions in Oxide Perovskites

Crystals of the oxide-perovskite family designated by the chemical formula ABO_3, where A = K, Sr, Ba, Ca, ... and B = Ta, Ti, Al, Pb and so forth, are rich in types of displacement that are attributed to the outset of collective motion of octahedral units BO_6^{2-} in the lattice. For titanium-oxide perovskites, substituting paramagnetic probes for the central Ti ions, the magnetic resonance spectra reflect the nature of the collective displacements in the critical region when the properties of crystals are not substantially modified at a low density of impurities. Such a model of the active group can be applied as a prototype to interpret structural transformations in other systems as well; hence, we summarize essential features of oxide perovskites in this section.

The structural change in $SrTiO_3$ crystals at $T_c = 105K$ can be recognized by a symmetry change from cubic to tetragonal, as the temperature is lowered through T_c, at which the unit cell is doubled in size. Three elastic domains are formed below T_c, which are characterized by tetragonal distortion along each of the symmetry axes. In the tetragonal domain, the phase is signified by unit cells in size of twice the original unit, indicating an alternate arrangement of oppositely rotated TiO_6^{2-} complexes, as illustrated for a $SrTiO_3$ crystal

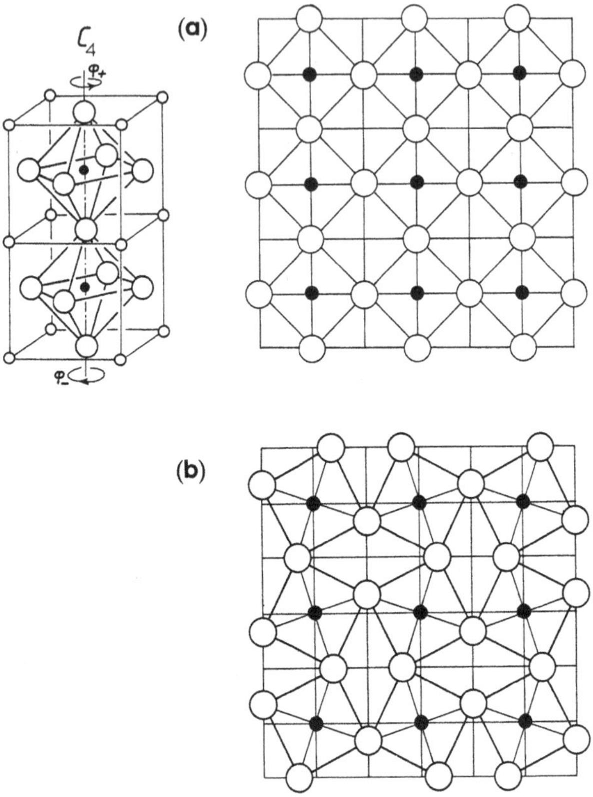

Fig. 10.1. Molecular arrangements in $SrTiO_3$ crystals. (a) Librational displacements $\pm\varphi$ (on the left insert) and the arrangement in the normal phase, (b) the phase below 105K.

in Fig. 10.1. From magnetic resonance spectra of Fe^{3+} ions substituted for Ti^{4+}, von Waldkirch, Müller and Berlinger [89] identified such an ordered state specified by a rotational angle φ of the TiO_6^{2+} around the C_4 axis, as indicated in the figure, where a tetragonal distortion along the c direction is shown specifically.

In fact, Fe^{3+} spectra in perovskites were complicated by an oxygen defect near the impurity site, which is usually designated Fe^{3+}-V_o. In such an Fe^{3+}-V_o complex, the charge cloud is distorted along the impurity-vacancy axis, which was identified by the unique direction characterized by the largest fine-structure splitting. Müller and his collaborators observed such spectra from Fe^{3+}-doped $SrTiO_3$ crystals at the resonance frequency 24GHz and obtained results interpreted by the angular displacement φ of TiO_6^{2-}, showing temperature dependences in various directions of B_o below T_c. Observed transition anomalies were very anisotropic as summarized in Fig. 10.2, which are characterized by considerable broadenings in two symmetry directions, while

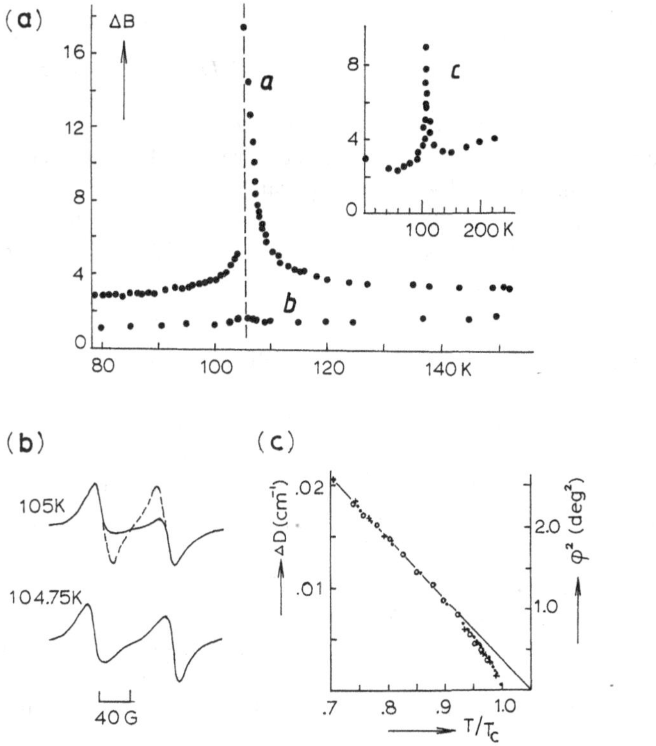

Fig. 10.2. Transition anomalies in Fe^{3+} spectra from $SrTiO_3$. (a) Linewidths, (b) an anomalous splitting for $B_o \parallel [101]$ at 105K, (c) temperature-dependence of $D_a \propto \varphi^2$.

virtually unchanged on the tetragonal axis. In a crystal tetragonal along the b direction, the anomalous widths in the a and c directions at 140K were nearly four times as large as those in the b direction (Fig. 10.2a). In addition, a Fe^{3+}-V_o line for $B_o \parallel [101]$ showed an anomalous splitting below T_c, which was appreciable at temperatures very close to T_c (Fig. 10.2b).

In the condensate model, such an anomaly can be attributed to binary fluctuations in phase mode between different wavevectors along symmetry axes, e.g. $\boldsymbol{q}_a$ and $\boldsymbol{q}_c$, so that the phase transition can be interpreted as two-dimensional in the b plane. On the other hand, the transition is characterized at $\frac{1}{2}\boldsymbol{G}$ in the reciprocal space, i.e. a point on the Brillouin-zone boundary; hence, it is not of the same character as in one-dimensional transitions observed at the zone center $\boldsymbol{G} = 0$. The anomalous line shown in Fig. 10.2b suggests that there should be clusters in two different types signified by wavevectors

$$\boldsymbol{q}_a = (\delta_a a^*, \tfrac{1}{2}b^*, 0) \quad \text{and} \quad \boldsymbol{q}_c = (0, \tfrac{1}{2}b, \delta_c c^*),$$

where δ_a and δ_c are incommensurate parameters along the a and c directions, respectively. In terms of the fine structure, we can consider binary fluctuations between

$$\Delta D_n(+\sigma) = a_n\sigma + b_n\sigma^2, \quad \text{and} \quad \Delta D_n(-\sigma) = -a_n\sigma + b_n\sigma^2$$

for the observed anomaly for $\boldsymbol{n} \parallel [101]$, leading to two domains separated by the $(10\bar{1})$ plane with decreasing temperature. In this specific direction of B_o, the resonance line begin to separate into two at T_c, while fluctuating between them. Hence, $D_n(+) = D_n(-)$ exactly at T_c, but the fluctuations between them start immediately below T_c. The binary fluctuations can thus be described by

$$\Delta D_n = \Delta D_n(+) - \Delta D_n(-) = 2a_n\sigma,$$

where σ is either in phase mode or amplitude mode across the crystal plane $(10\bar{1})$.

Noting that reported anomalous broadenings along the a and c axes are not equal, we can consider spontaneous stresses in the crystal, making these domain volumes slightly asymmetrical. In this case, the fluctuations are predominantly in phase mode, i.e. $\sigma = \sigma_o \cos\phi$, thereby explaining the observed anomalies in the [101] direction, provided that the factor a_n is nonzero.

For B_o applied in parallel to the a axis, on the other hand, the broadening can be due to either one of $\Delta D_n(+)$ or $\Delta D_n(-)$, where the term of $b_n\sigma^2$ can be significant, because the factor $a_{[100]}$ can vanish for rotational fluctuations around the principal axis parallel to the a axis. In the mean-field approximation, the order parameter is proportional to the rotational angle φ around a C_4 axis, and therefore we can write

$$\langle \Delta D_{[100]} \rangle \propto \langle \varphi^2 \rangle \propto T_o - T,$$

Figure 10.2c shows a plot of observed linewidths and estimated mean-field average of $\langle \varphi^2 \rangle$ in perovskites as a function of temperature, which are linear in a wide range close to T_c, although exhibiting a significant deviation in the critical region $T_c > T > 0.9T_c$.

On the basis of the above argument, the plane $(10\bar{1})$ in tetragonal $SrTiO_3$ crystals can be interpreted as the boundary between two domains, which are signified as binary order that is primarily considered as one-dimensional along the [100] and [001] directions. The unit-cell size is doubled in these domains as illustrated Fig. 10.1b, whereas on the $(10\bar{1})$ plane the angular displacements of TiO_6^{2-} octahedra around the b direction change directions of rotation all in phase, constituting the domain boundary. The order-variable is therefore a vector $\boldsymbol{\sigma} \parallel b$, signifying the domains by $\pm\boldsymbol{\sigma}$. Accordingly, the asymmetry in magnetic resonance linewidths between a and c directions in Fig. 10.2a can be attributed to such domain volumes in the sample crystals, although left unexplained in the original report. Nevertheless, the linear chains of pseudospins along the a and c axes are out-of-phase, constituting the basis of the binary order in two-dimension.

It is interesting that the above interpretation of critical anomalies in $SrTiO_3$ crystals is consistent with diffuse X-ray diffraction patterns (Fig. 6.3) from $NiNbO_3$ at 700°C reported by Comes et al. [19], although from the X-ray results, we can only substantiate that these one-dimensional fluctuations are primarily independent.

10.2 The Incommensurate Phase in β-Thorium Tetrabromide

Crystals of β-thorium tetrachloride, $ThBr_4$, exhibit a structural change from normal to incommensurate phases at $T_i \approx 95K$, which was discovered first with Raman studies [90], and then followed by neutron inelastic experiments. Bernard and his group [40] performed neutron experiments on β-$ThBr_4$ and found a well-defined soft mode from the phonon dispersion curve showing a dip at $Q_c = 0.310c^*$ in the $\boldsymbol{Q} \parallel \boldsymbol{c}$ scattering geometry, which was resolved in two branches at the bottom.

Emery, Hubert and Fayet [91] carried out magnetic resonance studies on β-$ThBr_4$ crystals doped with Gd^{3+} probes substituting for Th^{4+} ions, and they identified the active group as related to rotational angles $\pm\varphi$ around the $\bar{4}$ symmetry axis z. These authors reported various types of anomalous broadening of Gd^{3+} lines in the incommensurate phase, depending on the direction of the applied field. Furthermore, using Pa^{4+} probes in place of Th^{4+}, Zwanenburg and de Boer [92] observed anomalous g factor and hyperfine splitting in this phase. Evidenced by these results, the active group can be considered as related to two coupled adjacent Th^{4+} ions. These Th^{4+} ions are related by inversion along a direction perpendicular to the z axis, where each ion is surrounded by four Br^- ions in the shortest distances, and alternately arranged with two additional Br^- ions in between, as shown in Fig. 10.3a. We can assign two inversion-related pseudospins, σ_1 and σ_2 to Th^{4+} ions in the complex, signifying the chain of active groups along the $\bar{4}$ axis. Such a linear chain is characterized by 180° screw symmetry, where the nearest-neighbor interactions J_c between σ_1 and σ_2 can be competitive with correlations J'_c between two Br^- tetrahedral adjacent to each one of these pseudospins, as illustrated in Fig. 10.3b. Therefore, the phase below T_c can be incommensurate along the z axis, if signified by an irrational wavevector Q_c arising from competing correlations J_c and J'_c.

In their neutron inelastic scattering experiments, Bernard et al. [40] found that the phonon-dispersion curve was split into two branches at 81K, as shown in Fig. 4.9a, implying that fluctuations in the collective pseudospin motion are in phase and amplitude modes. For those pseudospins related by 180° screw symmetry, we write $\sigma_1 = e_q \exp i\phi$ and $\sigma_2 = e_q \exp i(\phi + \pi)$, where the phase angle $\phi = Q_c z - \omega_c.t$ is continuous in the long-wave approximation along the z direction taken on the $\bar{4}$ axis, and the short-range correlation energy can be expressed as

$$E = -J_c\sigma_1(0)\{\sigma_1(2Q_c c) + \sigma_1(-2Q_c c)\} - J'_c\sigma_1(0)\{\sigma_2(Q_c c) + \sigma_2(-Q_c c)\}$$

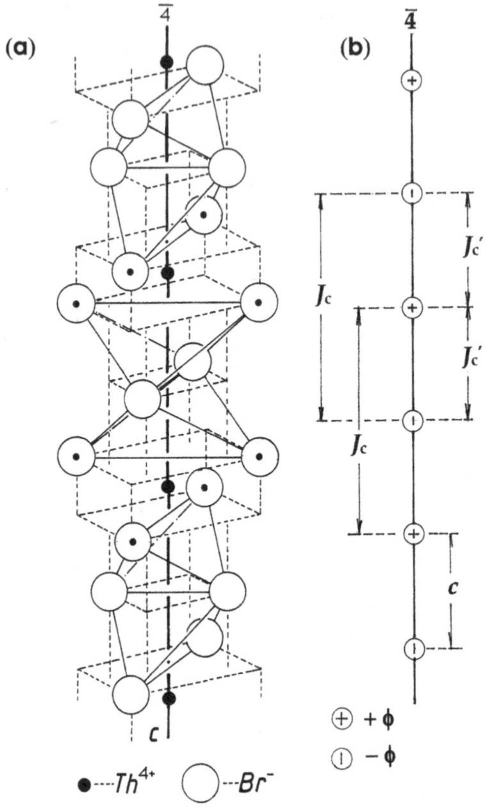

Fig. 10.3. (a) Molecular arrangement in β-$ThBr_4$ crystals. (b) The pseudospin arrangement for the short-range correlations, where ⊕ and • indicate opposite pseudospins in the cluster perpendicular to the c axis.

$$= -J(Q_c)e_{-Q}e_Q,$$

where

$$J(Q_c) = 4J_c\cos(2Q_c c) + 4J_c'\cos(Q_c c + \pi).$$

Here, we have ignored correlations in the a and b directions where pseudospins are arranged commensurately. Setting the differentiated $J(Q_c)$ with respect to Q_c equal to zero, we obtain the relation

$$-2J_c\sin(2Q_c c) + J_c'\sin(Q_c c) = 0,$$

which determines the specific real value of the incommensurate wavevector Q_c, i.e.

$$\cos(Q_c c) = J_c'/4J_c,$$

provided that $|J_c'/4J_c| < 1$. On the other hand, for a commensurate distribution, $Q_c = n\pi/2c$, where n is an integer, or $\cos(Q_c c) = 1$ and the nearest correlation $|J_c'|$ should be equal to $4|J_c|$.

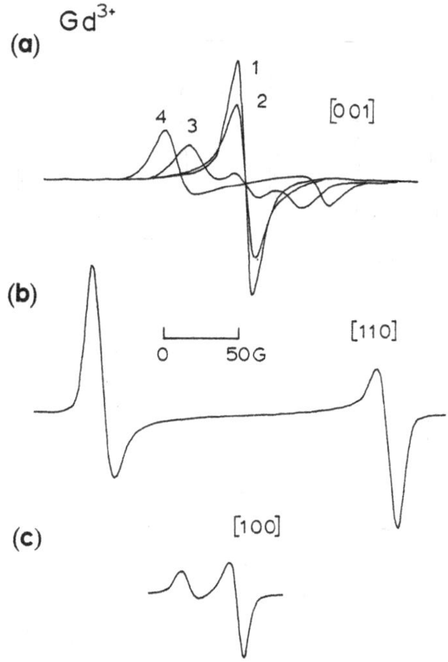

Fig. 10.4. Magnetic resonance anomalies in the incommensurate phase of $ThBr_4$ crystals. (a) $B_o \parallel [001]$, (b) $B_o \parallel [110]$ and (c) $B_o \parallel [100]$. (From J. Emery, S. Hubert and J. C. Fayet, J. Physique 46, 2099 (1985).)

In such a modulated lattice, the collective pseudospin mode can fluctuate as $\sigma_\pm = e_\pm \exp i\{(\pm Q_c \mp \Delta q)z - (\omega_c \pm \Delta\omega)t\}$, owing to the interaction with soft phonons. It is noted that $\sigma_+ = \sigma_-$ at $\Delta q = \pm Q_c$, and this degeneracy can be lifted if there is a periodic lattice potential $V = V_o \cos 2Q_c z$ that originates from the quartic anharmonicity in the chain structure. As the result, two independent modes $e_Q \cos\phi$ and $e_Q \sin\phi$ occur, where $\phi = \Delta q.z \pm \Delta\omega.t$ represents the fluctuations at Δq and $\Delta\omega$ in the vicinity of Q_c and ω_c.

Figure 10.4a shows the anomalies in Gd^{3+} spectra observed by Emery and his collaborators [91] for $B_o \parallel [001]$ in the critical region of β-$ThBr_4$ below 95°K. In these Gd^{3+} spectra, the two modes of fluctuations are evident, where the intensity of the sine mode diminishes with decreasing temperatures, as indicated by the sequence of 1-2-3-4 in the figure. The phase transition at 95°K is of a cell-doubling type, for which the nearest correlation J'_c should be positive. In the incommensurate phase of $ThBr_4$crystals, the long-range field is much weaker than in ferroelectric TSCC, as evident in the observed results; hence, the thermodynamic state below 95°K remains incommensurate. The spectrum for $B_o \parallel [110]$ in Fig. 10.4b shows that the cos mode dominates the incommensurate phase, where the fine-structure parameter is proportional to σ, whereas Fig. 10.4c shows the anomalous line in the [100] direction which is contributed by σ and σ^2 terms.

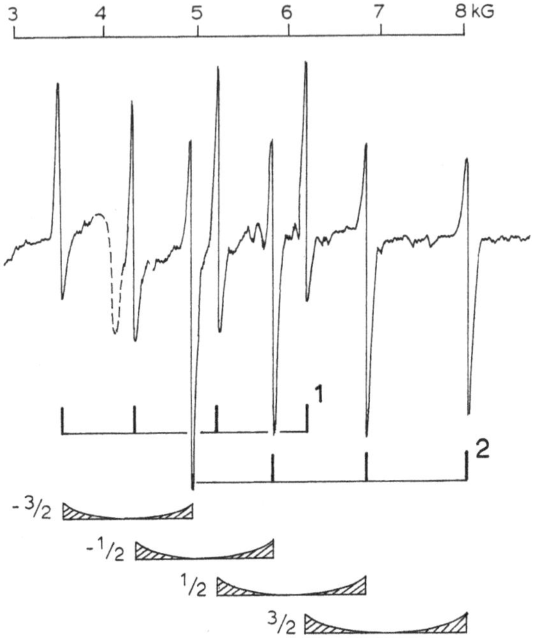

Fig. 10.5. Anomalous resonance lines of Pa^{4+} spectra in $ThBr_4$ at 4.2K. The line-shape of each hyperfine line can be explained in terms of Δg_n and ΔK_n, similar to VO^{2+} spectra in BCCD.

In contrast to trivalent Gd^{3+}, tetravalent $Pa^{4+}(5f^1, I(^{231}Pa) = 3/2)$ ions accompany no charge defects when substituted for Th^{4+}. Having only one electron in the $5f$-orbit, the magnetic resonance spectra are characterized by the Zeeman energy, and hyperfine and nuclear quadrupole interactions. Experiments on Pa-doped $ThBr_4$crystals were performed at liquid-helium temperatures, because the resonance lines were too broad to be resolved at temperatures above 20K. At 4.2°K, observed Pa^{4+} spectra showed anomalous splitting in hyperfine lines in all directions of B_o, which were assigned to "two impurity sites" in the unit cell in the ordered phase. Characterized by separated edges, the lattice is modulated approximately as proportional to $\sigma = \sigma_o \cos\phi$ $(0 < \phi < 2\pi)$ at 4.2K, as shown in Fig. 10.5, where forbidden lines are very weak in this specific direction. From observed Gd^{3+} spectra in the room-temperature phase, the impurity site is just one in the unit cell of $ThBr_4$, whereas two such anomalous Pa^{4+} sites in the low-temperature phase should be related by inversion symmetry between $\pm\sigma$ of incommensurate condensates. Zwanenburg et al. analyzed the Pa^{4+} spectra with the spin-Hamiltonian

$$\mathbf{H} = \beta\langle S|\mathbf{g}|B_o\rangle + \langle S|\mathbf{A}|I\rangle + \langle I|\mathbf{QT}|I\rangle - \gamma\langle I|B_o\rangle$$

in the strong-field approximation. Experimentally, the **g** and **A** tensors in $ThBr_4$crystals were found to have common principal axes, with the unique axis parallel to the crystallographic c axis. The principal values reported by the authors are

$$g_c = 1.76, \quad g_\perp = 1.05; \quad A_c = 695 \times 10^{-4}, \quad A_\perp = 430 \times 10^{-4}\mathrm{cm}^{-1},$$

and in such large anisotropic hyperfine tensors, incommensurate fluctuations are clearly evident, similar to VO^{2+} spectra in the incommensurate phase of BCCD.

10.3 Phase Transitions in Deuterated Biphenyl Crystals

Molecular crystals of biphenyl ($C_{12}H_{10}$) are utilized primarily as host crystals for phosphorescent naphthalene and phenanthrene molecules in triplet states. For a general reference on optical and magnetic experiments on triplet states, interested readers are referred to *Molecular Spectroscopy of Triplet States* by McGlynn et al. [93]. Huchison and his group [94] performed extensive magnetic resonance studies on triplet states of small aromatic molecules accommodated in various organic host crystals. Cullick and Gerkin [95] reported phase transitions in deuterated biphenyl crystals ($C_{12}D_{10}$) at 40K and 15K, which were first discovered in their magnetic resonance studies on phosphorescent triplet states of phenanthrene and naphthalene molecules as impurities.

According to Hirota and Hutchison [96], the phosphorescent time of uv-excited diphenyl crystals is of the order of $5 \sim 10$sec at 77K, and even longer at lower temperatures, which is sufficiently long to observe magnetic resonance of these excited molecules in triplet states. The triplet state of an aromatic molecule is due to two excited π-electrons that are characterized by the dipole-dipole interaction with parallel spins; hence the ground state is the triplet ^{3}S-state where $L = 0$ and $S = 1$ [93]. The dipole-dipole interaction can be expressed as a fine structure $\langle S|\mathbf{D}|S\rangle$ for the triplet spin $S = 1$, where $\mathbf{D}$ is a traceless tensor. Hutchison and Mangum [94] showed that the excited guest molecules are oriented in host crystals in the same way as the host molecule $C_{12}H_{10}$, being signified by two independent principal directions of the $\mathbf{D}$ tensor in the unit cell. The fine-structure energy can be written in the principal form as

$$\mathbf{H}_\mathrm{F} = D_\mathrm{x}S_\mathrm{x}{}^2 + D_\mathrm{y}S_\mathrm{y}{}^2 + D_\mathrm{z}S_\mathrm{z}{}^2 \quad \text{where} \quad D_\mathrm{x} + D_\mathrm{y} + D_\mathrm{z} = 0,$$

and the principal axes $(\mathrm{x}, \mathrm{y}, \mathrm{z})$ are taken to be consistent with molecular symmetry, as shown in Fig. 10.6b. The crystal structure of biphenyl is sketched in Fig. 10.6a. Using the traceless feature of $\mathbf{D}$, $\mathbf{H}_\mathrm{F}$ can be expressed in terms of two independent parameters D and E; that is,

$$\mathbf{H}_\mathrm{F} = DS_\mathrm{z}{}^2 + E(S_\mathrm{x}{}^2 - S_\mathrm{y}{}^2)$$

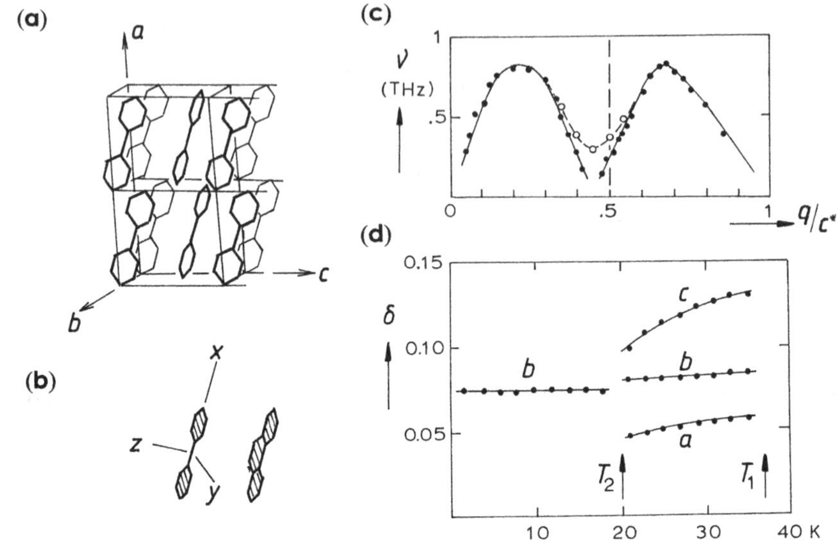

Fig. 10.6. (a) Molecular arrangement in biphenyl crystals. (b) The principal axes X, Y and Z can be set in common with the guest molecule phenantherene and the host molecule biphenyl. (c) The phonon-dispersion curve obtained by neutron inelastic scattering experiments. (d) Incommensuration parameters in the phases II and III of the host crystals. (From Cailleau, in *Incommensurate Phases in Dielectrics*, Vol. 2, pp. 72–100, ed. R. Blinc and A. P. Levanyuk (North-Holland, Amsterdam, 1986).)

where

$$D = (3/2)D_z \quad \text{and} \quad E = \tfrac{1}{2}(D_x - D_y).$$

Because of large values of D and E of the order of 0.1 and 0.01 cm^{-1} respectively, the fine structure plays a dominant role in a magnetic resonance experiment using a conventional laboratory magnet, and the spin-Hamiltonian is expressed as

$$\mathbf{H} = \beta\langle S|\mathbf{g}|B_o\rangle + \mathbf{H}_F + \mathbf{H}_{HF},$$

where the Zeeman energy is comparable with $\mathbf{H}_F$, and the last term $\mathbf{H}_{HF}$ represents hyperfine interactions with protons in the molecule $C_{12}H_{10}$, although negligible in deuterated $C_{12}D_{10}$. Because of large D and E, it is not quite adequate to consider $\mathbf{H}_F$ as a perturbation to the Zeeman energy in B_o. It is therefore logical to consider the Zeeman term together with the diagonal term of $\mathbf{H}_F$ for the unperturbed state, which is then perturbed by the off-diagonal elements of $\mathbf{H}_F$. In the zero field $B_o = 0$, the two separated energy levels, $e_{1,2}$ and e_3 due to D, called *zero-field splitting*, are signified by the electronic quantum numbers $M = \pm 1$ and 0. On the other hand, the degeneracy of $e_{1,2}$ for $M = \pm 1$ is lifted by an applied magnetic field B_o, resulting in three energy levels e_1, e_2 and e_3, among which transitions are determined by $e_i - e_j = h\nu$ (microwave quanta).

According to Hutchison and Mangum [94], these energy levels for B_o applied parallel to the principal directions are expressed as follows:

for $B_o \parallel$ z, $e_{1,2} = D + (\tan\alpha_z)E \pm g_z\beta B_o$ where $\tan 2\alpha_z = E/g_z\beta B_o$, and $e_3 = 0$,

for $B_o \parallel$ x, $e_{1,2} = \frac{1}{2}(D+E)(1-\tan\alpha_x) \pm g_x\beta B_o$ where $\tan 2\alpha_x = -\frac{1}{2}(D+E)/g_x\beta B_o$, and $e_3 = D - E$,

for $B_o \parallel$ y, $e_{1,2} = \frac{1}{2}(D-E)(1-\tan 2\alpha_y) \pm g_y\beta B_o$ where $\tan 2\alpha_y = \frac{1}{2}(D-E)/g_y\beta B_o$, and $e_3 = D + E$.

Figure 10.7a shows these energy levels plotted against B_o, where two allowed magnetic dipole transitions at a fixed microwave frequency are indicated by vertical arrows. Values of D, E and the principal g can be determined from the sets of observed transitions.

Cullick and Gerkin [95] studied triplet spectra from phenanthrene-d_{10} and naphthalene-d_8 molecules in host crystals of biphenyl-d_{10} in the temperature range between 5K and 80K, and found that these fine-structure parameters and zero-field splittings were temperature dependent, and that anomalous splitting and discontinuities were observed at about 42K and 15K, respectively. The temperature dependence is generally attributed to librational fluctuations around the molecular x axis, whereas discontinuous changes of the fine structure were considered as arising from structural changes in the host crystal. Cailleau et al. [98] carried out neutron inelastic scattering experiment on biphenyl crystals, and found soft phonons in phase and amplitude modes near the transition temperature $T_1 = 36$K. They also reported that the diffraction accompanied a strong first satellite reflection at 20K, indicating the presence of an intrinsic excitation. The soft mode is clear evidence for order variable condensates in $C_{12}H_{10}$ crystals, which should be responsible for the magnetic resonance anomaly from the triplet molecules in $C_{12}D_{10}$ at $T_1 = 42$K. Also noticeable is that line broadening in deuterated crystals observed by Cullick et al. appeared to be "anomalous" due to fluctuating σ, as seen from Fig. 10.7c. The anomaly in the zero-field splitting in biphenyl crystals at 42K can be interpreted as arising from a slow modulation, whereas the four resolved peaks between 42K and 15K in Fig. 10.7b were not identifiable by this model.

It is conceivable that phenyl molecules in librational motion are highly correlated in the crystal. Judging from the structure shown in Fig. 10.6a, correlations between similarly oriented molecules and those between dissimilar ones can be competitive along the c direction, resulting in incommensurate arrangements. From the neutron results, the incommensurability parameters can be expressed as

$$q_{II} = \pm(\delta_a \boldsymbol{a}^* - \delta_c \boldsymbol{c}^*) + \tfrac{1}{2}(1-\delta_b)\boldsymbol{b}^* \quad \text{in phase II (36K} \sim \text{20K)}$$

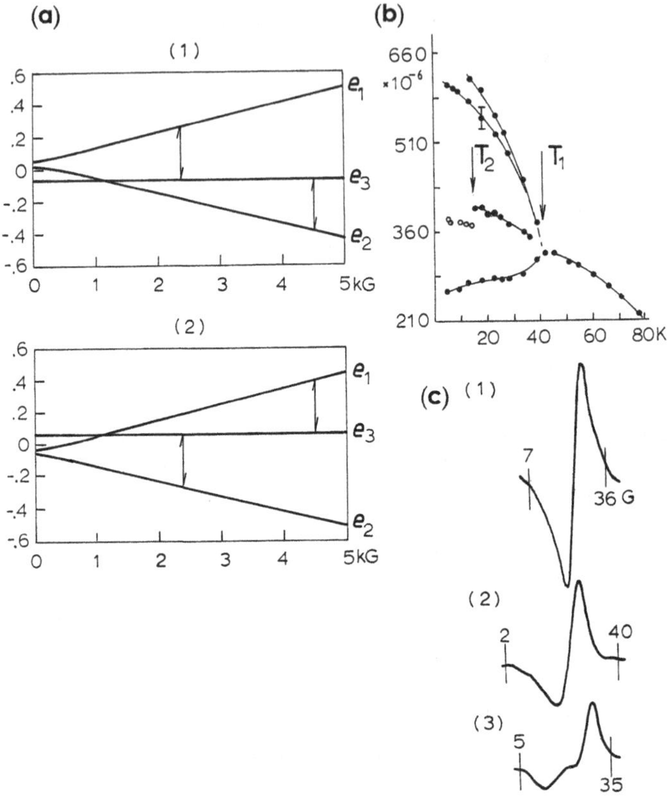

Fig. 10.7. (a) Energy levels of an excited triplet molecule in the magnetic field. Arrows indicate magnetic resonance transitions at a constant frequency. The diagrams (1) and (2) are drawn for different signs of the zero-field splitting constant. (b) Temperature dependence of the parameter $D - E$ observed from triplet phenantherene-d$_{10}$ in phase II of biphenyl crystals. (c) Magnetic resonance anomalies observed at microwave frequency 4419MHz, (1) at 42.3K, (2) at 41.5K and (3) at 40.5K. (From A. S. Cullick and R. E. Gerkin, Chem. Phys. 23, 217 (1977).)

and

$$q_{\text{III}} = \tfrac{1}{2}(1 - \delta_b)\boldsymbol{b}^* \quad \text{in phase III } (< 20\text{K}),$$

where these δ could be estimated from the short-range correlation parameters if known.

10.4 Successive Phase Transitions in A_2BX_4 Family Crystals

Crystals of the formula unit A_2BX_4 exhibit successive phase transitions from normal to incommensurate and then to commensurate phases, ending at an

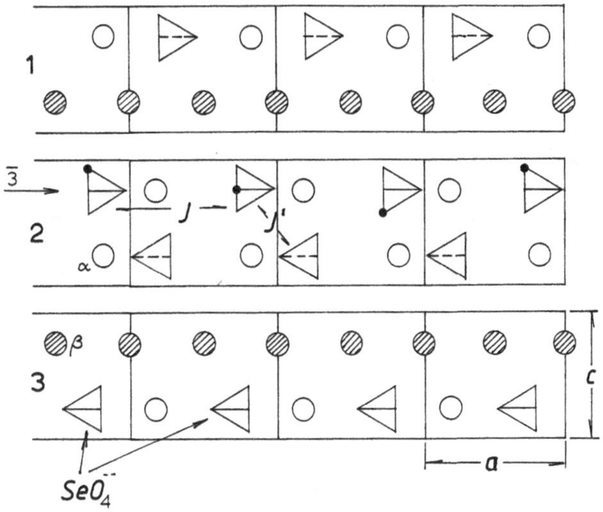

Fig. 10.8. Molecular arrangement in K_2SeO_4 in the ac plane, where the cell is divided into three layers 1, 2 and 3 perpendicular to the b axis, whose projections are shown for each layer. Here, open and shaded circles indicate crystallographically inequivalent K_α and K_β, respectively. Three units along the a direction are approximately equal to the wavelength of modulation. (From M. Iizumi, J. D. Axe and G. Shirane, Phys. Rev. B15, 4392 (1977).)

ordered phase that is known as ferroelectric in some systems. A variety of inorganic crystals belong to this category, including crystals where the ion A^+ can be K^+, Rb^+, NH_4^+, $N(CH_3)_4^+$, and so forth, and the tetragonal group BX_4^{2-} can be ${SO_4}^{2-}$, ${SeO_4}^{2-}$, ${ZnCl_4}^{2-}$, ${BeF_4}^{2-}$, and so forth. These crystals are isomorphous, showing commonly cascade phase transitions and "twinning" in some cases [99].

In $(NH_4)_2SO_4$ crystals, for example, the orthorhombic structure of the room–temperature phase is strained, resulting in a structure composed of three differently oriented ferroelastic domains, exhibiting a ferrielectric structural change at 223K [100]. On the other hand, K_2SeO_4 crystals are also strained at room temperature, but the phase below 129K is characterized as incommensurate. In these systems, pseudosymmetry is commonly found, as related to tetrahedral BX_4^{2-} ions and crystallographically inequivalent A^+ ions at α and β sites, as shown in Fig. 10.8 [101]. The twinned structure in $(NH_4)_2SO_4$ is signified by temperature-dependent pseudohexagonal symmetry along the a axis, as illustrated in Fig. 10.9 [99]. In each domain, tetrahedral BX_4^{2-} groups are related by $\bar{3}$ screw symmetry along the sliding direction parallel to the a axis, as shown in Fig. 10.8 for K_2SeO_4 [101]. It is considered that such pseudosymmetry plays a significant role for successive structural transformations in these systems.

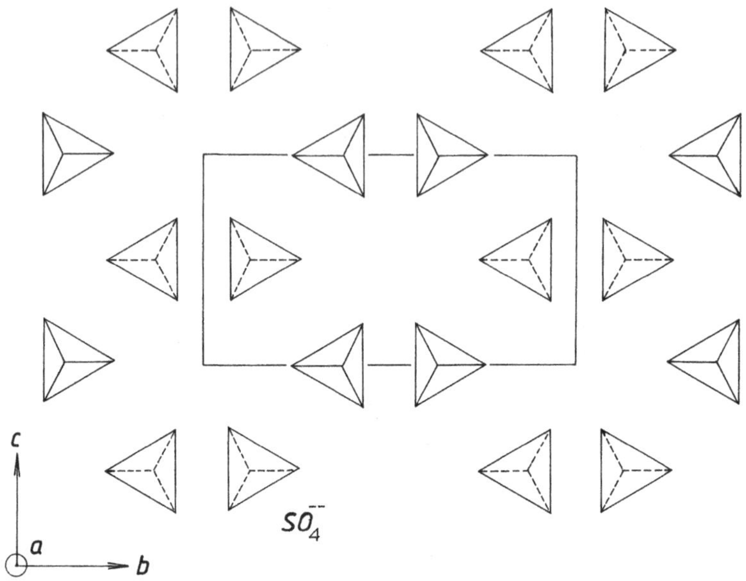

Fig. 10.9. Arrangement of SO_{42}-ions in $(NH_4)_2SO_4$ crystals in the *bc* plane in the room temperature phase. (From A. Sawada, Y. Makita and Y. Takagi, J. Phys. Soc. Japan 42, 1918 (1977).)

Iizumi and his co-workers [101] carried out extensive neutron inelastic scattering studies on structural changes in K_2SeO_4 crystals. The results showed that the transition at 129K is a normal-to-incommensurate structural change, which is not characterized by the loss of a macroscopic symmetry element, while the mirror symmetry on the *ac* plane is locally violated. Below $T_i = 129K$, K^+ ions are displaced out of the mirror plane, and tetrahedral $SeO_4{}^{2-}$ groups are rotated around the *c* axis, whereas in isomorphous crystals of K_2ZnCl_4, the rotational axis of $ZnCl_4{}^{2-}$ is slightly tilted from the *c* axis in the mirror *ac* plane. In K_2SeO_4, a soft mode was observed near $q \sim (1/3)a^*$ at about 130K, for which these authors showed that for one-phonon scattering the responsible normal mode can be characterized by librational fluctuations of $SeO_4{}^{2-}$ tetrahedrons around the *b* axis, accompanying linear displacements of K_α^+ and K_β^+ along the *c* direction. In the observed phonon dispersion curve shown in Fig. 4.5, the dip in the vicinity of $\boldsymbol{G}_i = 0.7\boldsymbol{a}^*$ implies the soft mode, whose incommensurate wavevector can be written as

$$\boldsymbol{q}_a = \tfrac{1}{3}(1 - \delta)\boldsymbol{a}^*.$$

Therefore, the order variable can be characterized by a phase variable $\phi = q_a x$ for spatial variation. The inelastic neutron scattering takes place primarily by the collective motion of heavy anions $SeO_4{}^{2-}$ accompanying displacements of K^+ ions. The crystal structure below 129K is characterized by the $\overline{3}$ screw symmetry along the *a* axis. Therefore, we can consider that the symmetric

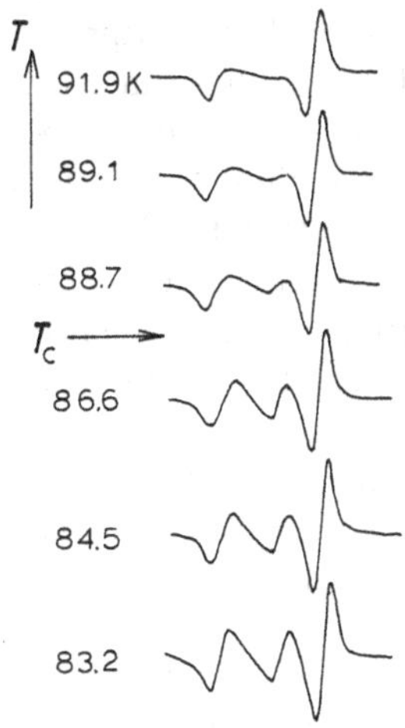

Fig. 10.10. Temperature variation of an anomalous resonance line of VO^{2+} spectra from a K_2SO_4 crystal. (From M. Fukui and R. Abe, J. Phys. Soc. Japan 12, 3942 (1982).)

mode of fluctuations $\sigma = \sigma_o \cos qx$ along the a direction can be pinned by the pseudopotential $V_3 \propto \cos(3q_a x)$, when $q \sim q_a$ (see Section 5.4). In fact, symmetry of V_3 is consistent with threefold ferroelastic domains that are quasi-trigonally related with respect to the a axis, along which the binary variable σ exhibits inversion symmetry.

While ferroelastic at room temperature, crystals of ammonium sulfate show a ferrielectric structural change at 223K, where the polar state is characterized by two unequal polar sublattices [100]. It is conceivable that reorientations of crystallographically inequivalent $(NH_4^+)_\alpha$ and $(NH_4^+)_\beta$ are responsible for the ferrielectricity in $(NH_4)_2SO_4$, as verified by Fujimoto et al. [102] in their magnetic resonance experiment.

Fukui and Abe [103] performed magnetic resonance studies on a VO^{2+}-doped K_2SeO_4 crystal, and observed anomalous lineshapes distinguishing the incommensurate phase from the following commensurate phase below 89K. These authors reported temperature-dependent anomalies in the ^{51}V hyperfine line at a lowest resonance field, as shown in Fig. 10.10, where the lineshape are interpretable with the formula $\Delta\nu = \nu_1\sigma + \nu_2\sigma^2$, where $\sigma = \sigma_o \cos\phi$, thereby showing that σ is the primary variable. The phases above and below T_i can be clearly distinguished by continuous and discrete angle ϕ in the range $0 \leq \phi \leq 2\pi$, respectively, indicating that the phase transition at 89K is a phase-locking phenomenon by the pseudopotential V_3. Numerically they simulated those lineshapes below 89K by discrete angles $\phi = 0, \pi/3$ and $2\pi/3$, as given by (5.18), using empirical value of the phase shift δa^* equivalent to 25°, substantiating the proposed mechanism.

The behavior of $BX_4{}^{2-}$ groups was studied in Rb_2ZnCl_4 and related crystals with paramagnetic Mn^{2+} and SeO_3^- probes substituted for B^{2+} and

BX_4^{2-} in A_2BX_4, respectively. Using Mn^{2+} probes, Fayet and his group [104-107] studied the modulated phase in Rb_2ZnCl_4 crystals, and Fukui et al. [108] and Kobayashi et al. [109] performed similar experiments on the $[N(CH_3)_4]_2ZnCl_4$ system, reporting results similar to Mn^{2+}anomalies from the incommensurate phase in BCCD. Among these reports, Pezeril et al. [104] identified the incommensurate phase from forbidden lines associated with $(-\frac{1}{2} \rightarrow +\frac{1}{2})$ transitions in Mn^{2+} spectra, and interpreted anomalies on $(3/2 \rightarrow 5/2)$ lines as due to frequency fluctuations described by $\Delta\nu = \nu_1\sigma + \nu_2\sigma^2$.

The incommensurate-commensurate transition is phase locking of $\sigma(\phi)$ by a pseudopotential, where we can expect discommensuration effects in the magnetic resonance spectra. Although no such evidence has been found in magnetic resonance studies, Unruh and his group [48] have reported such discommensuration lines at $T_i = 205$K in K_2ZnCl_4 crystals recorded by the transmission electron microscopy. The presence of an incommensurate phase in crystals of $A^+ = K^+$ and Rb^+ can be attributed to the absence of long-range order. On the other hand, ammonium sulfate is unique among others, in that the internal electric field makes no critical region at the transition.

Blinc and his co-workers [110] measured the electric field gradient tensor at ^{87}Rb nuclei in incommensurate phases of Rb_2ZnCl_4 and Rb_2ZnBr_4 crystals, showing that the lineshape of the $(\frac{1}{2} \rightarrow -\frac{1}{2})$ transition in ^{87}Rb-quadrupole resonance spectra was analyzable with the fluctuation formula $\Delta\nu = \nu_1\sigma + \nu_2\sigma^2$, depending on the direction of applied field B_o. Blinc et al. [83, 84] also reported a line due to what they called "floating" condensates, in addition to the pinned modulation near T_i of Rb_2ZnBr_4. Although observable in the NMR timescale in principle, such floating condensates can be identified as an amplitude mode.

10.5 Incommensurate Phases in $RbH_3(SeO_3)_2$ and Related Crystals

Orthorhombic crystals of $RbH_3(SeO_3)_2$ undergo a ferroelectric phase transition at about $T_c = 153$K, where the symmetry changes to monoclinic with the ferroelectric axis along the b direction. Many experimental studies [111] have been performed on this system since Levanyuk and Sannikov [112] predicted the presence of an incommensurate phase. Indeed, the incommensurate phase was later verified experimentally at temperatures below T_i that was just above T_c, although reported values of T_c depended apparently on the sample quality. Such transitions in sequence are considered as arising from primary and secondary order variables as in $(NH_4)_2SO_4$. In fact, the primary order variable was convincingly identified in these experimental studies, whereas the secondary variable can be considered from observed dielectric anomalies at T_c, below which the phase is only weakly polarized. For deuterated crystals $RbD_3(SeO_3)_2$, the incommensurate phase was confirmed by X-ray diffraction

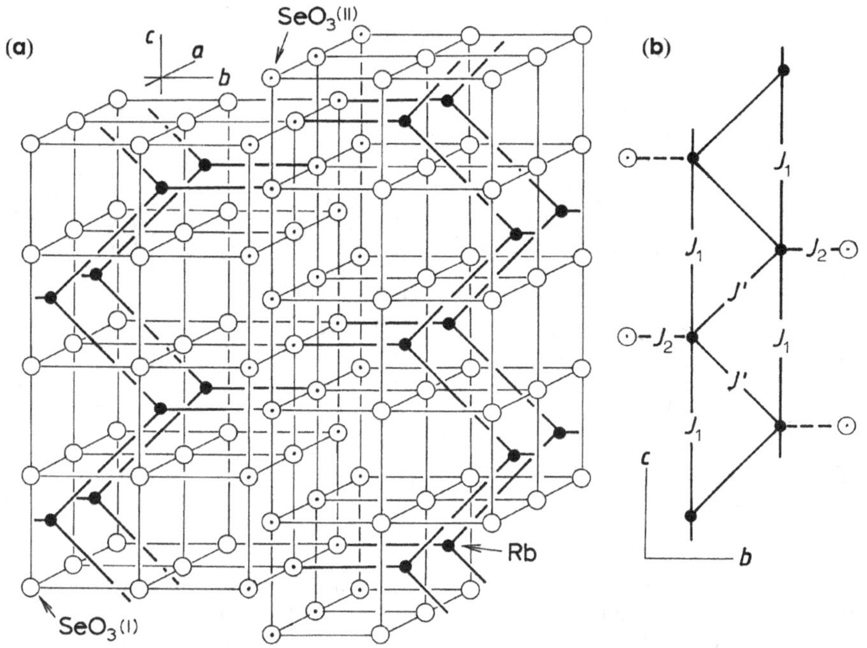

Fig. 10.11. (a) Molecular arrangement in the low temperature phase of $RbH_3(SeO_3)_2$ crystals. (From H. Grimm and W. J. Fitzgerald, Acta Cryst. A34, 268 (1978).) (b) A pseudospin chain model in $RbH_3(SeO_3)_2$.

studies by Makita and Suzuki [113], and the soft-mode vector was analyzed by Grimm and Fitzgerald [114] for the crystal structure below T_c.

Structures of $RbH_3(SeO_3)_2$ crystals at room temperature and in the ferroelectric phase were studied by a number of groups [115], showing that in the unit cell there are two $SeO_3{}^{2-}$ and three hydrogen bonds that are crystallographically inequivalent. Those $SeO_3{}^{2-}$ groups of one kind, denoted by SeO_3(I), are linked together via hydrogen bonds in one type, forming a chain along the c axis, whereas SeO_3(II) of the other kind form another chain along the a axis with hydrogen bonds of the second type. Parallel chains of SeO_3(I) and SeO_3(II) are in separate planes perpendicular to the b axis, being linked across via hydrogen bonds of the third kind. In this structure, each Rb^+ ion is coordinated by four SeO_3^+ ions in the ac planes and by a SeO_3(II) along the b direction. Figure 10.11a shows the crystal structure in the low-temperature phase sketched after the neutron work by Grimm et al., where a layer of SeO_3(II) ions is sandwiched between layers of SeO_3(I) and Rb ions. According to Shuvalov and his co-workers [116], the order variable is related to a reorientation of deformed $SeO_3{}^{2-}$ groups, for which the protons play no significant role.

In the crystal structure shown in Fig. 10.11a, each Rb^+ ion is at the center of a rectangle of four SeO_3(I) ions, coordinated by a SeO_3(II) at the apex of

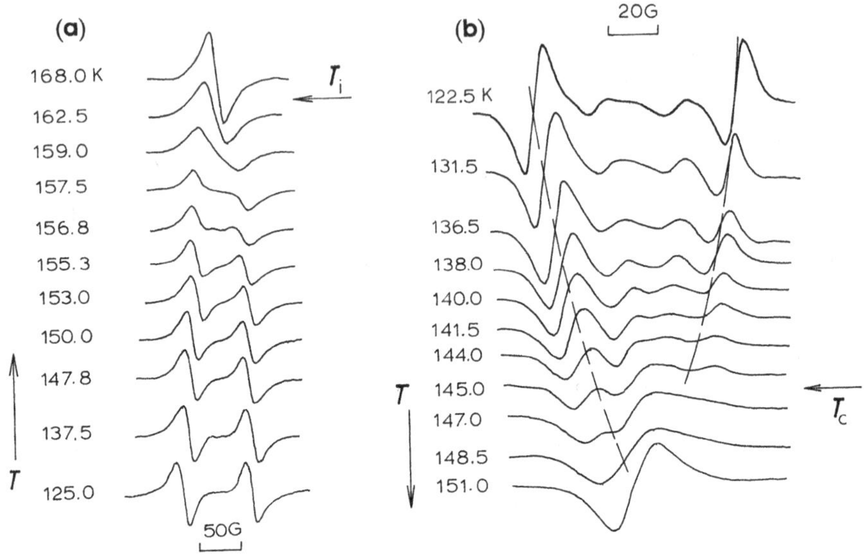

Fig. 10.12. Temperature changes of Cr^{3+} resonance lines from $RbH_3(SeO_3)_2$ crystals. (a) From M. Fukui, C. Takahashi and R. Abe, Ferroelectrics 36, 315 (1981). (b) From S. Waplack, S. Jerzak, J. Stankowski and L. A. Shuvalov, Physica 106B, 251 (1981).

a pyramid formed by these five $SeO_3{}^{2-}$ ions. Such Rb^+-$(SeO_3{}^{2-})_5$ complexes are arranged in a zigzag chain in the *bc* plane, being linked with the layer of SeO_3(II), as illustrated in Fig. 10.11b. Waplak et al. [117] and Fukui et al. [118] investigated independently the behavior of these pyramidal complexes by magnetic resonance studies using Cr^{3+} probes substituted for Rb^+ ions, and they identified the primary order variable for the incommensurate phase as related to the collective mode of these Rb complexes along the *c* axis. Although their results were not fully analyzed, the phase transitions at T_i and T_c were evident from the temperature-dependent Cr^{3+} spectra, which are shown in Figs. 10.12a and 10.12b, respectively.

In such a chain of Rb complexes as in Fig. 10.11b, Cr^{3+} probes were used to evaluate the short-range correlation scheme in $RbH_3(SeO_3)_2$ crystals, while the order variable $\sigma(\phi)$ can be defined from the observed **D**-tensor anomaly ΔD_n. As predicted, the phase ϕ describes the incommensure distribution of deformed $SeO_3{}^{2-}$ along the *c* direction. In the spectra of Fig. 10.12a, the anomaly can be explained as $\Delta D_c \propto A_c\sigma$, although the direction of the applied field was unspecified. The temperature variation indicates that the phase ϕ is continuous but showing discommensuration lines toward T_c, as seen from Fig. 10.12b.

For the incommensurate phase between T_i and T_c, two competing correlations along the *c* direction can be considered for the interaction scheme of Fig. 10.11b. Placing pseudospins σ at each Rb position, the short-range correlations between the nearest neighbors on the chain can be expressed as

$$E(\boldsymbol{q}) = -2{\sigma_o}^2 \boldsymbol{e}_q.\boldsymbol{e}_{-q} J_1(\boldsymbol{q}).$$

Considering the next nearest neighbors as in the *bc* plane, as indicated in Fig. 10.11b, we can include these additional interactions, i.e.

$$E(\boldsymbol{q}) = -4{\sigma_o}^2 \boldsymbol{e}_q.\boldsymbol{e}_{-q}[J_1 \cos(q_c c) + J' \cos(q_b b)\cos(\tfrac{1}{2}q_c c) + J_2 \cos(q_b b)],$$

which can be minimized with respect to $\boldsymbol{q}$ for the initial cluster for the phase transition at T_i. Disregarding all other short-range interactions, we set $q_b = 0$ and obtain

$$\cos(\tfrac{1}{2}q_c c) = -J'/4J_1,$$

which gives the incommensurate q_c along the *c* axis, provided that $|-J'/4J_1| < 1$.

Pending experimental results, the pseudospin can be expressed by a classical vector with components σ_b and $\sigma_c \propto \cos\phi$. In this case, σ_b can be considered as the secondary order variable, because its finite amplitude violates the mirror reflection symmetry of the SeO_3(II) plane. However, if the quasi mirror plane is not fully represented by σ_b, such a view may not be compatible with the soft layer mode proposed by Grimm et al. It is noted that the phase below T_c may not be necessarily ferroelectric if such a mirror plane is not verified.

10.6 Phase Transitions in $(NH_4)_2SO_4$ and NH_4AlF_4

Crystallographically, ammonium sulfate crystals belong to a group of A_2BX_4 type, sharing some features in common with others in the group, in that order variables are primarily related to orientation of heavy group ${BX_4}^{2-}$. However, $(NH_4)_2SO_4$ is unique among them, because pyramidal NH_4^+ ions exhibit another collective behavior below 223K, being associated with the secondary ordering in the ferrielectric phase.

Magnetic resonance studies on the structural change in $(NH_4)_2SO_4$ were performed using VO_2^+ impurities [73] and with NH_3^+ radicals in irradiated crystals [74][119]. In these works, the behavior of NH_4^+ ions was sampled by VO^{2+} and NH_3^+, whereas SO_3^- and SeO_3^- radicals in irradiated crystals were utilized to study the role of heavier SO_4^{2-} ions. However, the pseudosymmetry arising from near-trigonally oriented SO_4^- groups was not significantly reflected on the spectra of SO_3^- and SeO_3^- [120], although considered as responsible for breaking trigonal symmetry of the normal lattice [99]. On the other hand, NH_3^+ spectra from irradiated crystals were informative about NH_4^+ ions occupying distinct sites α and β. According to the normal-mode analysis for K_2SeO_4 by Iizumi et al. [101], the librational change of SeO_4^{2+} orientations around the *c* axis accompanies displacements of K_α^+ and K_β^+ out of the *ab*

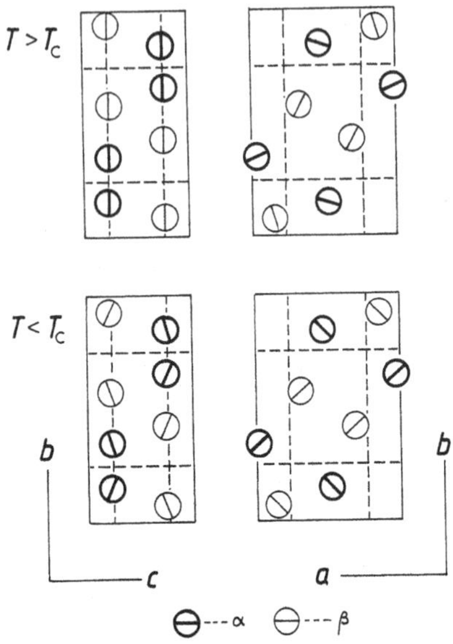

Fig. 10.13. Orientational arrangements of NH_4^+ ions in $(NH_4)_2SO_4$ crystals at above and below −50° C are compared in these diagrams, where SO_4^{2-} ions are omitted for clarity. The crystallographic inequivalence between $(NH_4)_\alpha$ and $(NH_4)_\beta$ is indicated by thick and thin circles. Arrows inside of these circles show direction of the ^{14}N-hyperfine tensor axis determined by magnetic resonance of NH_{3+} radicals. (From M. Fujimoto, L. A. Dressel and T. J. Yu, J. Phys. Chem. Solids 38, 97 (1977).)

plane; hence, by analogy, similar displacements of $(NH_4^+)_\alpha$ and $(NH_4^+)_\beta$ may be considered in $(NH_4)_2SO_4$. If this is the case, the orientation of a distorted NH_4^+ ion can be represented by the ^{14}N-hyperfine axis of a NH_3^+ probe, yielding information about NH_4^+ ordering in the ferrielectric phase. By virtue of different α and β sites, sublattices of $(NH_4^+)_\alpha$ and $(NH_4^+)_\beta$ can be considered as polar sublattices in the ferrielectric phase below 223K. The arrangement of NH_4^+ ions in such sublattices is sketched in Fig. 10.13 based on the magnetic resonance results of NH_3^+ [102].

The phase transition in NH_4AlF_4 crystals at 155K belongs categorically to a different group than $(NH_4)_2SO_4$; however, there is a common feature regarding the NH_4^+ order that is induced by the collective arrangement of heavier negative ions in the critical condition. Fayet carried out magnetic resonance studies [121] on NH_4AlF_4 crystals, using Fe^{3+} impurities, where the phase transition showed dual characters; namely order-disorder with respect to NH_4^+ ions and diplacive due to continuous displacements of AlF_4^- groups.

The crystal structure of NH_4AlF_4 at 155K consists of two-dimensional layers of NH_4^+ ions and of bipyramidal AlF_6 groups, which are alternately stacked up along the crystallographic c axis, as illustrated in Figs. 10.14a and 10.14b, when viewed in the ab plane and from a direction perpendicular to the c axis, respectively. In these figures, NH_4^+ ions indicated by triangles are

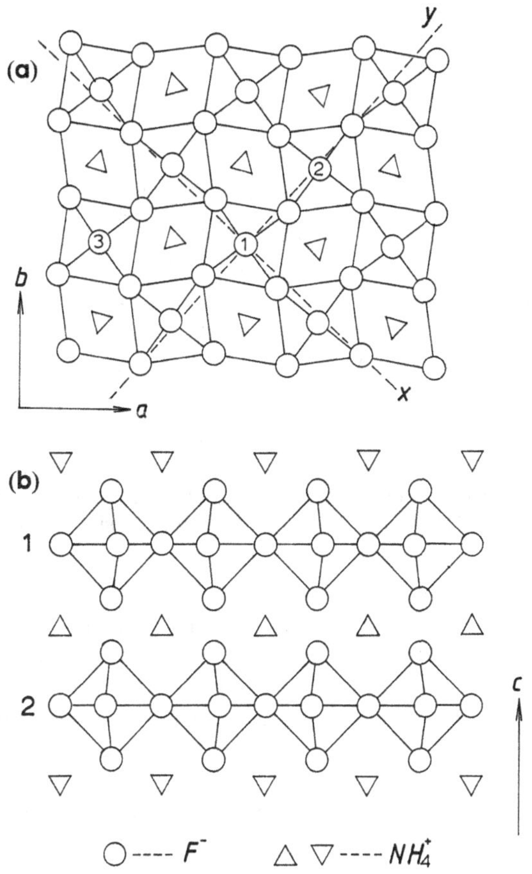

Fig. 10.14. Molecular arrangement in NH_4AlF_4 crystals. (a) the structure in a layer parallel to the *ab* plane. (b) alternate stack of layers of NH_4^+ and AlF_6^{3-} perpendicular to the *c* axis. (From J. C. Fayet, Helv. Physica Acta 58, 76 (1985).)

oriented in opposite directions parallel to the c axis, to which ± 1 states of a pseudospin can be assigned. On the other hand, each AlF_6 octahedron is coordinated by eight NH_4^+ in either state of + or −, and Fayet suggested that the ordered arrangement of eight NH_4^+ should be related to small angles of rotation $\pm\varphi$ of the AlF_6 octahedron around the c axis.

Magnetic resonance spectra of Fe^{3+} substituted for Al^{3+} ions were characterized by a uniaxial fine structure energy $\langle n|\mathbf{D}|n\rangle$ with respect to the direction $|n\rangle$ of the applied field B_o, where the unique tensor direction represents the axis of AlF_6 rotation. In the normal phase the Fe^{3+} spectra are characterized by one impurity site per unit cell while splitting into two components below T_c, as shown by Fig. 10.15. The linewidths were very broad due to unresolved superfine interactions with six ^{19}F nuclei ($I = \frac{1}{2}$) in the ligand, showing, however, a resolved structure in certain direction. In this direction, the spectrum

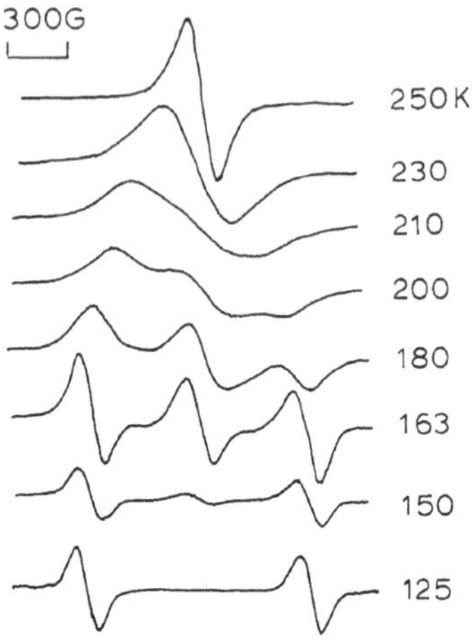

Fig. 10.15. Magnetic resonance anomalies in Fe^{3+} spectra from NH_4AlF_4 for $B_o \parallel [110]$ at and below 250K. (From J. C. Fayet, Helv. Physica Acta 58, 76 (1985).)

consists of a single line with diminishing intensity at a slow rate with decreasing temperature, on which a doublet in anomalous shape is superposed. Similar to the cell-doubling transition in $SrTiO_3$ at 105K, the result suggests that the phase transition in NH_4AlF_4 is binary in two-dimensional ordering. On the other hand, if the transition is first order, the single line may represent the normal phase.

The short-range cluster for such a displacive mechanism can be speculated as related to librational correlations between adjacent AlF_6 complexes, marked 1 and 2, along the x direction (and the same in the y direction), and between those marked 1 and 3 along the a axis (and the same along the b axis), as shown in Fig. 10.14a. Denoting these correlation parameters by J and J', respectively, we can express the condition for minimum correlations as

$$\cos(q_x \mathrm{d}) = \cos(q_y \mathrm{d}) = -(1 + J/2J'),$$

which are soluble for q_x and q_y if $0 > J/2J' > -1$. If this is the case, we can obtain order variables specified by wavevectors q_x and q_y in the ab plane, just as in perovskite crystals.

Noting that the observed spectrum is similar to the anomalous line from in $SrTiO_3$ (Fig. 10.2b), the anomalies in Fig. 10.15 can be described by $\Delta D \propto 2\sigma$ which is considered as related to binary fluctuations between σ_a and σ_b that are opposite in phase in this direction in the ab plane. In this case, alternate rotations of AlF_6 complexes along the x and y directions are related by inversion in the (110) plane, which can be considered as responsible for the anomalies in Fig. 10.15a. The normal-to-incommensurate phase transition in NH_4AlF_4 crystals should therefore be binary in two dimensions.

According to ref. [121], librational displacements of AlF_6 are coupled with NH_4^+ ordering at the transition temperature. Fayet's experiment was performed on AlF_6, giving direct support of a displacive view of the phase transition. Nevertheless, if magnetic probes sensitive to NH_4^+ ions were selected, the experiment might yield results in favor of order-disorder.

10.7 Proton Ordering in Hydrogen-Bonded Crystals

For classical pseudospins, the threshold of a phase transition is signified by the minimum correlation energy at T_c, where order variables emerge with nonzero amplitudes in a collective form. On the other hand, in a quantum-mechanical system as in hydrogen-bonded crystals, the ground state of a proton-bonded negative ionic group may be degenerate in relation to the proton arrangement, leading to a disordered state if they are not correlated spatially in the lattice [122]. In fact, such a case has not been found other than quantum liquids, besides these groups cannot be quite independent in crystalline states. If correlated at a slow rate, the order variable associated with the group may behave like a classical variable, resulting in significant strains in the lattice. Assuming that such "intersite" correlations are negligible, the proton interactions in each group may appear explicitly in the observed spectra, which are not, however, informative about the symmetry change in phase transitions.

In traditional arguments, the order-disorder phenomenon occurs thermally, independent of the hosting lattice; hence, in hydrogen-bonded crystals, phase transitions can be characterized by the absence of soft modes. Normally, thermal proton rearrangements can take place only as a part of the system, where the lattice structure may be considered to be intact. On the other hand, if lattice strains matter, protons in ordered arrangements should interact with an acoustic excitation. In this context, the traditional order-disorder theory cannot deal phase transitions characterized by a symmetry change.

Potassium dihydrogen phosphate KH_2PO_4 crystals, called KDP, exhibit a ferroelectric phase transition at 122K, for which Slater [123] and Takagi [124] proposed a simple model of proton ordering, although no hint was given at the spontaneous polarization below T_c. Nevertheless, their model was considered as a prototype of order-disorder because of its simple nature. On the other hand, in their Raman studies of KDP, Kaminov and Damen [125] found an intense temperature-dependent response overdamped in the low-frequency range, as shown in Fig. 10.16, where the observed lineshape was fitted to the damped harmonic oscillator equation with the characteristic frequency, $\omega_o \propto (T - T_c)^{1/2}$, and a constant damping factor, behaving like the Debye relaxation. It is not certain from their result whether such an overdamped soft mode can be coupled with proton ordering. Hence, the transition mechanism remains still controversial.

In Fig. 10.17, the active group in KDP crystals is sketched, for which Slater and Takagi proposed a model of thermal statistics for proton arrangements.

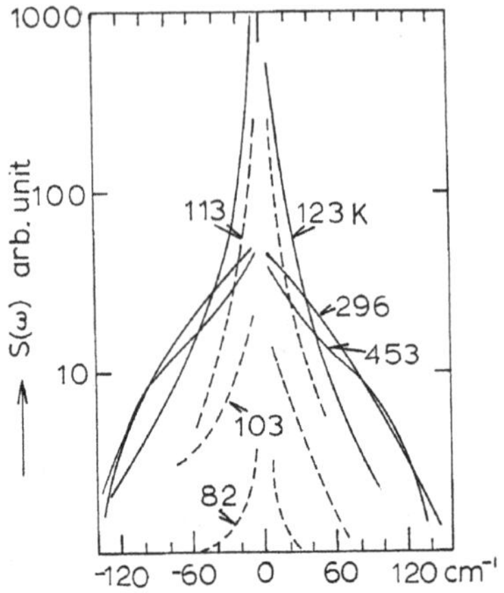

Fig. 10.16. Broad phonon spectra observed in KDP crystals. (From I. P. Kaminov and T. C. Damen, Phys. Rev. Lett. 20, 1105 (1968).)

In this model a PO_4^{3-} group is surrounded by four neighboring PO_4^{3-} ions, which are linked together via four hydrogen bonds. In Fig. 10.17b, we consider that four protons 1, 2, 3 and 4 are all close to the central PO_4^{3-} in the ground state, whereas 14 excited states are related to the locations of protons in these four hydrogen-bonds. Among the excited states, the first level ε_1 is due to four configurations where two of these protons, either in bonds (1, 3) or in bonds (2, 4), are near the central PO_4^{3-}, whereas the second level

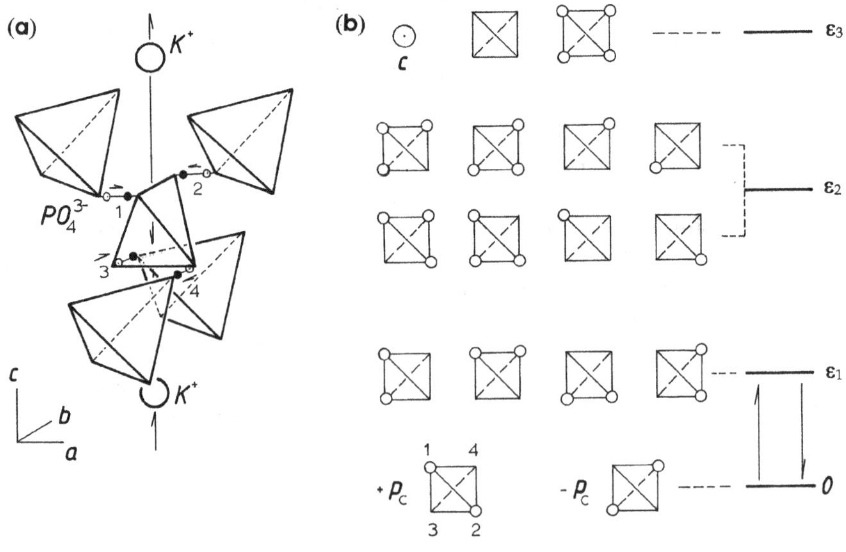

Fig. 10.17. (a) Proton configuration in the vicinity of PO_{43}-ion in KDP crystals. (b) The Slater-Takagi model.

ε_2 is related to three near protons, and the third one ε_3 corresponds to all four protons in the direct proximity or in distant locations, as illustrated in Fig. 10.17b. At $T_c = 122K$, the energy separations between proton states can be significantly greater than $k_B T_c$, so that the order variable associated with the cluster of PO_4^{3-} plus four protons is quantum-mechanical, according to the Landau criterion discussed in Section 3.2. In this case, these proton states are governed by Boltzmann probabilities, provided that these groups are uncorrelated in the crystal. In this model, the proton energies can, therefore, be considered thermally accessible, in contrast to the displacive crystals where the displacement is induced in a strained lattice.

Using the Slater-Takagi model, the statistical calculation shows that the transition temperature to the ferroelectric phase is determined by

$$k_B T_c = \varepsilon_1 \ln 2,$$

suggesting that clusters at the energy ε_1 for the proton configuration are significant at the transition threshold. It is noted that in this model the energy ε_1 is doubly degenerate with regard to inversion in the c direction; hence, the phase transition can be described by such a binary variable.

Although the actual transition mechanism cannot be found easily, it is believed that the proton ordering constitutes a part of the mechanism, as described by the Slater-Takagi model, for which there were some experimental studies. Magnetic resonance experiments on irradiated KH_2AsO_4 crystals, which are isomorphous to KDP, were carried out by two research groups led by Blinc [126] and McDowell [127], using $AsO_4{}^{4-}$ radicals as probes. Although these workers claimed that no significant change in the properties of KH_2AsO_4 was recognized by irradiation, in which paramagnetic $AsO_4{}^{4-}$ exhibited resonance spectra, showing the ferroelectric phase transition at 97K. The $AsO_4{}^{4-}$ spectra from irradiated KDP crystals may not exactly simulate the phase transition in KDP; however, the ^{75}As nucleus of spin $I = 3/2$ exhibited a large hyperfine splitting and additional small hyperfine splittings due to two protons, indirectly substantiating the Slater-Takagi model for KDP. It is noted that the proton hyperfine structure is sensitive for studying the proton arrangements. McDowell and his associates [127] performed ENDOR (electron-nuclear double resonance) experiments at 4.2K to study the model and determined the interaction tensors with two near protons and two far protons with greater accuracy than EPR spectra. In conventional EPR experiments, far proton interactions were too small to detect, while the temperature-dependence was not studied in the ENDOR measurement. Figure 10.18a shows a representative EPR spectrum from irradiated KH_2AsO_4 crystals reported in ref. [126], where the proton structure in 1:2:1 intensity ratio was barely resolved on the ^{75}As-hyperfine lines, although giving convincing evidence for two equivalent protons in the low temperature phase. In the ENDOR spectra, the structure was observed not only in better accuracy, but also a small additional structure caused by two far protons were revealed. Figure 10.18b shows the temperature dependence of the near proton lines reported in ref. [126], changing from

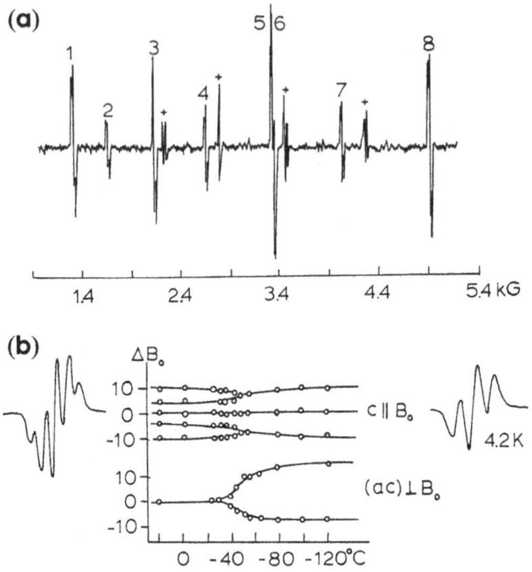

Fig. 10.18. (a) A magnetic resonance spectrrm of AsO_4^{4-} radicals at 4.2K from irradiated KH_2AsO_4 crystals. The lines marked 1, 2, ..., 8 are spectra of $AsO_4{}^{4-}$ at two inequivalent sites in the unit cell, whereas those marked + are due to unidentified species. (From N. S. Dalal, J. A. Hebden, D. E. Kennedy and C. A. McDowell, J. Chem. Phys. 66, 4425 (1977).) (c) Temperature dependence of the proton hyperfine structure in the spectra of AsO_4^{4-} in KH_2AsO_4. (From R. Blinc, P. Cevc and M. Schara, Phys. Rev. 159, 411 (1967).)

1:4:6:4:1 pattern to 1:2:1 near about 220K. This EPR result demonstrates that four protons in the group are equivalent in the high-temperature phase, while two at low temperature, substantiating the essential feature of the theoretical model together with the ENDOR result. It is realized that such a proton arrangement should be a slow process at low temperature, so that is timescale is a significant factor for observing the dynamics. Nevertheless, the model is substantiated, although the transition mechanism was not fully revealed in these magnetic resonance experiments. On the other hand, collective displacements of PO_4^{3-} ions appears to be a significant part of the transition mechanism, judging from the light-scattering results [125].

Epilogue

Macroscopically, structural phase transitions are characterized by a change in lattice symmetry. In second-order transitions, the loss of a symmetry element signifies the strained lattice, where a displacive mechanism is consequent on the symmetry change. I, therefore, discussed mainly displacive phase transitions in this book.

Although known theoretically, the Born-Huang principle came to my attention only after I witnessed the evidence in stressed and spontaneously modulated crystals. Guided by their principle, I discussed the critical region due to spontaneous strains, pending experimental evidence. The role played by a deformed lattice is explicit in observed anomalies, which are totally different from those arising from random fluctuations.

One of my objectives was to establish the physical origin for a collective classical pseudospins that can be expressed as $\sigma_o f(\phi)$. In fact, it was taken for granted in many theoretical works, leading to one-dimensional analyses of experimental results. Although considered for mathematical simplification, I felt it was a little too naïve to rely on such a formula for complex transition phenomena in three-dimensional crystals. On the other hand, anisotropic molecular correlations give rise to a quasi-one-dimensional pseudospin mode in the crystal as specified in the phase ϕ, which can, therefore, be justified in the limited applications.

Also, the presence of a unique axis was assumed in many works, which was found generally acceptable in anisotropic crystals. Consequently, in this book one-dimensional ordering was mainly discussed for practical applications. Nevertheless, critical anomalies are most explicit in one dimension, whereas less explicit in two and higher dimensions, which I should have discussed with appropriate examples. In fact, we came across two-dimensional fluctuations in crystals undergoing cell-doubling phase transitions as witnessed in the tetragonal phase of $SrTiO_3$, and perhaps in NH_2AlF_4 as well. I could have discussed it in Chapter 3 using the incommensurate condition ${\boldsymbol{e}_{\pm q}}^2 = 0$ for displacement vector $\boldsymbol{e}_{\pm q}$, in which vectors $(e_{qx} = \pm i e_{qy}, e_{qz})$ and $(\mp i e_{-qx} = e_{-qy}, -e_{-qz})$ can be assigned for breaking tetragonal symmetry along the z direction. I believe that such an interpretation paves the way to further studies on the structural transformation mechanism beyond one-dimensional arguments.

We have postulated that the critical temperature T_c is determined by minimum correlations of displacive variables. Such correlated variables may, however, be associated with a precursory process above T_c, as we have seen among practical systems. The ferroelectric phase transition in $BaTiO_3$ crystals, for example, is still controversial, since anomalies observed by different experiments showed results incompatible with one another, indicating that such transitions may not be purely displacive. Different timescales may be involved in observed anomalies; however, it is not surprising that such a precursory process as in $BaTiO_3$ may exist in general, for which further investigations are necessary, particularly, at lower frequencies than the characteristic fluctuations.

The soliton potential for pseudospin propagation is an acceptable concept for relating it with the deformed crystal structure. With the condensate model, we can explain its finite amplitude σ_o, for which the long-range order is responsible. However, the long-range interactions should be partly involed in a random mechanism of phase transitions, in order to account for the temperature dependence.

Appendix

The Adiabatic Approximation

For critical anomalies at structural phase transitions of second order, we have considered the interaction between order variables and the lattice, as implied by the lattice dynamical theory in Born-Huang's monograph [18]. We have in fact utilized their physical implication, while the theory was too formal to be modified for the problem of structural transformations. Nevertheless, the theory showed that anharmonicity arises from the lattice if perturbed adiabatically by fluctuating order variables, whose states are modulated along with nuclear displacements. On the other hand, in the non-adiabatic case, the order variable is primarily independent of the nuclear motion, but subjected to random agitation by thermal phonons. Although dynamically defined in the Born theory, the word "adiabatic" can be literally applied to thermodynamics of crystalline states. As the nuclear system represents the lattice, states of the order variables can be determined statistically if the interaction is non-adiabatic, otherwise representing classical displacements from lattice points under the adiabatic condition. The lattice structure should be unchanged under a constant volume condition. Hence, thermodynamically, a lattice structure is primarily deformable by an adiabatic interaction with order variables or external stresses. While interested readers can be referred to their original monograph, we review in this appendix the Born–Huang theory, as modified for a spontaneous structural change between crystalline phases, to show that anharminic potentials emerge as a consequence of condensate fluctuations.

For displacive ordering, we have considered a variable associated with a displacement in an active complex, which is an adequate model at least for phase transitions in perovskite crystals. Such displacive order variables are considered as primarily independent of the lattice of heavier active groups, provided that the correlation energies are negligible compared with the thermal energy k_BT. However, if characterized by discrete energy levels with a gap greater than k_BT, order variables are "quantum-mechanical" according to Landau's categories. In a binary system, such an order variable can be described in terms of probabilities in the Boltzmann statistics, which are expressed effectively by ± 1 states of an Ising spin. On the other hand, arriving at the minimum correlation energy at a transition threshold T_c with decreasing temperature, a lattice distortion begins to occur adiabatically.

Dynamically, the order variable and the active complex are distinguishable in terms of their masses and, hence, their interactions can be asymptotically evaluated in terms of the parameter κ defined as

$$\kappa = (m/M)^{1/4}, \tag{A.1}$$

where m and M are effective masses. Assuming that their interaction U emerges at the threshold temperature T_c, we analyse it with the approximation, known as the Born-Oppenheimer approximation. In this case, the composite system can be described primarily by fluctuating kinetic energies, namely

$$K_\sigma = -(\hbar^2/2m)\sum\nolimits_i(\partial^2/\partial x_i^2) \quad \text{and} \quad K_L = -(\hbar^2/2M)\sum\nolimits_i(\partial^2/\partial X_i^2),$$

and the interaction potential U between them is expressed as a function of x_i and X_i. Here, we consider that the interaction U takes place in one-dimension along x_i and X_i direction. As discussed in Chapter 3, this assumption is generally acceptable for anisotropic systems. Further, if $K_\sigma > K_L$, the Hamiltonian of the order variable system $\mathbf{H}_o = K_\sigma + U$ is regarded as perturbed by K_L, that is

$$\mathbf{H} = \mathbf{H}_o + K_L, \quad \text{where} \quad \mathbf{H}_o = \mathbf{H}_o(x, \partial/\partial x, X). \tag{A.2}$$

To solve this perturbation problem, we can use asymptotic expansions with respect to the small parameter κ introduced by (A.1), and write

$$K_L = \kappa^4\mathbf{H}_1(\partial/\partial X), \quad \text{where} \quad \mathbf{H}_1(\partial/\partial X) = -\sum\nolimits_i(\hbar^2/2m)(\partial^2/\partial X_i^2).$$

The Hamiltonian $\mathbf{H}$ in (A.2) can then be expressed as

$$\mathbf{H} = \mathbf{H}_o + \kappa^4\mathbf{H}_1 \tag{A.3}$$

for which the Schrödinger equation is

$$(\mathbf{H} - E(X))\psi(x, X) = 0, \tag{A.4}$$

considering the nuclear coordinate X as a parameter. We assume that the equation (A.4) is solved for the wave function of $\varphi(x, X)$ at an arbitrary fixed position of the active group X, i.e.,

$$(\mathbf{H}_o - \varepsilon_o(X))\varphi_n(x, X) = 0, \tag{A.5}$$

where n is the quantum number of the order variable. Assuming that the function $\varepsilon_o(X)$ has been obtained at a certain configuration X_o, the equation (A.4) can be written in terms of a small displacement κu of an active group from X_o, i.e.,

$$X = X_o + \kappa u. \tag{A.6}$$

Expanding $\varphi_n(x,X)$ asymptotically into a power series

$$\varphi_n(x,X)=\varphi_n(x,X_o+\kappa u)=\varphi_n^{(0)}+\kappa\varphi_n^{(1)}+\kappa^2\varphi_n^{(2)}+\dots, \tag{A.7}$$

and the eigenvalues are also expressed as

$$\varepsilon_n(X)=\varepsilon_n(X_o+\kappa u)=\varepsilon_n^{(0)}+\kappa\varepsilon_n^{(1)}+\kappa^2\varepsilon_n^{(2)}+\dots, \tag{A.8}$$

where the coefficients $\{\varphi_n^{(r)},\varepsilon_n^{(r)}\}$, $(r=1,2,\dots)$ are proportional to u^r. We can also write

$$\begin{aligned}\mathbf{H}_o(x,\partial/\partial x,X)&=\mathbf{H}_o(x,\partial/\partial x,X_o+\kappa u)\\&=\mathbf{H}_o^{(0)}+\kappa\mathbf{H}_o^{(1)}+\kappa^2\mathbf{H}_o^{(2)}+\dots\end{aligned} \tag{A.9}$$

where these $\mathbf{H}_o^{(r)}$ are operators with respect to x, and homogeneous functions of u^r. Substituting these expansions into (A.5), we obtain the following equations from each power of κ^r, i.e.

$$(\mathbf{H}_o^{(0)}-\varepsilon_n^{(0)})\varphi_n^{(0)}=0, \tag{A.10a}$$

$$(\mathbf{H}_o^{(0)}-\varepsilon_n^{(0)})\varphi_n^{(1)}=-(\mathbf{H}_o^{(1)}-\varepsilon_n^{(1)})\varphi_n^{(0)}, \tag{A.10b}$$

$$(\mathbf{H}_o^{(0)}-\varepsilon_n^{(0)})\varphi_n^{(2)}=-(\mathbf{H}_o^{(1)}-\varepsilon_n^{(1)})\varphi_n^{(1)}=-(\mathbf{H}_o^{(2)}-\varepsilon_n^{(2)})\varphi_n^{(0)}, \tag{A.10c}$$

$\dots\dots$

Since $\partial/\partial X=(1/\kappa)(\partial/\partial u)$, the kinetic energy K_L can be expressed as

$$K_L=\kappa^4\mathbf{H}_1=\kappa^2\mathbf{H}_1^{(2)}\quad\text{where}\quad\mathbf{H}_1^{(2)}=-\sum(\hbar^2/2m)(\partial^2/\partial u^2).$$

Combining this with the unperturbed expansion $\mathbf{H}_o$ of (A.9), the perturbed Hamiltonian $\mathbf{H}$ can be written as

$$\mathbf{H}=\mathbf{H}_o^{(0)}+\kappa\mathbf{H}_o^{(1)}+\kappa^2(\mathbf{H}_o^{(2)}+\mathbf{H}_1^{(2)})+\kappa^3\mathbf{H}_o^{(3)}+\dots. \tag{A.11}$$

Here, we consider that coefficients from different expansions in these terms of $\kappa^{(r)}$ are the same order of magnitude in the region where $\psi(x,u)$ does not vanish, and solve (A4) by the usual perturbation method. Thus, we write

$$E=\varepsilon_n^{(0)}+\kappa E_n^{(1)}+\kappa^2E_n^{(2)}+\dots. \tag{A.12}$$

and

$$\psi=\psi_n^{(0)}+\kappa\psi_n^{(1)}+\kappa_n^2\psi_n^{(2)}+\dots, \tag{A.13}$$

where $E_n^{(r)}$ for the energy eigenvalue E should be contant, being independent of u, unlike $\varphi_n^{(r)}$. Using (A.19) and (A.11) into (A.4), we obtain the following successive equations

$$(\mathbf{H}_o^{(0)}-\varepsilon_n^{(0)})\psi_n^{(0)}=0, \tag{A.14a}$$

$$(\mathbf{H}_o^{(0)}-\varepsilon_n^{(0)})\psi_n^{(1)}=-(\mathbf{H}_o^{(1)}-E_n^{(1)})\psi_n^{(0)}, \tag{A.14b}$$

$$\begin{aligned}(\mathbf{H}_o^{(0)}-\varepsilon_n^{(0)})\psi_n^{(2)}=&-(\mathbf{H}_o^{(1)}-E_n^{(1)})\psi_n^{(1)}\\&-(\mathbf{H}_o^{(2)}+\mathbf{H}_1^{(2)}-E_n^{(2)})\psi_n^{(0)},\end{aligned} \tag{A.14c}$$

$\dots\dots$

From the homogeneous differential equation (A.14a), we have $\psi_{\rm n}^{(0)} \sim \chi^{(0)}(X_{\rm o})\varphi_{\rm n}^{(0)}(x, X_{\rm o})$ as the zero-order solution, where $\chi^{(0)}(X_{\rm o})$ can be an arbitrary factor that depends on $X_{\rm o}$. However, for the perturbed wave function of the zeroth order, we can write

$$\psi_{\rm n}^{(0)}(x,u) = \chi^{(0)}(u)\varphi_{\rm n}^{(0)}(x), \tag{A.15}$$

where $\chi^{(0)}(u)$ is a function of u, to be determined in the following by higher order calculations. The inhomogeneous equation (A.14b) can be solved if the right side is orthogonal to $\varphi_{\rm n}^{(0)}$, i.e.

$$\int \psi_{\rm n}^{(0)}(x)(\mathbf{H}_{\rm o}^{(1)} - E_{\rm n}^{(1)})\psi_{\rm o}^{(0)}(x,u)\mathrm{d}x$$
$$= \chi^{(0)} \int \varphi_{\rm n}^{(0)}(x)(\mathbf{H}_{\rm o}^{(1)} - E_{\rm n}^{(1)})\varphi_{n}^{(0)}(x)\mathrm{d}x = 0,$$

whereas from the inhomogeneous (A.10b), we obtain

$$\int \varphi_{\rm n}^{(0)}(x)(\mathbf{H}_{\rm o}^{(1)} - \varepsilon_{\rm n}^{(1)})\varphi_{\rm n}^{(0)}\mathrm{d}x = 0.$$

Therefore, we have $\varepsilon_{\rm n}^{(1)} = E_{\rm n}^{(1)}$. By definition of the Taylor expansion, we can express $\varepsilon_{\rm n}^{(1)} = (\partial\varepsilon_{\rm n}/\partial X_{\rm o})u$, which should be equal to zero, if $X_{\rm o}$ represents the equilibrium configuration of complexes. The eigenvalue E and, hence, $E_n^{(1)}$ must be constant and independent of u, and therefore we obtain $E_{\rm n}^{(1)} = 0$. Using this result from (A.14b) and (A.10b), we find that a particular solution of the inhomgeneous equation is given by $\chi^{(0)}(u)\varphi_{\rm n}^{(1)}(u)$. To this, we may add any solution of the corresponding homogeneous equation, so that the general solution for $r = 1$ can be expressed as

$$\psi_{\rm n}^{(1)}(x,u) = \chi^{(0)}(u)\varphi_{\rm n}^{(1)}(x,u) + \chi^{(1)}(u)\varphi_{\rm n}^{(0)}(x). \tag{A.16}$$

Using this expression for $\psi_{\rm n}^{(1)}$ and $E_{\rm n}^{(1)} = 0$ in (A.14c),

$$(\mathbf{H}_{\rm o}^{(0)} - \varepsilon_{\rm n}^{(0)})\psi_{\rm n}^{(2)} = -\mathbf{H}_{\rm o}^{(1)}\chi^{(0)}\varphi_{\rm n}^{(1)} - (\mathbf{H}_{\rm o}^{(2)} + \mathbf{H}_{1}^{(1)} - E_{\rm n}^{(2)})\chi^{(0)}\varphi_{\rm n}^{(0)} - \mathbf{H}_{\rm o}^{(1)}\chi^{(1)}\varphi_{\rm n}^{(0)}. \tag{A.17}$$

The third and first terms on the right side can be replaced, respectively, by $\chi^{(1)}$ times $\mathbf{H}_{\rm o}\varphi_{\rm n}^{(0)}$ in (A.10b) for $E_n^{(1)} = 0$ and by $\chi^{(0)}$ times $\mathbf{H}_{\rm o}^{(1)}\varphi_{\rm n}^{(1)}$ derived from (A.10c) likewise, resulting in the expression

$$(\mathbf{H}_{\rm o}^{(0)} - \varepsilon_{\rm n}^{(0)})(\varphi_{\rm n}^{(2)} - \chi^{(0)}\varphi_{\rm n}^{(2)} - \chi^{(1)}\varphi_{\rm n}^{(1)}) = -(\mathbf{H}_{1}^{(2)} + \varepsilon_{\rm n}^{(2)} - E_{\rm n}^{(2)})\chi^{(0)}\varphi_{\rm n}^{(2)}.$$

This equation is soluble if

$$\int \varphi_{\rm n}^{(0)}(\mathbf{H}_{1}^{(2)} + \varepsilon_{\rm n}^{(2)} - E_{n}^{(2)})\chi^{(0)}\varphi_{\rm n}^{(0)}\mathrm{d}x = 0,$$

which is nevertheless satisfied because the integrand is independent of x, and hence

$$(\mathbf{H}_1^{(2)} + \varepsilon_n^{(2)} - E_n^{(2)})\chi^{(0)}(u) = 0. \tag{A.18}$$

It is noted that (A.18) is the equation determining the nuclear motion, for which $\kappa^2\mathbf{H}_1$ and $\kappa^2\varepsilon_n^{(2)}(u)$ represent the kinetic and potential energies in this approximation. Since $\varepsilon_n^{(2)}(u)$ is a quadratic function of u, (A.18) describes harmonic vibrations of the lattice in accuracy up to the first-order of κ, which is known as the *harmonic approximation.*

The authors showed the mathematical procedure to obtain solutions in higher-order accuracy. However, here, we only quote results of the second-order calculation that gives the adiabatic approximation. For the detail, interested readers are referred to the original literature. They showed that the second-order term of the wave function has the form

$$\psi_n^{(2)}(x,u) = \chi^{(0)}(u)\varphi_n^{(2)}(x,u) + \chi^{(1)}(u)\varphi_n^{(1)}(x,u) + \chi^{(2)}(u)\varphi_n^{(0)}(x), \tag{A.19}$$

where the functions $\chi^{(1)}$ and $\chi^{(2)}$ satisfy the differential equations:

$$(\mathbf{H}_1^{(2)} + \varepsilon_n^{(2)} - E_n^{(2)})\chi^{(1)}(u) = -(\varepsilon_n^{(3)} - E_n^{(3)})\chi^{(0)}(u)$$

and

$$(\mathbf{H}_1^{(2)} + \varepsilon_n^{(2)} - E_n^{(2)})\chi^{(2)}(u) = -(\varepsilon_n^{(3)} - E_n^{(3)})\chi^{(1)}(u) - (\varepsilon_n^{(4)} + C - E_n^{(4)})\chi^{(0)}(u),$$

where C is a constant. Using (A.15), (A.16) and (A.19), for the approximation up to κ^2, we can write

$$\begin{aligned}\psi_n(x,u) &= \psi_n^{(0)} + \kappa\psi_n^{(1)} + \kappa^2\psi_n^{(2)}\\ &= \chi^{(0)}\{\varphi_n^{(0)}(x) + \kappa\varphi_n^{(1)}(x,u) + \kappa^2\varphi_n^{(2)}(x,u)\}\\ &\quad +\kappa\chi^{(1)}(u)\{\varphi_n^{(0)}(x) + \kappa\varphi_n^{(1)}(x,u)\} + \kappa^2\chi^{(2)}(u)\varphi_n^{(0)}(x)\\ &= \{\chi^{(0)}(u) + \kappa\chi^{(1)}(u) + \kappa^2\chi^{(2)}(u)\}\varphi_n(x,X).\end{aligned} \tag{A.20}$$

In this approximation, the wavefunction (A.20) has a simple interpretation, indicating that the order variable and the nucleus are independent. We say that these are interacting *adiabatically*, and this approximation to the order of κ^2 is called the adiabatic approximation. Applying this to a periodic lattice of identical complexes, the order variable may be modulated along with nuclear displacements, for which we should have a responsible agent.

In the adiabatic approximation, the effective potential for nuclear motion can be derived from the effective Hamiltionan

$$\kappa^2\mathbf{H}_1^{(2)} + \kappa^2\varepsilon_n^{(2)}(u) + \kappa^3\varepsilon_n^{(3)}(u) + \kappa^4[\varepsilon_n^{(4)}(u) + C], \tag{A.21}$$

which was taken up to κ^4 or to the fourth power of the nuclear displacement for discussions in Chapter 4.

The asymptotic analysis is discussed in detail in ref. [18], however the results are complicated and no physical significance was implemented beyond harmonic and adiabatic approximations. Therefore, in this appendix, no further discussion is continued to higher-order calculations.

The above theory is purely mechanical, while the temperature-dependence must be sought from an additional thermally accessible mechanism. In fact, for a soft mode, Cowley [35] considered phonon scattering processes by the terms $\varepsilon_n^{(3)}(u)$ and $\varepsilon_n^{(4)}(u)$, for which the phonon densities were considered statistically with the high-temperature approximation. Experimentally, such processes can be described as thermal relaxation from the interaction categorized dynamically in the adiabatic approximation.

Further, it is noted that the condition $\varepsilon_n^{(1)} = 0$ assumed in the above argument is related to symmetric variations of u. However, such displacement cannot be symmetrical if there are anti-symmetric potentials in a crystal; either applied externally or due to an internal origin. For the latter case, we have considered an internal field of long-range correlations, as discussed in Chapters 4 and 5. In any case, an acoustic excitation $u(x,t)$ at a long wavelength should occur adiabatically, which is an essential excitation for maintaining thermodynamic stability of the lattice, as proposed originally by Born and Huang.

References

[1] C.J.Adkins, *Equilibrium Thermodynamics* (McGraw-Hill, London 1968).

[2] M. W. Zemansky, *Heat and Thermodynamics*, 5th ed. (McGraw-Hill, New York, 1957).

[3] A. B. Pippard, *Classical Thermodynamics* (Cambridge Univ. Press, London, 1964).

[4] C. Kittel and H. Kroemer, *Thermal Physics*, 2nd ed. (W. Freeman, San Francisco, 1980).

[5] H. D. Megaw, *Crystal Structures: A Working Approach*, (W. B. Saunders, Philiadelphia, 1973).

[6] See Chapter 9 of ref. [3].

[7] L. D. Landau and E. M. Lifshitz, *Statistical Physics*, trans. by E. Peierls and R. F. Peierls (Pergamon Press, London, 1958).

[8] C. Kittel, *Introduction to Solid State Physics*, 6th ed. pp 633-635 (J. Wiley, New York, 1956).

[9] H. E. Stanley, *Introduction to Phase Transitions and Critical Phenomena* (Oxford Univ. Press, New York, 1971).

[10] R. A. Cowley and A. D. Bruce, *Structural Phase Transitions* (Taylor and Francis, London, 1981).

[11] F. C. Nix and W. Shockley, Rev. Mod. Phys. 10, 1 (1938); T. Muto and Y. Takagi, *Solid State Physics* 1, 194 (1955).

[12] R. Blinc and B. Zeks, *Soft Modes in Ferroelectric and Antiferroelectrics*, Chapter 5 (North Holland, Amsterdam, 1974).

[13] L. Onsager, Phys. Rev. 65, 117 (1944).

[14] R. Becker, Z. Angew. Phys. 6, 23 (1954); *Theory of Heat*, trans. by G. Leibfried, 2nd ed. (Springer, New York, 1967).

[15] H. Gränicher and K. A. Müller, Mat. Res. Bull. 6, 977 (1971).

[16] G. Wannier, *Statistical Physics*, Wiley, New York, 1966, p. 330.

[17] L. D. Landau and E. M. Lifshitz, *Quantum Mechanics*, trans. by E. Peierls and R. F. Peierls, 46 (Pergamon Press, London, 1958).

[18] M. Born and K. Huang, *Dynamical Theory of Crystal Lattices* (Oxford Univ. Press, Oxford, 1954).

[19] R. Comes, R. Currat, F. Desnoyer, M. Lambert and A. M. Quittet, Ferroelectrics 12, 3 (1976).
[20] K. A. Müller, W. Berlinger and F. Waldner, Phys. Rev. Lett. 21, 814 (1968).
[21] M. Fujimoto, S. Jerzak and W. Windsch, Phys. Rev. B34, 1668 (1986).
[22] A. Yoshimori, J. Phys. Soc. Japan, 14, 807 (1959).
[23] M. Fujimoto, Ferroelectrics 47, 177 (1983).
[24] T. Ashida, S. Bando and M. Kakudo, Acta Cryst. B28, 1131 (1972).
[25] E. Nakamura, K. Itoh, K. Deguchi and N. Mishima, Jpn. J. Appl. Phys. Supp. 24-2, 393 (1985).
[26] S. Jerzak and M. Fujimoto, Can. J. Phys. 63, 377 (1985).
[27] P. A. Lee, T. M. Rice and P. W. Anderson, Solid State Comm. 14, 703 (1974).
[28] C. Kittel, *Introduction to Solid State Physics*, 3rd ed. J. Wiley, New York, 1966, p. 486.
[29] J. M. Ziman, *Models of Disorder*, (Cambridge University Press, Cambridge, (1979), p. 23.
[30] W. Cochran, Adv. Phys. 9, 387 (1960); ibid. 10, 401 (1961).
[31] P. W. Anderson, in *Fizika Dielectrikov*, ed. by G. I. Shanavi, Moskow, 1960.
[32] J. Pryzystava, *Physics of Modern Materials*, IAEA, Vienna, 1980, vol. 2.
[33] R. F. Peierls, *Quantum Theory of Solids*, Oxford Univ. Press, London, 1955, p. 108.
[34] R. J. Elliott and A. F. Gibson, *An Introduction to Solid State Physics and its Applications*, MacMillan, London, 1976, Chap. 3.
[35] R. A. Cowley, Pep. Prog. Phys. 31, 123 (1968).
[36] T. Riste, E. J. Samuelsen, K. Otnes and J. Feder, Solid State Comm. 9, 1445 (1971).
[37] S. M. Shapiro, J. D. Axe, G. Shirane and T. Riste, Phys. Rev. B6, 4332 (1972).
[38] A. Sawada and M. Horioka, Jpn. J. Appl. Phys. Supp. 24-2, 390 (1985).
[39] K. A. Müller, *Lecture Notes in Physics*, vol. 124, p209, ed. by C. P. Enz (Springer, Heidelberg, 1971).
[40] L. Bernard, R. Currat, P. Dalamoye, C. M. E. Zeyen, S. Hubert and R. de Kouchkovsky, J. Phys. C16, 433 (1983).
[41] M. Wada, H. Uwe, A. Sawada, Y. Ishibashi, Y. Takagi and T. Sakudo, J. Phys. Soc. Japan, 43, 544 (1977).
[42] M. J. Rice, in *Solitons and Condensed Matter Physics*, vol. 8, p 246, ed. by A. R. Bishop and T. Schneider (Springer, Berlin, 1978).
[43] Cz. Pawlaczyk, H. –G. Unruh and J. Petzelt, Phys. Stat. Sol. (b)136, 435 (1986).
[44] M. Fujimoto, Cz. Pawlaczyk and H. –G. Unruh, Phil. Mag. 69, 919 (1989).
[45] J. G. Kirkwood and I. Oppenheim, *Chemical Thermodynamics*, McGraw-Hill, New York, 1961, Chap. 1.

[46] F. C. Frank and J. H. van der Merwe, Proc. Roy. Soc. London, A198, 205 (1949).
[47] H. Böttger, *Principles of the Theory of Lattice Dynamics*, Physik Verlag, Weiheim, 1983.
[48] Xiaoquin Pan and H. -G. Unruh, J. Phys. Cond. Matter 2, 323 (1990).
[49] J. A. Krumshansl and J. R. Schrieffer, Phys. Rev. B11, 3535 (1835).
[50] S. Aubry, J. Chem. Phys. 64, 3392 (1976).
[51] G. L. Lamb, Jr., *Elements of Soliton Theory*, (J. Wiley, New York, 1980).
[52] T. Toda, *Daenkansu Nyumon* (*Introduction to Elliptic Functions* Nippon Hyoronsha, Tokyo 2001 (in Japanese).
[53] P. M. Morse and H. Feshbach, *Methods of Theoretical Physics*, p 1651 (McGraw-Hill, NewYork, 1953).
[54] P. M. de Wolff, Acta Cryst. A30, 777 (1974); ibid. A33, 493 (1977),
A. Janner and T. Janssen, Phys. Rev. B15, 643 (1977),
T. Janssen and A. Janner, Physica 126A, 163 (1984),
T. Janssen, Phys. Rep. 168, 55 (1988).
[55] M. P. Schulhof, P. Heller, R. Nathans and A. Linz, Phys. Rev. B1, 2403 (1970).
[56] R. Pinn and B. E. F. Fender, Physics Today 38, 47 (1985).
[57] See Chapter 4, p 204 in ref. [15].
[58] P. S. Peercy, J. F. Scott and P. M. Bridenbaugh, Bull. Am. Phys. Soc. 21, 337 (1976);
J. C. Toledano, G. Errandonea and J. P. Jaguin, Solid State Comm. 20, 905 (1976).
[59] E. B. Wilson, J. C. Decius and P. C. Cross, *Molecular Vibrations* (McGraw-Hill, New York 1955).
[60] J. F. Scott, Rev. Mod. Phys. 46, 83 (1974).
[61] J. F. Scott, *Spectroscopy of Structural Phase Transitions* in *Light Scattering near Phase Transitions*, ed. by H. Z. Cummins and A. P. Levanyuk (North Holland, Amsterdam, 1983).
[62] T. Hikita, P. Schnackenberg and V. Higo Schmidt, Phys. Rev. B31, 299 (1985).
[63] G. V. Kozlov, A. A. Volkov, J. F. Scott, G. E. Feldkamp and J. Petzelt, Phys. Rev. B28, 225 (1983).
[64] K. Deguchi, N. Aramaki, E. Nakamura and K. Tanaka, J. Phys. Soc. Japan, 52 1897 (1983).
[65] J. Petzelt, G. V. Kozlov and A. A. Volkov, Ferroelectrics 73, 101 (1987).
[66] Cz. Pawlaczyk, H. -G. Unruh, Phys. Stat. Sol. (b)136, 435 (1986).
[67] J. Petersson, Z. Naturforsch. (a)34, 538 (1979).
[68] F. Bloch, Phys. Rev. 70, 460 (1946).
[69] A. Abragam and M. H. L. Pryce, Proc. Roy. Soc. A205, 135; ibid. A206, 135, 173 (1951).
[70] A. Abragam and B. Bleaney, *Electron Paramagnetic Resonance of Transition Ions* (Clarendon Press, Oxford, 1970).

[71] M. Fujimoto and L. A. Dressel, Ferroelectrics 8, 611 (1974); ibid. 13, 449 (1976).
[72] M. Fujimoto, T. J. Yu and K. Furukawa, J. Phys. Chem. Solids 39, 345 (1978).
[73] M. Fujimoto, K. Furukawa and T. J. Yu, J. Phys. Chem. Solids 40, 101 (1979).
[74] M. Fujimoto, Ferroelectrics 47, 177 (1983).
[75] H. J. Rother, J. Albers and A. Klöpperpieper, Ferroelectrics 54, 107 (1984).
[76] W. Brill and K. H. Ehses, Jpn. J. Appl. Phys. 24 (Suppl. 24-2), 826 (1985).
[77] A. A. Volkov, Yu G. Goncharov, G. V. Kozlov, J.Albers and J. Petzelt, JETP 44, 606 (1986).
[78] R. Ao and G. Schaack, Ind. J. Pure Appl. Phys., Raman Diamond Jubilee (1988).
[79] W. Brill, W. Schilkamp and J. Spilker, Z. Kristallogr. 172, 281 (1985).
[80] M. Fujimoto and Y. Kotake, J. Chem. Phys. 90, 532 (1989).
[81] M. Fujimoto and Y. Kotake, J. Chem. Phys. 91, 6671 (1989).
[82] R. N. Rogers and G. E. Pake, J. Chem. Phys. 33, 1107 (1960).
[83] B. W. van Beest, A. Janner and R. Blinc, J. Phys. C16, 5409 (1983).
[84] R. Blinc, D. C. Allion, P. Prelovsek and V. Rutar, Phys. Rev. Lett. 50, 67 (1983).
[85] R. Blinc, F. Milia, B. Topic and S. Zumer, Phys. Rev B29, 4173 (1984).
[86] R. Blinc, S. Zuzinic, V. Rutar, J. Seliger and S. Zumer, Phys. Rev. Lett. 44, 609 (1980).
[87] R. Blinc, Phys. Rep. 79, 331 (1981).
[88] P. Segransan, A. Janossy, C. Berthier, J. Mercus and P. Butaud, Phys. Rev. Lett. 56, 1854 (1986).
[89] Th. von Waldkirch, K. A. Müller and W. Berlinger, Phys. Rev. B5, 4324 (1972); ibid. 1052 (1973).
[90] S. Hubert, P. Dalanoye, S. Lefrant, M. Lepostollec and M. Hussonios, J. Solid State Chem. 36, 36 (1981).
[91] J. Emery, S. Hubert and J. C. Fayet, J. Physique Lett. 45, 693 (1983); J. Pjysique 46, 2099 (1985).
[92] G. Zwanenburg, *Thesis* Catholic Univ. Nijmegen 1990
[93] S. P. McGlynn, T. Azumi and M. Kinoshita, *Molecular Spectroscopy of Triplet States* (Prentice Hall, Englewood Cliffs, 1969).
[94] C. A. Hutchison, Jr. and B. W. Mangum, J. Chem. Phys. 29, 952 (1958); ibid 34, 908 (1961):
R. W. Brandon, R. E. Gerkin and C. A. Hutchison, Jr., J. Chem. Phys. 41, 3717 (1964):
R. W. Brandon, G. L. Cross, C. E. Davoust, C. A. Hutuchison, Jr., B. E. Kohler and R. Sibley, J. Chem. Phys. 43, 2006 (1965).
[95] A. S. Cullick and R. E. Gerkin, Chem. Phys. 23, 217 (1977).
[96] N. Hirota and C. A. Hutchison, Jr., J. Chem. Phys. 42, 2869 (1965).

[97] A. W. Hornig and J. S. Hyde, Mol. Phys. 6, 33 (1963).
[98] H. Cailleau, J. C. Messager, F. Moussa, F. Bugant, C. M. E. Zeyen and C. Vettier, Ferroelectrics 67, 3 (1986);
H. Cailleau, F. Moussa, C. M. E. Zeyen and J. Bouillot, J. Physique Coll. 42, 704 (1981).
[99] Y. Makita, A. Sawada and Y. Takagi, J. Phys. Soc. Japan 41, 167 (1976).
[100] H. -G. Unruh, Solid State Comm., 8 1951 (1970).
[101] M. Iizumi, J. D, Axe and G. Shirane, Phys. Rev. B15, 4392 (1977).
[102] M. Fujimoto, L. A. Dressel and T. J. Yu, J. Phys. Chem. Solids, 38, 97 (1977).
[103] M. Fukui and R. Abe, J. Phys. Soc. Japan, 51, 3942 (1982).
[104] M. Pezeril, J. Emery and J. C. Fayet, J. Physique Lett. 41, 499 (1980).
[105] M. Pezeril and J. C. Fayet, J. Physique Lett. 43, 267 (1982).
[106] A. Kaziba, M. Pezeril, J. Emery and J. C. Fayet, J. Physique Lett. 46, 387 (1985).
[107] A. Kaziba and J. C. Fayet, J. Physique 47, 239 (1986).
[108] M. Fukui and R. Abe, Jpn. J. Appl. Phys. 20, L-533 (1981); I. Suzuki, K. Tsuchida, M. Fukui and R. Abe, ibid. 20, L-840 (1981).
[109] T. Kobayashi, M. Suhara and M. Machida, Phase Trans. 4, 281 (1984).
[110] R. Blinc, D. C. Ailion, J. Dolinsek and S. Zumer, Phys. Rev. Lett. 50, 67 (1983).
[111] L. Aa. Shuvalov, N. R. Ivanov, N. V. Gordeyeva and L. F. Kirpichnikova, Soviet Phys. Crystallgr. 14, 554 (1970);
K. Gesi, K. Ozawa and Y. Makita, Jpn. J. Appl. Phys. 12, 1963 (1973).
[112] A. P. Levanyuk and D. G. Sannikov, Soviet Phys. Solid State 12, 1418 (1971).
[113] Y. Makita and S. Suzuki, J. Phys. Soc. Japan, 36, 1215 (1974).
[114] H. Grimm and W. J. Fitzgerald, Acta Cryst. A34, 268 (1978).
[115] A. B. Tovbis, T. S. Davydova and V. I. Simonov, Soviet Phys. Crystrallogr. 17,81 (1972).
R. Tellgren, D. Armed and R. Luminga, J. Solid State Chem. 6, 250 (1975).
[116] L. A. Shuvalov, N. R. Ivanov, N. V. Gordeyeva and L. F. Kirpichnikova, Phys. Lett. A33, 490 (1970).
[117] S. Waplak, S. Jerzak, J. Stankowski and L. A. Shuvalov, Physica 106B, 251 (1981).
[118] M. Fukui, C. Takahashi and R. Abe, Ferroelectrics 36, 315 (1981).
[119] N. Shibata, R. Abe and I. Suzuki, J. Phys. Soc. Japan 41, 2011 (1976);
R. Abe and N. Shibata, J. Phys. Soc. Japan 43, 1308 (1977).
[120] S. Jerzak, private communication
[121] J. C. Fayet, Helv. Physica Acta 58, 76 (1985).
[122] R. M. Stratt, J. Chem. Phys. 84, 2315 (1985).
[123] J. C. Slater, J. Chem. Phys. 9, 16 (1941).
[124] Y. Takagi, J. Phys. Soc. Japan, 3, 271 (1948).
[125] I. P. Kaminov and T. C. Damen, Phys. Rev. Lett. 20, 1105 (1968).

[126] R. Blinc, P. Cvec and M. Schara, Phys. Rev. 159, 411 (1967).
[127] N. S. Dalal, C. A. McDowell and R. Srinivasan, Chem. Phys. Lett. 4, 97 (1969); Phys. Rev. Lett. 25, 823 (1970); Mol. Phys. 24, 1051 (1972).
N. S. Dalal and C. A. McDowell, Phys. Rev. B5, 1074 (1972).
N. S. Dalal, J. A. Hebden and C. A. McDowell, J. Chem. Phys. 62, 4404 (1974).

Index